Ecotoxicological Risk Assessment of the Chlorinated Organic Chemicals

Other titles from the Society of Environmental Toxicology and Chemistry (SETAC):

Principles and Processes for Evaluating Endocrine Disruption in Wildlife
R.J. Kendall, R.L. Dickerson, J.P. Giesy, W.P. Suk, editors
1998

Chemical Ranking and Scoring: Guidelines for Relative Assessments of Chemicals
M.B. Swanson, A.B. Socha, editors
1997

Chemically Induced Alterations in Functional Development and Reproduction of Fishes
R.M. Rolland, M. Gilbertson, R.E. Peterson, editors
1997

Atmospheric Deposition of Contaminants to the Great Lakes and Coastal Waters
J.E. Baker, editor
1997

Ecological Risk Assessment of Contaminated Sediments
C.G. Ingersoll, T. Dillon, G.R. Biddinger, editors
1997

Reassessment of Metals Criteria for Aquatic Life Protection: Priorities for Research and Implementation
H.L. Bergman and E.J. Dorward-King, editors
1997

Whole Effluent Toxicity Testing: An Evaluation of Methods and Prediction of Receiving System Impacts
D.R. Grothe, K.L. Dickson, D.K. Reed-Judkins, editors
1996

The Multi-Media Fate Model: A Vital Tool for Predicting the Fate of Chemicals
C.E. Cowan, D. Mackay, T.C.J. Feijtel, D. Van de Meent, A. DiGuardo, J. Davies, N. Mackay, editors
1995

Aquatic Dialogue Group: Pesticide Risk Assessment and Mitigation
J.L. Baker, A.C. Barefoot, L.E. Beasley, L. Burns, P. Caulkins, J. Clark, R.L. Feulner, J.P. Giesy, R.L. Graney, R. Griggs, H. Jacoby, D. Laskowski, A. Maciorowski, E. Mihaich, H. Nelson, R. Parrish, R.E. Siefert, K.R. Solomon, W. van der Schalie, editors
1994

For information about any SETAC publication, including SETAC's international journal *Environmental Toxicology and Chemistry*,
contact the SETAC Office, 1010 N. 12th Avenue, Pensacola, Florida, USA 32501-3367
T 850 469 1500 F 850 469 9778 E setac@setac.org http://www.setac.org

Ecotoxicological Risk Assessment of the Chlorinated Organic Chemicals

Edited by
John Carey
Philip Cook
John Giesy
Peter Hodson
Derek Muir
J. William Owens
Keith Solomon

Proceedings from the Pellston Workshop
on Environmental Risk Assessment for Organochlorine Compounds
24–29 July 1994
Aliston, Ontario, Canada

SETAC Special Publications Series

SETAC Liaison
Greg Schiefer
Society of Environmental Toxicology and Chemistry (SETAC)

Current Coordinating Editor of SETAC Books
C. G. Ingersoll
U.S. Geological Survey

Publication sponsored by the Society of Environmental Toxicology
and Chemistry (SETAC) and the SETAC Foundation for Environmental Education

SETAC PRESS

Cover by Michael Kenney Graphic Design and Advertising
Editing and typesetting by Wordsmiths Unlimited
Indexing by Nanette Cardon

Library of Congress Cataloging-in-Publication Data

Ecotoxicological risk assessment of the chlorinated organic chemicals / edited by John Carey ... [et al.].
p. cm. -- (SETAC special publications series)
"Proceedings from the Pellston Workshop on Environmental Risk Assessment for Organochlorine Compounds, 24–29 July 1994, Aliston, Ontario, Canada."
Includes bibliographical references and index.
ISBN 1-880611-07-4
1. Organochlorine compounds--Environmental aspects--Congresses. 2. Organochlorine compounds--Toxicology--Congresses. 3. Ecological risk assessment--Congresses. I. Carey, John H. II. Pellston Workshop on Environmental Risk Assessment for Organochlorine Compounds (1994 : Ontario) III. SETAC (Society) IV. Series.
QH545.O72E36 1998
571.9'57--dc21 98-4840
CIP

Information in this book was obtained from individual experts and highly regarded sources. It is the publisher's intent to print accurate and reliable information, and numerous references are cited; however, the authors, editors, and publisher cannot be responsible for the validity of all information presented here or for the consequences of its use. Information contained herein does not necessarily reflect the policy or views of the Society of Environmental Toxicology and Chemistry (SETAC).

International Standard Book Number 1-880611-07-4
Printed in the United States of America
05 04 03 02 01 00 99 98 10 9 8 7 6 5 4 3 2 1

(∞) The paper used in this publication meets the minimum requirements of the American National Standard for Information Sciences—Permanence of Paper for Printed Library Materials, ANSI Z39.48-1984.

Reference Listing: Carey J, Cook P, Giesy J, Hodson P, Muir D, Owens J, Solomon K, editors. 1998. Ecotoxicological risk assessment of the chlorinated organic chemicals. SETAC Pellston Workshop on Environmental Risk Assessment for Organochlorine Compounds; 1994 Jul 24–29; Aliston, Ontario, Canada. Published by the Society of Environmental Toxicology and Chemistry (SETAC), Pensacola, Florida, USA. 397 p.

The SETAC Special Publications Series

The SETAC Special Publications Series was established by the Society of Environmental Toxicology and Chemistry (SETAC) to provide in-depth reviews and critical appraisals on scientific subjects relevant to understanding the impacts of chemicals and technology on the environment. The series consists of single- and multiple-authored or edited books on topics reviewed and recommended by the SETAC Board of Directors and approved by the Publications Advisory Council for their importance, timeliness, and contribution to multidisciplinary approaches to solving environmental problems. The diversity and breadth of subjects covered in the series reflect the wide range of disciplines encompassed by environmental toxicology, environmental chemistry, and hazard and risk assessment. Despite this diversity, the goals of these volumes are similar; they are to present the reader with authoritative coverage of the literature, as well as paradigms, methodologies and controversies, research needs, and new developments specific to the featured topics. All books in the series are peer reviewed for SETAC by acknowledged experts.

The SETAC Special Publications are useful to environmental scientists in research, research management, chemical manufacturing, regulation, and education, as well as to students considering careers in these areas. The series provides information for keeping abreast of recent developments in familiar subject areas and for rapid introduction to principles and approaches in new subject areas.

SETAC would like to recognize the past SETAC Special Publications Series editors:

T. W. La Point, The Institute for Environmental and Human Health,
Texas Tech University, Lubbock TX

B. T. Walton, U.S. Environmental Protection Agency,
Research Triangle Park NC

C. H. Ward, Department of Environmental Sciences and Engineering,
Rice University, Houston TX

Contents

List of Figures

List of Tables

Preface and Dedication

This book is dedicated to the memory of Ulf Ahlborg and Eva Voldner. Both participated in the workshop, and both made significant contributions to our understanding of the behavior and effects of persistent and bioaccumulative substances in the environment. Their willingness to participate in the workshop that resulted in this book is further evidence of their interest in the field and their dedication to the advancement of science.

Acknowledgments

The Society of Environmental Toxicology and Chemistry (SETAC), the organizing committee of the Workshop on Environmental Risk Assessment for Organochlorine Compounds, and the editors of this book wish to thank the workshop participants for their input and hard work. The chapters were written by workgroups, whose members are acknowledged below as contributing authors.

Ulf Ahlborg, David Allen, Karlheinz Ballschmiter, Terry Bidleman, Robert Boethling, John Carey, Jerry Clark, Philip Cook, Joseph DePinto, Gary Dickson, Russell Erickson, Heidelore Fiedler, Michel Fournier, John Giesy, Michael Gilbertson, Frank Gobas, Trevor Green, Anders Grimvall, Peter Hodson, Phil Howard, Bo Jansson, Paul Jones, Richard Kimerle, Don Mackay, Foster Mayer, Lynn McCarty, Derek Muir, Kelly Munkittrick, Ross Norstrom, Mats Olsson, J. William Owens, Rodney Parrish, Richard Peterson, Christoffer Rappe, Richard Reitz, Mirja Salkinoja-Salonen, Wayne Schmidt, Dick Sijm, Keith Solomon, John Stein, Kenneth Stromborg, Shinsuke Tanabe, Dorien ten Hulscher, Peter Thomas, Donald Tillitt, Martin van den Berg, Glen Van Der Kraak, Gillman Veith, Eva Voldner, Phil Watanabe, Randall Wentsel, Maurice Zeeman

Contributors

The Pellston Workshop on Environmental Risk Assessment for Organochlorine Compounds was financially supported by:

American Forest and Paper Association
Chemical Manufacturers Association
Ciba-Geigy Corporation
Environment Canada
Euro Chlor
U.S. Environmental Protection Agency

This publication does not necessarily reflect the position or the policy of these organizations. The workshop organizers graciously thank Rod Parrish, Linda Longsworth, Allison Bough, and the other SETAC/SETAC Foundation staff for their support, Mimi Meredith for her editing and layout skills, and Gary Klecka, Dow Chemical Company and Gary Heinz, Patuxent Wildlife Research Center, for their timely reviews.

About the Editors

John Carey is the director of Environment Canada's largest research facility, the Rivers Research Branch, National Water Research Institute, in Burlington, Ontario. His research interests lie in environmental dynamics of organic pollutants in surface waters, photochemical processes in natural waters, advanced oxidation processes as methods for destroying organics in wastewater, and fate and effects of bleached kraft mill effluent. Dr. Carey holds a B.Sc. in inorganic chemistry and an M.Sc. in chemistry from the University of Windsor, and earned a Ph.D. in chemistry from Carleton University in 1974.

Philip Cook is a research chemist and senior USEPA ERL-Duluth scientific expert on risk assessment methods for toxic organic chemicals in freshwater ecosystems, including the Great Lakes. He received a B.S. in chemistry from Tufts University in 1961, an M.S. in geochemistry from the Colorado School of Mines in 1968, and a Ph.D. in chemistry from the University of Wisconsin-Madison in 1972. He has recently completed reports on dioxin in Lake Ontario associated with the Hyde Park Superfund case and on assessment of TCDD risk to aquatic life and wildlife, and he has formulated a comprehensive strategy for USEPA research toward a bioaccumulation prediction capability that meets future risk assessment needs of regulatory programs.

John Giesy is professor of zoology at Michigan State University. He received a B.S. degree in biology and chemistry from Alma College, Alma, Michigan in 1970 and earned master's and doctorate degrees in limnology from Michigan State University in 1971 and 1974. Professor Giesy has done extensive research in metal speciation, multispecies toxicity testing, biochemical indicators of stress in aquatic organisms, fate and effects of polycyclic aromatic hydrocarbons, and photo-enhanced toxicity of organic compounds. He is a charter member of SETAC, receiving its Founders Award in 1995, and was the Society's president in 1991.

Peter Hodson is the director of the School of Environmental Studies at Queen's University, Kingston, Ontario. His special interests include toxicological consequences of MFO induction in fish and causes of the decline of the American eel. In 1968, he received a B.Sc. in physiology from McGill University, followed by an M.Sc. in biology from the University of New Brunswick in 1970 and a Ph.D. in zoology from the University of Guelph in 1974. He was the 1989 Stephenson Lecturer, Canadian Conference for Fisheries Research, served as the president of SETAC in 1994, and is an aquatic toxicology editor for *Environmental Toxicology and Chemistry*.

Derek Muir is a research scientist for Environment Canada in Burlington, Ontario. He received a Ph.D. in agricultural chemistry in 1977, has been an adjunct professor at the University of Manitoba and the University of Guelph, and is highly committed to the training of graduate students in environmental chemistry. He is especially interested in persistent organic contaminants in the Arctic. In 1992, Dr. Muir received the F.W. Karasek Award in Environmental Chemistry from the University of Waterloo. He has twice served on the editorial board of *Environmental Toxicology and Chemistry*, and he is a charter member of SETAC.

J. William Owens is principal scientist with Procter & Gamble in Cincinnati, Ohio. He is a National Science Foundation fellow and earned a B.S. in biology in 1967 and a Ph.D. in molecular biology in 1976. Of particular interest to him are the environmental impacts of manufacturing facilities and the application and integration of life-cycle assessment with ecological risk assessment.

Keith Solomon is the director of the Centre for Toxicology and professor in the Department of Environmental Biology, University of Guelph. He holds B.Sc. degrees in zoology and chemistry, a B.Sc. and an M.Sc. in zoology from Rhodes University, and M.S. and Ph.D. degrees from the University of Illinois, Urbana-Champaign. He received the 1993 SETAC-ABC Laboratories Award for Environmental Education and is an active member of SETAC, the Entomological Society of America, and the Toxicology Forum. He directs an active program of research into the fate and effects of pesticides in the environment as well as exposure of humans to pesticides.

List of Attendees and Participants*

Ulf G. Ahlborg (deceased)
Institute of Environmental Medicine
Karolinska Institute
Stockholm, Sweden

David Allen
University of California-Los Angeles
Los Angeles CA

Karlheinz Ballschmiter
University of Ulm
Analytical Chemistry and
Environmental Chemistry
Ulm, Germany

Terry F. Bidleman
Atmospheric Environment Service
ARQP
Downsview ON, Canada

Robert Boethling
USEPA
Washington DC

John Carey
Environment Canada
National Water Research Institute
Burlington ON, Canada

Jerry E. Clark
National Fish and Wildlife Foundation
Washington DC

Philip Cook
USEPA
Environmental Research Laboratory
Duluth MN

Joseph V. DePinto
University at Buffalo
Great Lakes Program
Buffalo NY

Gary Dickson
Ciba Plant Protection
Ciba-Geigy Corporation
Greensboro NC

Russell Erickson
USEPA
Environmental Research Laboratory
Duluth MN

Heidelore Fiedler
University of Bayreuth
Ecological Chemistry and Geochemistry
Bayreuth, Germany

Michel Fournier
University of Quebec-Montreal
Department of Biological Science
Montreal, Canada

John P. Giesy
Michigan State University
Department of Fisheries and Wildlife
East Lansing MI

Michael Gilbertson
International Joint Commission Great
Lakes Regional Office
Windsor ON, Canada

Frank Gobas
Simon Fraser University
School of Resource and
Environmental Management
Burnaby BC, Canada

Trevor Green
Zeneca Central Toxicology Laboratory
Alderley Park
Cheshire, United Kingdom

* These affiliations were correct at the time of the workshop.

Anders Grimvall
Department of Water and
Environmental Studies
Linkoping University
Linkoping, Sweden

Peter Hodson
Canada Centre for Inland Waters
Environment Canada
Burlington ON, Canada

Phil Howard
Syracuse Research Corporation
Chemical Hazard Assessment Division
Syracuse NY

Bo Jansson
Institute of Applied Environmental
Research
Stockholm University
Solna, Sweden

Paul Jones
ESR: Environmental Gracefield
Research Centre
Lower Hutt, New Zealand

Richard A. Kimerle
Monsanto Corporation
St. Louis MO

Don Mackay
University of Toronto
Institute for Environmental Studies
Department of Chemical Engineering
Toronto ON, Canada

Foster Mayer
USEPA
Environmental Research Laboratory
Gulf Breeze FL

Lynn S. McCarty
L.S. McCarty Scientific Research
and Consulting
Oakville ON, Canada

Derek Muir
Freshwater Institute
Department of Fisheries and Oceans
Winnipeg MB, Canada

Kelly Munkittrick
GLLFAS, Department of Fisheries
and Oceans
Environment Canada
Burlington ON, Canada

Ross Norstrom
Environment Canada CWS/NWRC
Ottawa ON, Canada

Mats Olsson
Swedish Museum of Natural History
Contaminant Research Group
Stockholm, Sweden

J. William Owens
The Procter and Gamble Company
Environmental Sciences Department
Cincinnati OH

Rodney Parrish
SETAC/SETAC Foundation
Pensacola FL

Richard E. Peterson
University of Wisconsin-Madison
School of Pharmacy
Madison WI

Christoffer Rappe
Umeå University
Institute for Environmental Chemistry
Umeå, Sweden

Richard Reitz
McLaren/Hart
Midland MI

Mirja Salkinoja-Salonen
University of Helsinki
College of Agriculture and Forestry
Department of Applied Chemistry
and Microbiology
Helsinki, Finland

Wayne A. Schmidt
National Wildlife Federation
Great Lakes National Resources Center
Ann Arbor MI

Dick T.H.M. Sijm
Fahrenheitstraat 44
Amersfoort, Netherlands

Keith R. Solomon
Canadian Centre for Toxicology
University of Guelph
Guelph ON, Canada

John E. Stein
NOAA/National Marine Fisheries Service
Environmental Conservation Division
Seattle WA

Kenneth L. Stromborg
U.S. Fish and Wildlife Service
Green Bay WI

Shinsuke Tanabe
Ehime University
College of Agriculture
Department of Environment Conservation
Matsuyama, Japan

Th.E.M. (Dorien) ten Hulscher
RIZA, Environmental Chemistry
Lelystad, Netherlands

Peter Thomas
University of Texas
Marine Science Institute
Port Aransas TX

Donald Tillitt
National Biological Survey
National Fisheries Contaminant Research Center
Columbia MO

Martin van den Berg
University of Utrecht
Research Institute of Toxicology
Utrecht, Netherlands

Glen Van Der Kraak
University of Guelph
Department of Zoology
Guelph ON, Canada

Gilman D. Veith
USEPA
Environmental Research Laboratory
Duluth MN

Eva Voldner (deceased)
Atmospheric Environment Service
AROI
Downsview ON, Canada

Phil Watanabe
The Dow Chemical Company
Health and Environmental Sciences
Midland MI

Randall S. Wentsel
U.S. Senate Committee on Environment and Public Works
Washington DC

Maurice Zeeman
USEPA
OPPT
Washington DC

Executive Summary

The Society of Environmental Toxicology and Chemistry (SETAC) held a Workshop on Environmental Risk Assessment for Organochlorine Compounds in Aliston, Ontario in July 1994, in which 50 scientists from the United States, Canada, Germany, Sweden, Finland, the Netherlands, Great Britain, Japan, and New Zealand participated. The scientists noted that chlorine-containing substances do not have sufficiently common chemical and biological properties for them to be represented as a single class of substances. Therefore, calls to ban all uses of chlorine to protect the environment are not scientifically valid or supportable.

The scientists agreed that the chemical and biological properties of concern for any substance, whether or not it contains chlorine, are a combination of

- environmental persistence;
- bioaccumulation in organisms, in which a special concern is highly bioaccumulative substances that biomagnify in food chains; and
- high toxic potency, in which relatively small doses cause effects for significant endpoints of concern, such as reproduction and development.

This combination of properties can cause apparently small releases of a substance to accumulate in the environment over time and to reach harmful concentrations. Therefore, it is this combination of properties that should be the focus of societal concern and management efforts. A very small number of chlorine containing substances, such as polychlorinated biphenyls (PCBs), dichlorodiphenyltrichloroethane (DDT), polychlorinated dibenzo-*p*-dioxin and dibenzofurans (PCDD/PCDFs), dieldrin, aldrin, chlordane, endrin, heptachlor, hexachlorobenzene (HCB), mirex, and toxaphene are known to have this combination of properties. Additionally, such nonchlorinated substances as mercury and the polyaromatic hydrocarbons exhibit these properties.

The important findings from this SETAC-sponsored international workshop are these:

- The current approaches to estimating the environmental risks are adequate for most chlorine-containing substances.
- Steps taken to manage the risks of chlorine-containing substances have already improved environmental quality in some nations and regions. However, additional controls at a global level on a select number of highly toxic, persistent, and bioaccumulative substances are needed.
- A total ban on chlorine use is not supported by the evidence. Most chlorine-containing substances (> 80%) can be produced and used safely for beneficial purposes.

Chapter 1 of the workshop proceedings offers a review of the current production levels and uses of chlorine-containing substances and the byproducts of this production. Natural production of chlorine-containing substances is also discussed. One sign of progress is that, during the past 10 to 20 years, regulations have been effective in curtailing important sources of both direct emissions of persistent and bioaccumulative chlorine-containing substances, such as PCBs and DDT, and byproducts, such as PCDD/PCDFs.

The pattern of distribution and the ultimate fate of a substance according to its physicochemical properties are discussed in Chapter 2. The compounds of concern are those with intermediate molecular weights ($200 < MW < 500$) and that are semivolatile, lipophilic, and persistent in the environment. Examples include such highly polychlorinated aromatic substances as DDT, chlordane, PCBs, polychlorinated benzenes (PCBzs), and PCDD/PCDFs, whose properties give rise to atmospheric transport that results in global redistribution to remote regions. Scientists now understand these fundamental properties and can, with relative confidence, model and predict environmental distribution of substances. Scientists have lesser confidence when predicting physical, chemical, and biological degradation processes and rates, since these are very dependent upon local conditions.

Bioaccumulation is discussed at length in Chapter 3 and occurs both via direct uptake (bioconcentration) and transfer through food chains (biomagnification). The most important chemical properties affecting bioaccumulation are

- the extent to which a chemical partitions into the lipids in an organism from external media and
- the structural characteristics that control the persistence of a substance within an organism, such as its rate of metabolic transformation.

Other properties are also important in this complex process, including the bioavailability (the proportion of the total amount of a substance that is available for uptake by an organism), the characteristics of the organism and its metabolic capability, the physicochemical characteristics of the ecosystem, and the structure of the food chain. The role of these characteristics are understood and can be modeled with increasing accuracy.

Fate models establish a quantitative relationship between releases of a substance and its different concentrations in air, water, sediments, soil, and biota as a function of known chemical and environmental properties. Models are crucial to risk assessment and management of substances in the environment, as they enable integration of the fate, distribution, and bioaccumulation of substances and of the characteristics of ecosystems into an assessment of management options. These fate models apply to all substances. Chlorine-containing substances do not possess properties that make their environmental fate unique. The fate and distribution of PCBs and chlorinated pesticides have been modeled successfully on a site-specific basis (in small watersheds, harbors, bays, and lakes $< 100 \text{ km}^2$), and models are now being extended to regional and global scales. Most importantly, models provide a predictive capability to assess potential exposures and doses as discussed in Chapter 4. Models in the aquatic environment are further developed than those for the terrestrial environment. Continued improvement in the models requires better emission data, improved physicochemical and reactivity data, and more monitoring to confirm predictions, validate the models, and identify research needs.

Chlorine-containing substances can cause a spectrum of toxic effects, as discussed in Chapter 5. The nature of their effects and potencies overlap those of nonchlorinated substances. The mechanisms of toxicity or modes of action range from nonspecific general narcosis at high exposures (called "baseline toxicity") to specific mechanisms such as enzyme inhibition, metabolic activation of reactive intermediates, endocrine interference, immune suppression, and neurological and behavioral effects. Once more, the specific structure and physicochemical properties of the molecule, regardless of the presence or absence of chlorine, determines its action and toxicity. At this time, effect models can be used to predict general narcosis and are useful to assess possible toxicity for some known mechanisms. However, models are not sufficient to replace experimental observations. The toxicity or potential effects of substances on organisms has typically been based on the results of laboratory experiments and, in some circumstances, field data. Laboratory studies are available for characterizing many effects of substances on aquatic biota but are less developed for wildlife.

Further methodological development is needed, particularly in the case of neurotoxic, immunotoxic, reproductive, and endocrine effects. There is growing evidence that some natural and synthetic substances, including some chlorine-containing substances, may have the potential to disrupt reproductive and endocrine function. These effects and their ecological risks are poorly characterized, and there is an urgent need to evaluate these hazards using existing methods and to develop new assessment methods appropriate for these effects. In addition, successful application of laboratory studies requires that exposure methods and test endpoints be selected to suit the characteristics of the substances and ecosystems of concern.

There have been adverse effects on wildlife (terrestrial and aquatic organisms) from chlorine-containing substances displaying the combined properties of persistence, bioaccumulation, and toxicity. Identified substances include PCBs, PCDD/PCDFs, and several chlorinated pesticides, including DDT, dieldrin, and toxaphene. This evidence is discussed in Chapter 6. It is important to note that modern pesticides, including those containing chlorine, do not have the persistence, bioaccumulation potential, or broad-spectrum toxicity typical of these older pesticides.

The history of effects of substances released into the environment on wildlife demonstrates the following lessons:

- Adequate hazard evaluations and risk assessments are required before substances are released to the environment.
- Risk assessments and their degree of complexity should be specific to the use and expected production volumes of the substance.
- Field observations and biological monitoring provide invaluable indicators of unexpected effects.
- Managing risk requires institutional frameworks for reevaluating the use and release of substances, if new evidence of adverse effects is discovered.

The most encouraging sign of progress in reducing effects is that populations of several species of birds, fish, and mammals have apparently recovered from the effects of persistent chlorinated substances, such as PCBs and DDT, and their recovery parallels declines in observed environmental concentrations of such chemicals. However, at some sites, environmental concentrations may still be sufficient to cause adverse effects.

Field biomonitoring of wildlife provides a critical check on the success of regulations and can serve as an early warning for unanticipated effects. While laboratory methods for hazard assessment are relatively well-developed, methods to characterize effects in natural ecosystems are less so. Routine biomonitoring requires an improved capability to detect effects and to establish cause-effect relationships. Improved integration of chemical, biochemical, and population- and ecosystem-level measurements is needed as well as better sampling designs to address spatial and temporal considerations.

Risk assessment combines the available information on chemical properties, environmental levels, environmental models, laboratory toxicity studies, field biomonitoring, etc., to protect ecosystems as discussed in Chapter 7. The basic scientific procedures for ecological risk assessment have now been developed in North America and Europe, and these procedures are being incorporated into regulatory frameworks and guidelines. These guidelines utilize a tiered assessment process where chemical, biological, and production volume criteria are considered. Ecotoxicological risk assessment is broadly applicable to estimate risk from both chlorinated and nonchlorinated substances.

Persistent and bioaccumulative substances are a significant challenge due to their complex behavior. However, the results of recently developed predictive models for PCBs in Green Bay and TCDD and related substances in Lake Ontario (validated by field monitoring) indicate that environmental behavior can be successfully addressed by ecological risk assessment tools. As a result, workshop attendees concluded that the approaches and frameworks currently used by regulators to assess the environmental risks of chlorine-containing substances were applicable, even in the case of toxic, persistent and bioaccumulative substances like PCDD/PCDFs.

Despite the progress in developing ecological risk assessment approaches, further improvements are needed. Risk assessment procedures rely upon the availability of accurate scientific data and must cover multiple endpoints. Important improvements are needed:

- Generation and collection of data on chemical behavior and toxicity, such as a minimum or base dataset for substances, must be increased.
- Additional important endpoints such as development and endocrine disruption must be included.
- Toxicity testing procedures for chronic exposures and reproductive effects must be developed further.

Risks to humans and the environment are managed based on the information and the perspective provided by risk assessment, along with a broad set of technical, economic, societal, and policy inputs. Risk management for chlorine-containing substances is not inherently different from that for nonchlorinated substances, as discussed in Chapter 8. The standard approach of discharge criteria and improved technology to limit emissions applies in almost all cases. Use restrictions and bans for the substances that display the greatest persistence, bioaccumulation, and toxicity are already in place in many nations. The effectiveness of risk management strategies and decisions should be continually appraised in order to improve the process and the results. Important sources of information are the level and trends of environmental concentrations and biomonitoring of the health of previously affected and sentinel populations. The latter ensures that remediation is effective and that potentially unrecognized effects are quickly discovered. These monitoring outcomes must be clearly communicated to the public to build their understanding of complex issues and to maintain their confidence in the risk management process.

Several factors combine to make risk management for chlorinated substances more difficult than for nonchlorinated substances. All chlorinated substances are often mistakenly associated with a few persistent and bioaccumulative substances, such as DDT, PCBs, and PCDD/PCDFs. The long depuration times for contaminated sites, extraordinary clean up costs, disposal and landfill issues, and residual effects in wildlife continually focus attention on these problem chlorine-containing substances. This leads to the misperception that risks from all chlorinated substances may be increasing rather than decreasing and makes risk management and risk communication more difficult.

For persistent, bioaccumulative, and toxic substances, full international coordination is lacking despite the recent efforts by the United Nations Economic Commission for Europe and the United Nations Commission on Sustainable Development. At the national level, some countries have developed comprehensive frameworks to address risks from substances and have generated significant progress in managing and reducing risks within their jurisdictions. Other countries need assistance in creating regulatory frameworks that do not drain scarce resources or hamper their economic development. In cases where these substances are used to protect human life or for other critical beneficial uses, less hazardous alternatives need to be identified and made available. Persistent, bioaccumulative, and toxic substances may pose particular cost and management challenges for developing economies. These economies are faced simultaneously with controlling chemical production and release, substituting other substances, remediating contaminated sites, and disposing of existing stocks.

To resolve the global issue that is presented by the persistent, bioaccumulative, and mobile organochlorines (OCs), it is suggested that an international agreement, similar to the Montreal Protocol on chlorofluorocarbons (CFCs), is needed to deal with the select group of chlorine-containing substances that have the greatest persistence, bioaccumulation, and toxicity. Such an agreement should contain provisions for the necessary educational, technical, and financial assistance to implement the agreement.

Introduction

The Need for Solutions

Since the nineteenth century, chlorine and chlorinated compounds have been produced in industrial processes. Chlorine chemistry has been used for public health benefits to disinfect potable water supplies and treated sewage, and in industry, chlorine has become a critical component of numerous chemical processes. However, certain chlorinated compounds have been both persistent in the environment and bioaccumulative, for example, polychlorinated dioxins and furans (PCDD/PCDFs) polychlorinated biphenyls (PCBs), mirex, toxaphene, dichlorodiphenyltrichloroethane (DDT) and its metabolites,. Environmental scientists have accumulated evidence for adverse effects from these particular compounds. The regulatory agencies of most nations have now banned the deliberate production of several chlorinated compounds and have regulated others using risk assessment principles. Simultaneously, industrial and other human activities are being modified to reduce environmental releases of chlorinated materials. As a result, environmental concentrations of many chlorinated substances have declined since peaking in the 1970s, and previously affected species such as the bald eagle, osprey, and herring gull have begun to recover.

Despite the downward trends in chemical concentrations and the apparent initiation of recovery for several species, the International Commission for the Great Lakes (IJC) considers chlorinated organic chemicals as a single class of similarly behaving compounds and recommends the complete phase-out or banning of chlorine chemistry. This recommendation comes despite growing evidence that a wide variety of chlorinated compounds are also produced naturally by both biotic and abiotic processes and are presumed to be assimilated by the environment. The expressed rationale of the IJC is that

1) the discharge of chlorinated chemicals as a class of compounds continues to pose significant environmental risks,
2) environmental protection cannot be adequately assured through current regulatory and risk assessment mechanisms, and

Introduction Preview

35.4527
A3 tet
-100.79^{TP}
-34.05
3.17
3.16
$[Ne]3s^2p^5$
Chlorine
17
±1,3,5,7
Cl

3) chlorinated compounds cannot be controlled by no-observed-effect concentrations, and therefore, a strict and absolute zero-discharge philosophy is necessary.

The IJC phaseout recommendation departs from standard practices of risk assessment and risk management. Regulatory bodies have attempted to employ a more data-based approach on a compound-by-compound basis, such as the Great Lakes Initiative. A review of the environmental risks of chlorine and chlorinated substances and the applicability of risk assessment and management methods by an independent scientific body is needed.

Workshop Focus

The Workshop on Environmental Risk Assessment for Organochlorine Compounds was held to review the environmental effects of the spectrum of organochlorine (OC) compounds produced by human activity and to evaluate those risk assessment and risk management approaches currently in place and those proposed, relative to chlorine chemistry. Specifically, the workshop participants sought to accomplish the following:

- Identify historical and current environmental concentrations of chlorinated compounds in the environment, including those compounds identified to be persistent and bioaccumulative such as DDT, PCBs, and dioxins.
- Determine environmental fate and effects of various chlorinated compounds in the atmosphere, water, sediments, and various trophic levels of organisms.
- Assess the hypothesis that all OC compounds possess the characteristics of a single, general chemical class.
- Evaluate scientific predictive models used to determine the fate of chlorinated compounds and concentration trends, especially in fish, wildlife, terrestrial animals, and plants, with respect to environmental risk assessment, risk management, and determining remediation strategies.
- Evaluate current regulatory and scientific environmental risk assessment methodologies used to protect the environment.
- Identify current methods that are applicable to chlorinated compounds, and develop improved risk and hazard assessment processes and risk management strategies for chlorinated compounds.
- Identify regions and aquatic and/or wildlife species in which these risk assessment methodologies can be further tested and validated, and identify research needs to validate these methodologies for chlorinated chemicals.

The workshop followed the format of previous Pellston workshops and was conducted under the auspices of the Society of Environmental Toxicology and Chemistry (SETAC), an international professional society, and the SETAC Foundation for Environmental Education. SETAC purposely brings together environmental scientists and engineers involved in the use, protection, and management of the environment from diverse backgrounds in academia, government, industry, and public-interest groups. The SETAC Foundation is a nonprofit organization that supports publications of scientific value, training and education, and information on environmental toxicology and chemistry.

Workshop Process

Key scientists from the world scientific community were invited to attend a workshop in Aliston, Ontario, to develop a science-based assessment of issues regarding chlorine and chlorinated compounds in the environment. The specific objectives of the workshop were to

- assess and characterize the risks from OC compounds in the environment,
- identify risk management options that will address the issues of OC compounds in the environment,
- develop strategies and identify research needs that address critical issues in the ecotoxicology of OC compounds, and
- communicate the results of the workshop.

To clarify the major objectives of the workshop, several key presentations were given by invited speakers. The participants were then divided into several workgroups, where the topics in the following chapters were addressed. The tasks of the workgroups were to prepare outlines for the chapters and expand these into more comprehensive documents. During the workshop, summary workgroup presentations were made to the entire group in plenary session, and comments from the participants were incorporated into the developing document.

After the workshop, a small group of editors combined the individual sections into the chapters that follow. The task of the editors was to arrange the information in a logical sequence while preserving the essence of consensus reached at the workshop. The editors gratefully acknowledge the contributing authors for their helpful and constructive comments and suggestions.

CHAPTER 1

Sources, History, and Uses of Chlorine

Although anthropogenic sources of organochlorines (OCs) have received the bulk of scientific attention, OCs are not produced solely by human activity. A diverse array of OCs is produced and released into the environment by natural processes or organisms in both aquatic and terrestrial settings. Natural production has occurred for millions of years with no apparent adverse effects nor evidence of high persistence. Many of these natural compounds are identical or structurally similar to anthropogenic OC compounds.

Humans have used chlorine as chloride (salt) for thousands of years but have used molecular chlorine only since the late 1700s. The use of chlorine grew rapidly from the late 19th century to well into the mid-20th century. Chlorine is now used worldwide in 4 broad categories: 1) as a powerful oxidant, 2) in the synthesis of chlorinated organic compounds, 3) as an intermediate in the production of nonchlorinated organic chemicals, and 4) in the production of inorganic chemicals.

Chlorine gas production exceeds 28 million metric tons per year worldwide, resulting in a spectrum of inorganic products such as titanium and titanium dioxide, hydrochloric acid, polysiloxanes, bromine, sodium and calcium hypochlorites not covered as bleaches, and both aluminum and ferric chloride salts.

Chlorine's oxidizing power has major applications as a bleach for pulp and paper and as a disinfectant for water and sewage. The chemical industry produces a significant number and variety of chlorinated organics, ranging from polyvinylchloride (PVC) to solvents to specialty chemicals and pharmaceuticals. Significant quantities of chlorine are used as critical intermediates for other organic chemicals, including phosgene, epichlorohydrin, chlorohydrin for propylene oxide, and critical synthetic agents such as Grignard reagents.

Industrial chemical production inevitably leads to the formation of byproducts and the release of some emissions of both byproduct and intended product. During the past decade, increasing regulations have accelerated the decline in direct OC and other emissions. Important sources of various persistent and bioaccumulative chlorinated byproducts have been identified, and most have been addressed by either process alternations or production bans on the parent compound.

CHAPTER PREVIEW

Natural Sources of Organochlorines

Although anthropogenic sources of organochlorines (OCs) have received the bulk of scientific attention, OCs are not produced solely by human activity. A diverse array of OCs is produced and released into the environment by natural processes or organisms in both aquatic and terrestrial settings. This production has occurred for millions of years without apparent adverse effects. There has been no observed accumulation of these compounds in environmental compartments, suggesting that they are not highly persistent. Many of these compounds are either identical to or structurally similar to those OC compounds produced by humans.

Because sources of OC compounds include fungi and bacteria, the natural production of OCs apparently has ancient origins. All 4 of the major halogens—fluorine, chlorine, bromine, and iodine—can occur in natural organic forms, and the structure of the compounds produced can range from simple alkanes to complex macromolecules (Gribble 1992, 1994). The identified natural OCs now exceed 1500 compounds and include a number of compounds that are polychlorinated (Gribble 1994). A diversity of species in terrestrial and aquatic environments are capable of producing these halogenated compounds in small quantities (µg/kg).

Several natural organohalogens, however, are produced in significant amounts. Chloromethane, whose atmospheric concentration is greater than any other organohalogen, is almost exclusively of natural origin (Harper 1993). Chlorinated humic substances are also naturally produced in large quantities (Asplund and Grimvall 1991). On a global scale, organically bound chlorine detected as part of total organic halogens in soil and sediments is almost exclusively of natural origin, and nonspecific chlorination processes involving the formation of active chlorine are responsible for this large-scale chlorination of organic matter (Asplund et al. 1994). Chlorophenolic structures that are similar to compounds released from pulp mills occur in natural humus (Dahlman et al. 1993), and there is strong evidence that 2,4,6-trichlorophenol and other chlorinated phenols are naturally produced (Hodin et al. 1991; Grimvall et al. 1994). Even in heavily industrialized watersheds such as the Rhine River, a considerable percentage of the detectable adsorbable organic halogen (AOX) appears to be from natural sources (Van Loon, Boon, de Jong, and de Groot 1993). The natural production mechanism appears to involve haloperoxidase enzymes (Neidleman and Geigert 1986). First discovered in the late 1950s (Shaw and Hager 1959), these enzymes are involved in the production of a variety of volatile and phenolic OC products (Walter and Ballschmiter 1991, 1992). No significant accumulation of these natural OCs has been observed in the environment, leading to a presumption that these compounds are degraded in the environment and metabolized by organisms.

Natural production of well-known environmental toxicants may also occur. Measurable background concentrations of polychlorinated dibenzo-*p*-dioxins (PCDDs) and polychlorinated dibenzofurans (PCDFs) have been observed in soil samples that were collected and sealed in the 19th century, i.e., before the onset of the large-scale anthropogenic production of chlorine and chlorinated organic compounds. But during the 20th century, the environmental levels increased dramatically above background levels, indicating the overwhelming importance of anthropogenic sources. The issue of whether polychlorinated biphenyls (PCBs) are naturally produced is still unresolved, but if they are, their impact is negligible compared to that of anthropogenic production.

Anthropogenic History and Use of Chlorine

Humans have used chlorine as chloride (salt) for thousands of years, but have used molecular chlorine only since the late 1700s. The use of chlorine grew rapidly from the late 19th century to well into the mid-20th century. Chlorine is now used worldwide in 4 broad categories:

1) as a powerful oxidant such as in water purification,
2) to synthesize chlorinated organic compounds such as polyvinylchloride (PVC) plastics,
3) as an intermediate in the production of hundreds of nonchlorinated organic chemicals such as plastics and pharmaceuticals, and
4) to produce inorganic chemicals such as titanium dioxide.

The evolution of chlorine use has been based on economic and environmental needs.

The use of sodium chloride precedes recorded history, and the use of solutions known to contain hydrochloric acid are recorded in Roman and Arabic texts. Karl Scheele (Sweden) is credited with the discovery of chlorine gas in 1774. Subsequently, Humphrey Davy (England) determined that chlorine was an element and suggested its name from the Greek *chloros* for "green." Since then, numerous applications have been developed for chlorine in industrial chemistry, medicine, and public health. Berthollet (France) first combined chlorine and a solution of potassium hydroxide in 1785 to produce a bleaching solution for cloth, with later applications for plant and wood fibers. The medical use of chlorine began in 1826 with chlorine water in delivery wards to prevent puerperal fever, and chloroform was first used for medical anesthesia in 1847. In the late 19th century, the production of chlorine gas by electrolysis began. Chlorine was first used for water disinfection in 1896 at Pola in Austria-Hungary and in 1897 at Maidstone in England during typhoid epidemics. In 1913, chlorine was first used for the treatment of sewage at Altoona PA. The growth of U.S. production of chlorine was steady and rapid until about 1970 (Figure 1-1), after which production plateaued. World production in industrialized countries has followed a similar pattern. Since 1985, an especially sharp rate of growth has occurred in Asia, driven by PVC production.

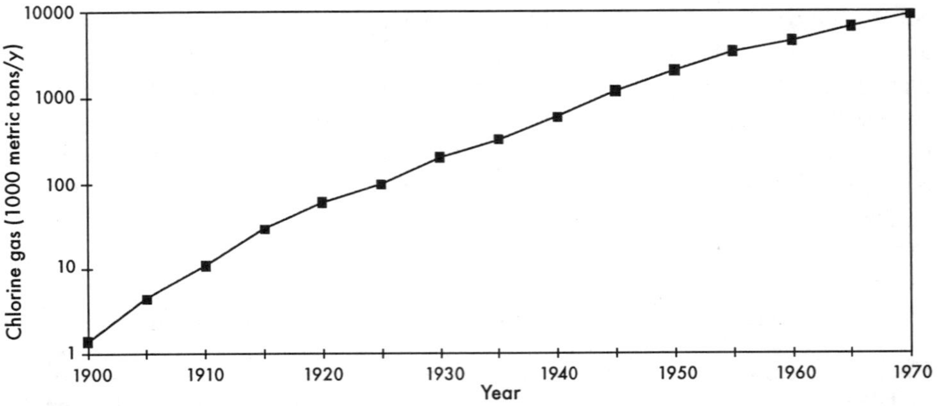

Figure 1-1 History of production volumes of chlorine gas in the U.S.

The primary uses of chlorine have changed over time. Originally confined to wood pulp bleaching and sanitation, chlorine's use has grown to dominate the chemical industry today. In the 1920s and 1930s, the first large-scale production techniques for solvents, vinyl chloride, PCBs, silicones, and other materials were developed. After World War II, the uses of chlorine in the chemical industry shifted to these new products, resulting in steady increases in chlorine use for several decades (Figure 1-2).

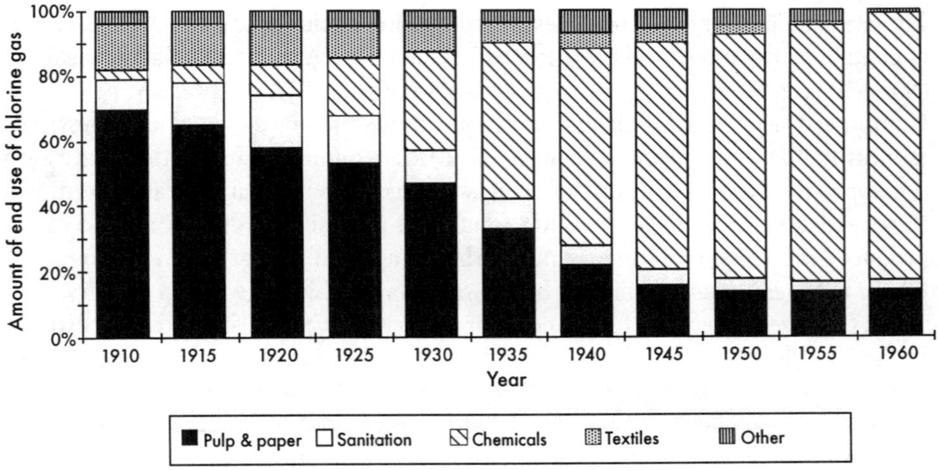

Figure 1-2 History of end use of chlorine gas in the U.S., showing changes in applications over time

In the 1960s and 1970s, the recognition of certain chlorinated chemicals as persistent, bioaccumulative, toxic substances led to production bans in most industrialized countries for PCBs, dichlorodiphenyltrichloroethane (DDT), dieldrin, mirex, toxaphene, and other chlorinated pesticides. In the late 1980s, with the signing of the Montreal Protocol to protect stratospheric ozone, the phase out of chlorofluorocarbons (CFCs) and other volatile and stable chlorine and bromine compounds began. There currently are 4 major anthropogenic applications of chlorine:

1) Chlorine is contained in the final organic product such as PVC and chlorinated solvents.
2) Chlorine is used for its oxidizing potential, and the chlorinated reaction products can be released to the environment, such as in pulp bleaching, water disinfection, and sewage treatment.
3) Chlorine chemistry is used as a process intermediate in the synthesis of non-chlorinated materials such as urethanes, polycarbonates, and other organic chemicals.
4) Chlorine chemistry is used in the production of inorganic chemicals such as titanium dioxide, silicones, and ferric chloride.

Overall, chlorine chemistry has literally thousands of individual uses and applications.

Natural and anthropogenic production of chlorine gas

Chlorine gas is produced naturally and anthropogenically. Human-made production exceeds 28 million metric tons per year worldwide. Production is currently concentrated in North America, Western Europe, and Japan (see Figure 1-3 for 1992 estimates). There are multiple anthropogenic end uses and applications for chlorine.

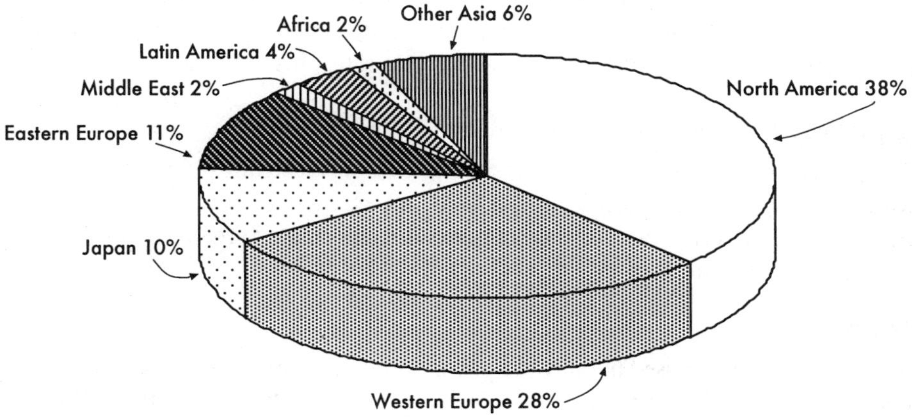

Figure 1-3 Worldwide production of chlorine gas in 1992. Total 34 million metric tons. Redrawn with permission from Curlin and Bommaraju 1993.

Elemental chlorine results from both natural and anthropogenic production. At the ocean–atmosphere interface, trace amounts of chlorine gas are produced naturally as the result of the reaction between tropospheric ozone and chloride ion in sea spray (Keene et al. 1993; Pzenny et al. 1993). Anthropogenically, the electrolysis of brine produces chlorine gas, sodium hydroxide, and hydrogen gas:

$$2\,NaCl + 2\,H_2O \overset{energy}{\rightarrow} 2\,NaOH + Cl_2 + H_2 \qquad \text{(Equation 1-1).}$$

Three processes are now in use to make chlorine gas: 1) the mercury cell process (~ 15% of U.S. capacity), 2) the membrane process (~ 5% of capacity), and 3) the diaphragm process (~ 80% of capacity). Both the mercury cell and diaphragm processes were developed and in use prior to 1900. In contrast to the early dominance of the diaphragm process in North America, chlorine production in Europe and Japan relied primarily on the mercury cell process. For new production, current economics and the need to reduce mercury emissions favor the membrane process. In the 1970s, all 3 production processes switched from graphite electrodes to ruthenium or titanium anodes because of improved energy efficiencies and to prevent chlorine–carbon contact that would produce chlorinated organics as impurities. Significant chlorine production is captive, meaning that the chlorine is used at the site where it was manufactured. However, chlorine can be liquefied and shipped short distances in pipelines or long distances in railroad tank cars or other pressurized containers.

Chlorine gas is used in numerous processes on every continent (for end uses in the U.S. during 1992, see Figure 1-4). In sanitation, chlorine use for sewage treatment has expanded with the growth of municipal treatment systems that control the release of coliform bacteria. In the pulp and paper industry, chlorine gas is being replaced by chlorine dioxide and other bleaching agents. Within the chemical industry, the direct use in CFC refrigerants is rapidly decreasing, and the use of chlorinated solvents is in a slow, gradual decline. Simultaneously, new applications of chlorine emerge, such as in CFC substitutes that degrade before reaching the stratosphere.

Application of chlorine as an oxidant

Chlorine's oxidizing power has 3 major applications: 1) bleaching pulp and paper, 2) disinfecting water, and 3) disinfecting sewage. In each case, the reactant organic material (wood or microbial cells) is complex, and chlorinated organic byproducts may reach the environment.

Pulp and paper

Chlorine gas has been used in pulp and paper bleaching to remove residual lignin from fiber after the initial kraft and sulfite "cooking" processes. Although > 90% of the chlorine is typically converted to chloride ion within the process, many OC compounds are created as a result of the complex organic materials in trees. Almost all

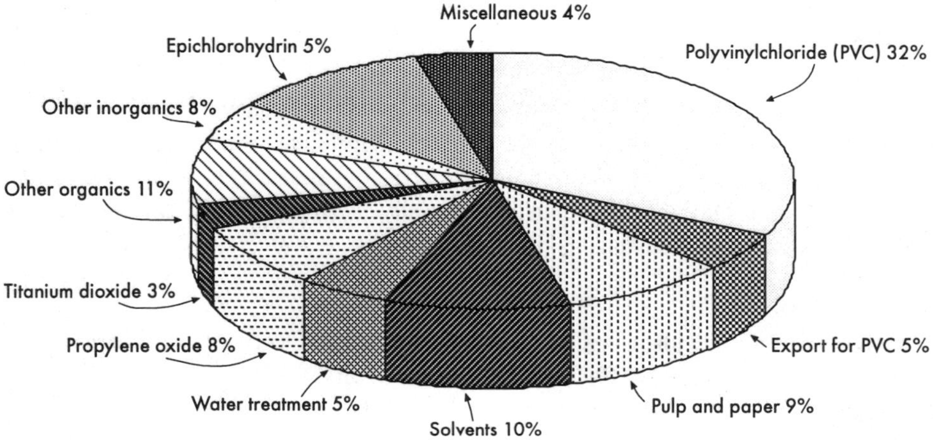

Figure 1-4 End use and applications of chlorine gas in the U.S.

aromatic rings in the residual lignin are destroyed during chlorine bleaching, and a significant number of carboxyl groups are created, conferring water solubility on the reaction products. Less than 1% of the OC material in pulp effluents appears to be hydrophobic (< 0.1% has a log K_{OW} > 3). The use of chlorine in pulp and paper bleaching has declined steadily for these reasons:

1) improved delignification "cooking" processes and the additional step of oxygen delignification reduce the amount of residual lignin before bleaching, thereby reducing the amounts of bleaching chemicals needed, and
2) alternative bleaching chemicals are used to avoid the production of any measurable traces of 2,3,7,8-tetrachlorodibenzo-*p*-dioxin (2,3,7,8-TCDD) and 2,3,7,8-tetrachlorodibenzofuran discovered in bleach plant effluents in the late 1980s.

Water and sewage disinfection

Chlorine has major public health applications for potable water disinfection and sewage disinfection. OC formation in these processes is reduced by water filtration and by the settling or removal of biosolids before the addition of chlorine. In the case of water purification, natural humic acids appear to be the primary source of OC byproducts. Economy, broad efficacy, ease of use, and, especially, the ability to deliver an effective disinfecting residual within the distribution system compared to chemical alternatives all favor chlorine use in both the developing world and industrialized nations.

Additional chlorine oxidant applications include household and commercial laundry bleach (sodium hypochlorite), bleaching applications in fiber and textile processing (cotton bleaching), surface disinfectants (hospitals and food processing), and swimming pool treatments.

Application of chlorine to produce chlorinated final products

A significant number and variety of chlorinated organics are produced in the chemical industry. They range from PVC (millions of metric tons per year) to solvents (several hundreds of thousands of metric tons) to specialty chemicals and pharmaceuticals (from thousands of metric tons to kilograms).

Polyvinylchloride

The dominant worldwide use of chlorine is for the synthesis of vinyl chloride monomer (VCM) used to produce PVC plastics. Because of its stability and processing flexibility, PVC is one of the most versatile thermoplastics available. PVC use accounts for about 28% of the world's annual chlorine gas consumption (U.S. use is greater, at 32%), and its 1993 production on all continents approached nearly 19 million metric tons .

In the 1930s, the first commercial production of VCM was based on the reaction of HCl with acetylene, but production was constrained by the price of acetylene. With the availability of large volumes of an alternative, cheaper ethylene feedstock after World War II, VCM production rapidly increased. VCM production is based on 3 separate process steps using 2 different ways to synthesize 1,2-dichloroethane or ethylene dichloride (EDC). In the first method, EDC is produced through the direct chlorination of ethylene. In the second method, EDC is produced by an oxychlorination reaction of ethylene with hydrochloric acid and oxygen. Finally, the production of VCM from EDC is achieved by pyrolysis, which releases hydrochloric acid. These 3 steps are combined in industrial production. The HCl acid released from pyrolysis is reused in a "balanced" process, in which half of the EDC comes from an oxychlorination step using the HCl and the other half of the EDC is from direct chlorination:

Direct chlorination	$CH_2{=}CH_2 + Cl_2$	$\rightarrow$	$ClCH_2CH_2Cl$
Oxychlorination	$CH_2{=}CH_2 + 2\ HCl + 1/2\ O_2$	$\rightarrow$	$ClCH_2CH_2Cl + H_2O$
Pyrolysis	$2\ ClCH_2CH_2Cl$	$\rightarrow$	$2\ CH_2{=}CHCl + 2\ HCl$
Balanced VCM	$2\ CH_2{=}CH_2 + Cl_2 + 1/2\ O_2$	$\rightarrow$	$2\ CH_2{=}CHCl + H_2O$

The PVC process has been continually modified for economic, environmental, and safety reasons. Of particular importance is the reduction of worker exposure to VCM. Examples include the use of oxygen streams rather than direct air in the oxychlorination step, which 1) significantly reduces total air emissions (by > 20 X) and 2) reduces the energy required to strip this waste stream. Other examples include hydrogenation of acetylene in the HCl recovery stream (< 0.2 mol. %), which eliminates a major pollution source of aromatic chlorinated tars from the heavy ends recovery step.

Overall, the modern focus is on capture and use of such byproducts as 1,1,2-trichloroethane and on incineration or oxidation of both gaseous and liquid wastes to eliminate OC emissions. In some PVC production facilities, a nearly complete elimination of OC emissions has been achieved, as these residuals are subjected to a fluidized bed, catalytic oxidation resulting in the formation of HCl, H_2O, N_2O, O_2, and CO_2 gases (Kirk-Othmer 1993). It should be emphasized that no 2 plants have identical processes, controls, or environmental equipment, so releases and byproducts vary among PVC production sites.

The wide application of PVC as a major industrial commodity is due not only to its chemical stability but also to the broad range of properties in its finished products, resulting from 1) the availability of several polymerization processes, 2) variables within the polymerization conditions of each process, 3) numerous compound agents to modify properties, and 4) unique post-polymerization chemical modifications. Hence, PVC can be processed into almost any shape, ranging from highly flexible wire insulators to rigid pipe. The construction industry is the largest end user in North America, using almost 60% of PVC production in pipe, siding, flooring, window and gutter extrusions, and other building materials. A growing area of PVC use is in automobile production, where light weight and durability are critical to fuel economy. The proportion of these PVC applications is shown in Figure 1-5.

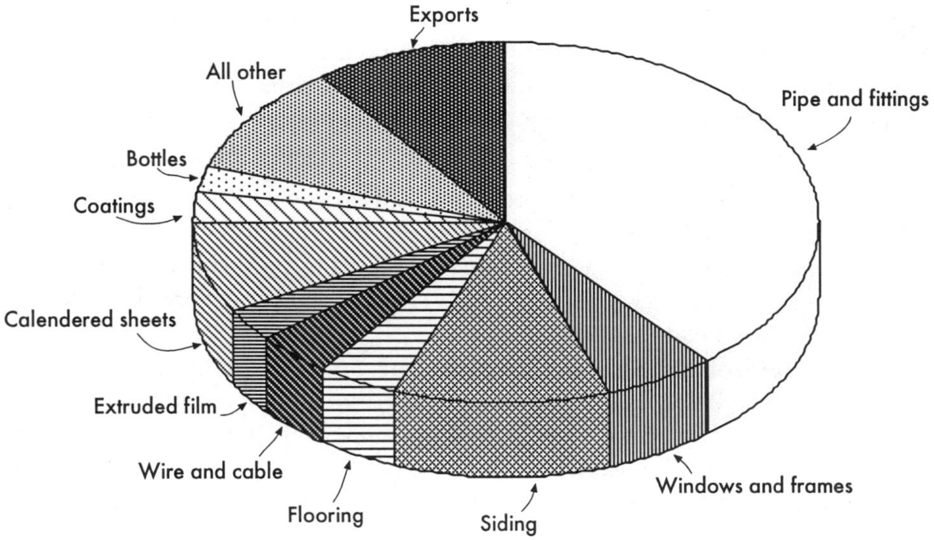

Figure 1-5 End uses and applications of PVC in the U.S. for 1993

Polyvinylidenechloride

A second polymer, polyvinylidenechloride, combines chemical stability with low permeability to most gases and vapors. About 150,000 metric tons of the 1,1-dichloroethylene monomer are produced globally each year. Polyvinylidenechloride end uses range from a copolymer with PVC for certain specialty resins and pipes to special packaging films that preserve labile foods and pharmaceuticals, and industrial coatings where chemical resistance and barrier properties are important.

Neoprene or chlorobutyl rubber

A third chlorinated polymer is neoprene rubber. This polymer is produced from chloro-1,3-butadiene monomer, and its special properties — thermal stability, oil resistance, and flexibility — are superior to other rubber materials and elastomers. A major end use is an isoprene-isobutylene copolymer for the inner lining of radial tires, where its low air permeability and heat and flexural resistance add significantly to a tire's life. Overall, nearly 400,000 metric tons of isoprene are produced globally each year.

Solvents

The production of chlorinated solvents is normally an integrated process in which the byproducts of one reaction are purified to become either a final co-product of the overall scheme or an intermediate for a subsequent reaction. In all cases, there are a significant number of site-to-site variations in the actual chemical process or reaction steps such as reactor design and conditions. This is due to the economics, technical and engineering experience of particular manufacturers, patents for specific reaction designs, and successive technical improvements in newer plants versus older ones.

The choice of a chlorinated solvent depends upon intended application, desired combination of chemical properties, economy, and environmental safety considerations. Chlorinated solvents provide a range of needed properties such as greater solvent power, low flammability, and energy efficiency in removal of chlorine from process streams and in recycling and recovery steps. Briefly, the major chlorinated solvents and their end uses are these:

- **tetrachloroethylene** or **perchlorethylene**, primarily for dry cleaning of clothing and for limited metal cleaning and degreasing applications. Recently, demand has declined as recycling and emissions containment are implemented. Perchlorethylene provides a nearly universal, high solvent power. It is also nonflammable and is energy efficient to use and to recover, compared to water or other substitutes.
- **carbon tetrachloride**, primarily (> 90%) as feedstock for refrigerant single-carbon (C_1) CFCs, demand for which is rapidly declining as a result of the phaseouts negotiated under the Montreal Protocol. Carbon tetrachloride was removed from U.S. consumer products in the early 1970s. Some specialty uses exist as a process solvent.

- **trichloroethane**, as a major industrial solvent and degreasing substance. 1,1,1-trichloroethane combines low toxicity, low reactivity, minimum flammability hazard, and excellent solvent capabilities. Under the Montreal Protocol (United Nations Environmental Program [UNEP] 1987), production and use of trichloroethane will be restricted.
- **methylene chloride**, in paint stripping formulations and as both a process solvent and an extractant. The latter use includes processing food spices, certain plastics, and pharmaceuticals. Methylene chloride is also used in metal cleaning and degreasing applications. Use as an aerosol propellant has been eliminated.
- **methyl chloride**, as a key intermediate in the production of other C_1 chlorinated hydrocarbons. The largest single application is in silicone chemistry, and methyl chloride has significant use as a chemical methylating agent. Most methyl chloride production is captive, with on-site use.
- **trichloroethylene**, as the favored solvent in vapor phase degreasing and cleaning applications because of its solvent properties, lack of flammability, and high volatility. Minor solvent and specialty process applications also exist.
- **chloroform**, whose primary use (> 80%) was for the production of C_1 CFCs, which are now being phased out. Chloroform is also used for synthesizing fluorochemicals and fluoropolymers and as a specialty solvent, often in pharmaceutical applications.
- **chlorobenzenes**, as sources of particular chlorinated anilines, which are then intermediates for the production of herbicides, bacteriostatic substances, and dyes. Currently, the primary chlorobenzenes produced are the *mono*, the *ortho*-, and the *para*-dichloro. Monochlorobenzene is a feedstock for nitrochlorobenzene, which is used in dye intermediates. Chlorobenzenes have limited specialty solvent applications. In the case of *p*-dichlorobenzene, it is an intermediate for the production of specialty polyphenylene sulfide resins.
- **chlorotoluenes** and **benzylchloride** (ring-chlorinated and methyl-chlorinated toluenes), as intermediates for other feedstock chemicals, including benzyl alcohol and benzaldehyde. The ultimate end products include a large array of perfumes, flavorings, benzyl peroxide, UV-stabilizers, pharmaceuticals, bactericides, and other organic materials.
- **chlorinated paraffins** (CPs), as the last major group of relatively high-molecular-weight chlorinated hydrocarbons deliberately produced as final, commercial products. A variety of CPs may be produced, depending upon the hydrocarbon chain length and the chlorine substitution. The alkyl chain lengths vary from C_{10} to over C_{30}, and the chlorine substitution varies from 35 to 70%. Moderately important applications of CPs as plasticizers exist in plastics and paint formulations, metal-working lubricants, adhesive formulas, and fire-retardant formulas.

- **refrigerants**, as replacements for CFC, produced from chlorine-containing feedstocks in which hydrofluoric acid is used to replace chlorine. These refrigerants are designed to degrade rapidly in the environment. One or more hydrogen atoms are left attached to the carbon skeleton, providing reactivity with the hydroxyl radical to reduce atmospheric persistence. Fluorine is not reactive with ozone in the stratosphere.

Application of chlorine to produce intermediates for nonchlorinated final products

Significant quantities of chlorine are used as critical intermediates for other organic chemicals. Examples include phosgene, the initial intermediate to produce isocyanate feedstocks for both polyurethanes and polycarbonates, epichlorohydrin for resins, chlorohydrin for propylene oxide used to produce polypropylene, specialty chemicals and pharmaceuticals, and critical synthetic agents such as Grignard reagents.

Phosgene

The combination of carbon monoxide and chlorine yields phosgene ($COCl_2$), which is a highly versatile material by virtue of its reactions with a range of organic and inorganic compounds. Phosgene is manufactured and used exclusively on-site because of its toxicity and reactivity. The primary end use of phosgene (> 80%) is in reaction with amines to produce isocyanates:

$$RNH_2 + COCl_2 \rightarrow RNCO + 2HCl \qquad \text{(Equation 1-2).}$$

The isocyanate is itself a chemical intermediate that is reactive with active hydrogens (H_A) on other organic compounds to produce diisocyanates:

$$R_1NCO + H_AR_2 \rightarrow R_1NHC = OR_2 \qquad \text{(Equation 1-3).}$$

Such aliphatic and aromatic diisocyanates form the basis for polyurethane production in which a diisocyanate reacts with the terminal hydroxyl groups of polyether and polyester materials to form linear and branched polymers. Such materials, when combined with process and compounding aids, result in flexible foams, rigid foams, coatings, adhesives, and elastomers. Global production of urethanes is well over 2 million metric tons.

About 10% of phosgene is used for the production of polycarbonate plastics, which are polyester-based polymers:

$$n\,OHROH + n\,COCl_2 \rightarrow OH[RCO]_n + 2n\,HCl \qquad \text{(Equation 1-4).}$$

The hardness, transparency, thermal stability, and other engineering properties of these plastics have led to polycarbonate applications, from break-resistant windows to goggles and face shields to automotive bumpers. Global production of polycarbonates approaches 200,000 metric tons/y. The remaining 10% of the total phosgene is used in the production of numerous individual chemicals, including herbicides, pharmaceuticals, and specialty chemicals.

Chlorohydrins

An organic epoxide structure can be created by removal of a chlorine and a hydroxyl hydrogen when the chlorine and the hydroxyl are on adjacent carbons. The resulting epoxide can then be used as an intermediate for the synthesis of other chemicals. The 2 largest end uses for the chlorohydrin reaction are 1) the production of propylene oxide that is used primarily for polypropylene oxide polymers and 2) the synthesis of epichlorohydrin that is used in the manufacture of numerous resins and polymers, as it is a bifunctional reagent (the epoxide and the chlorine). Numerous minor commercial applications exist for synthesizing specialty chemicals using chlorohydrins such as alkanolamines.

Cellulose derivatives

Derivatized cellulose chemicals are manufactured primarily through chlorine chemistry. Methyl and ethyl cellulose are synthesized by methyl and ethyl chloride reaction with hydroxyl groups on the cellulose. Carboxymethylcellulose derivatives are synthesized using chloroacetic acid. Hydroxypropylcellulose is produced from propylene oxide that is derived from chlorohydrin reactions. Variations in both cellulose polymer length and the degree of substitution result in the production of an array of materials, from cellulose acetate to viscosity control agents for aqueous and solvent solutions.

Nylon-6,6

Nylon-6,6 is a copolymer of adipic acid and hexamethylenediamine and is the primary nylon used for carpeting and numerous other applications. Significant hexamethylenediamine is derived from 1,4-dichloro-2-butene (from the chlorination of 1,3-butadiene).

Fluoropolymers

Production of almost all fluoropolymers begins with chlorinated materials. Polytetrafluoroethylene (PTFE or Teflon) is produced from chloroform by reaction with hydrogen fluoride followed by condensation to form tetrafluoroethylene for polymerization. The inert nature of PTFE has led to its extensive use in seals, gaskets, packings, and reactor and pipe coatings, where chemical resistance is at a premium. Global PTFE production exceeds 50,000 metric tons/y.

Grignard reagents and Friedel-Crafts reactions

The Grignard reaction is a widely used synthetic reaction for producing pharmaceuticals and other critical chemical intermediates. Basically, an organohalogen is reacted with magnesium, usually in an anhydrous ether solvent. Although reactivity decreases in the order I > Br > Cl (fluorine is inert under most conditions), the OCs usually are the commercial compounds of choice, based on their cost. The synthesized Grignard reagent is reactive with a spectrum of organohalogens, carbonyls, carbon–carbon double bonds, nitrites, oxygen and sulfur, epoxides, and other groups.

The Friedel-Crafts-type reaction, using an aluminum, ferric, or zinc chloride catalyst, may be the most versatile synthetic process in organic chemistry. The original reaction involves the alkylation of an aromatic nucleus with alkyl halides and alcohols. This has been expanded to include a number of condensation reactions between various aromatic and aliphatic compounds, isomerization reactions, polymerizations, and acylation of aromatics. Additional derivatizations and additions are literally too numerous to mention. The Friedel-Crafts reactions comprise a cornerstone of the organic chemical industry from large-scale process to specialty chemicals to pharmaceuticals. In fact, over 85% of all pharmaceuticals and agricultural chemicals utilize chlorine chemistry somewhere in their synthetic pathway.

Application of chlorine to produce inorganic substances as final products

Chlorine gas is used to produce a spectrum of inorganic products such as titanium and titanium dioxide, hydrochloric acid, polysiloxanes, bromine, sodium and calcium hypochlorites not covered as bleaches, and both aluminum and ferric chloride salts.

Titanium dioxide

Titanium dioxide, which provides paints with their base white color and covering ability and is also the primary agent to color transparent plastics, is produced using chlorine. This is also the production source for most purified titanium metal. The basis is the selective fractionation of titanium from its metal ores, using chlorine and carbon in a fluidized bed reactor:

$$TiO_2 \text{ (as ore)} + 2\,Cl_2 + C \rightarrow TiCl_4 + CO_2 \qquad \text{(Equation 1-5).}$$

The volatile titanium chloride is collected and purified by distillation. The titanium tetrachloride is then subjected to oxidation to yield pure titanium dioxide, and the chlorine gas is recycled. In the case of the purified metal, magnesium or sodium metal is used as a reductant. Well over 1 million metric tons of titanium dioxide are produced globally each year, with ~ 50% produced using chlorine. Approximately 150,000 metric tons of titanium metal are produced using the chlorine process.

Hydrochloric acid

Hydrochloric acid (HCl) is generated as a byproduct from several reactions. Production of HCl from the replacement of chlorine with fluorine in the production of CFCs is decreasing. The largest end use of HCl is the surface treatment of steel and iron metal products, referred to as "pickling." Hydrochloric acid is a major reactant in chlorine chemistry, with several structures containing carbon–carbon double and triple bonds.

Other uses include pH adjustment, ion exchange regeneration, production of high-fructose corn syrup, and enhancing petroleum recovery from carbonate rock. Global production exceeds 5 million metric tons/y.

Siloxanes

Silicone chemistry is based upon the reaction of methyl chloride with silicon. The methylchlorosilane materials then are used to produce siloxanes directly. Both linear and cyclic products can be produced. The polymerization reaction may be characterized as

$$R_3SiCl + H_2O \rightarrow R_3SiOH + HCl \qquad \text{(Equation 1-6)},$$

and, where one R is Cl, then

$$R_3SiOH + R_3SiCl \rightarrow R_3SiOSiR_3 \qquad \text{(Equation 1-7)},$$

with the process of hydrolysis and polymerization then repeating if one R is Cl (a broad functional range for R exists, but methyl groups are most common).

Nonmethyl organic side chains that produce siloxanes with an additional spectrum of properties can be formed either by direct reaction of the organic with the reactive chlorine–silicone bond or by Grignard reagents. Additional organosilicone materials may result from attaching polyethers through the terminal hydroxyl group. In all cases, polymer cross-links and branches can be introduced to create a diverse family of elastomeric materials. Silicones are used in fluid lubricants, hydraulic fluids, heat exchange fluids, construction sealants, medical products, fluid antifoaming agents, water repellents, electrical insulators, certain tubing and gasket products, adhesives, release agents, and other uses. Global production approaches 400,000 metric tons/y.

Aluminum

Aluminum chloride is manufactured by several processes, including the direct reaction of chlorine gas with molten aluminum metal or aluminum in a fluidized bed reactor. Aluminum chloride hydrates are used in textile finishes and as the starting material for specialized aluminum refractory materials. The basic aluminum chlorides, a combination of hydroxide and chloride salts, are used in deodorants and antiperspirants and as part of flocculating agent formulas for water and sewage treatments.

Ferric compounds

Relatively water-soluble ferric chlorides are key components in the tertiary treatment of sewage to remove inorganic phosphorus, which promotes eutrophication when released to surface waters. When ferric compounds are combined with flocculating agents, phosphorus concentrations in effluents of less than 1 mg/L can be achieved. Ferric chloride sources range from the recycling of metal surface treatment waste streams to reactions of chlorine gas with metal filings.

Phosphates

Combining phosphorus and chlorine to produce phosphorus trichloride is the initial step in several pathways to synthesize organophosphorus compounds, including detergents, dispersants, and corrosion inhibitors, based on linear alkyl and also aryl groups such as in phosphites and phosphonates. Further conversion of the trichloride to phosphorus oxychloride produces additional organophosphorus esters and amides.

Sulfur

Thionyl chloride ($SOCl_2$) and sulfuryl chloride (SO_2Cl_2) are used to produce acid chlorides, alkyl chlorides, and other organic intermediates for synthetic schemes, as well as sulphonated specialty chemicals. The combination of lithium and thionyl chloride is the basis for small, high-energy batteries.

Releases of Anthropogenic Organochlorines to the Environment

Industrial chemical production inevitably leads to the formation of byproducts and the release of some emissions of both byproduct and intended product. During the past decade, increasing regulation has accelerated the decline in direct OC and other emissions. In addition, important sources of various persistent and bioaccumulative chlorinated byproducts have been identified, and most of their emissions have been addressed by either process alterations or production bans on the parent compound.

Anthropogenic input to the environment

Manufactured chemicals such as OC compounds can reach the environment in several ways. One significant group of chemicals is produced in large volumes for deliberate release into the environment: agricultural chemicals such as insecticides and herbicides. Some of these compounds contain chlorine. Other intentionally produced chemicals include manufactured intermediates and end products that are released at manufacturing sites in a manner dependent upon the process and the engineering controls. Certain processes such as pulp manufacture are relatively open and result in significant releases, while other processes are increasingly tightly closed. In the past decade, the formation and emission of unwanted byproducts from the chemical industry in general has decreased. For example, in the U.S., atmospheric releases of Toxic Release Inventory (TRI) compounds have fallen by nearly one-third since the program began in 1988 (USEPA 1994a). One example of the TRI emissions data trend is shown in Figure 1-6. "Fugitive" emissions are created by leakage from liquids that are pumped through pipes and valves with the use or displacement of air streams and during the filling of tanks and shipping containers. Final disposal of nonrecyclable manufacturing waste streams is moving toward such on-site processes as incineration

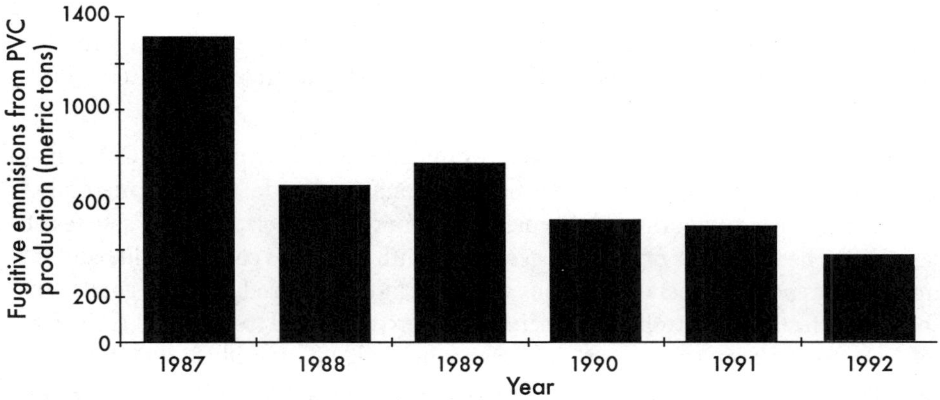

Figure 1-6 Toxics Reduction Inventory (TRI) emission trends for a major U.S. PVC producer in 1000s of kilograms. The TRI occurs under the Superfund Amendments and Reauthorization Act (SARA 1986) legislation in the U.S.

or direct chemical oxidation, which destroy the byproduct, rather than transferring the waste streams to landfills or deep wells.

Manufacturing, however, is not the only point of release for anthropogenic OCs. Three other important stages potentially lead to environmental releases in the life-cycle of a compound: distribution, use, and disposal after use. In the distribution stage, both storage of chemicals and their transportation by pipeline, rail, road, water, and air are potential sources of fugitive and accidental releases, and they are increasingly regulated. The use and the disposal of manufactured chemicals are important sources of environmental releases. Certain end products such as the chlorinated solvents can be released or lost in significant quantities during the use stage; therefore, proper use and handling practices are important to minimizing releases. The disposal stage is increasingly controlled for all chemicals, encouraging their destruction at well-designed, licensed facilities.

Industrial chemistry is rarely as selective as biological enzymatic processes. Chemical reactions may have several minor and even major side reactions that result in the formation of byproducts, and such formation is usually compounded in the scale-up from laboratory to industrial production. When groups of unwanted byproducts result from incomplete or competing side reactions that occur during production, improved process control and/or process modification is necessary to eliminate or minimize byproduct formation and to recover and dispose of the byproduct. These challenges are magnified when existing equipment must be modified because a retrofit situation usually involves some design and engineering compromises. Thus, the environmental safety and economic challenges for technical process design and engineering are to to minimize the formation of byproducts and their release to the environment.

The environmental inputs of OCs cover an extremely varied group of chemicals. They range from single compounds such as the highly volatile C_1–C_2 chlorinated solvents and water-soluble phenols and pyridines to partially characterized complex mixtures such as pulp bleaching discharges and incineration streams.

Finally, apart from deliberate manufacture or manufacturing byproducts, OCs can also be released in combustion processes, including municipal, industrial, and medical incineration, where chlorinated precursors and inorganic chloride are the sources of chlorine. The sources of organohalogens differ with respect to controllability and temporal trends. In general, emissions related to industrial production of goods can be more effectively controlled than can other sources, and the trend is that emissions of OCs are decreasing and in certain cases are negligible. This is illustrated by the dramatic reduction in OC discharge from the pulp and paper industry during the past few years. Improved delignification and new bleaching technologies have not only reduced the emission of OCs measured as AOX, but the full substitution of chlorine gas for bleaching has virtually eliminated the discharge of toxic PCDD/PCDFs and polychlorinated phenols. Reductions in emissions of PCDD/PCDFs from waste incineration provides another example of successful new technical solutions that will gradually decrease the anthropogenic input of OCs to the environment (Hinton and Lane 1991).

Heterologous emissions of mixtures

The application of chlorine as a general oxidant in the pulp and paper, water purification, and sewage treatment industries results in the creation of mixtures that are emitted to the environment. The reactions of chlorine with lignin and humic acids are, therefore, discussed separately.

Chlorolignin

The product of pulp bleaching, chlorolignin, is a heterogeneous group of heavily oxidized organohalogens. Chlorolignins are produced from oxidation and chlorination of lignin and have far less aromatic character than lignin alone. These products are poorly soluble in nonpolar solvents (Banerjee et al. 1984). The main portion of chlorolignin is water soluble at pH > 8 (to ~ 500 mg/L), but below pH 7, it forms colloids and micelles; at lower pH, it precipitates and becomes increasingly soluble into nonpolar solvents (up to 5% in heptane at pH 2), while approximately 70 to 100% is soluble in tetrahydrofuran at pH < 4. Less than 3% of the total organically bound halogens in these effluents are known low-molecular-weight compounds, such as chlorophenols, chloroguaiacols, and chlorinated resin or fatty acids. Less than 1% is volatile organic halogen. Because the quantitative chemical composition and the chemical structure of the molecules are unknown, chlorolignin and other components are usually quantified as AOX-Cl, i.e., active carbon-adsorbable organic chlorine. In the few circumstances where bleached pulp is still produced with elemental chlorine, chlorolignins are likely to represent the largest emission of organic halogens into the environment. The chlorolignins adsorb to (but do not greatly accumulate in) sediment and microbial biofilms in the aqueous ecosystem.

A small part of the chlorolignin discharge (1 to 5%) is soluble in nonpolar solvents; < 1% is formed of volatile halogen compounds, the main part of the AOX being represented by rather hydrophilic compounds and approximately 70 to 100% being soluble in tetrahydrofuran. Of the chlorolignins discharged by bleached pulp mills, 30 to 70% is biodegradable by biological wastewater treatment. The remainder is very persistent, with a half life of > 1 year in water ecosystems (Saski et al. 1996).

In the older literature, the molecular size of chlorolignins from bleached kraft pulp production was estimated to be in the 800 to 10,000 g/mol range. It was assumed that molecules in this size range were too large to be bioavailable. Recently, however, these size ranges have been questioned. For example, the molecular size of chlorolignins from some bleached kraft mills was found to range from 100 g/mol to 5000 g/mol and up to 25,000 g/mol from bleached sulfite pulp mills (Pellinen et al. 1985; Jokela et al. 1992; Van Loon, Boon, and de Groot 1993). The molecular weight (MW) averages from 200 to 600 g/mol for bleached kraft pulp chlorolignin from different mills. Approximately 95% of the material has a molecular weight between 100 and 2000 g/mol (Jokela et al. 1992). The chlorine-to-carbon ratios of chlorolignins vary from 1:15 to 1:40 (mol/mol) (Bryant et al. 1988; Jokela et al. 1992, 1993) for kraft pulp bleached with elemental chlorine but are lower for pulp bleached with chlorine dioxide, where they vary from 1:100 to 1:200 (Lindström and Österberg 1984; Österberg and Lindström 1985a, 1985b; Solomon et al. 1994).

Water purification and sewage treatment

The organohalogen substance formed during chlorine disinfection of humic-rich raw water is called "chlorohumus." This substance is similar to chlorolignin, and no way of differentiating between the two has been reported for situations where both contribute to the organohalogen content of drinking water (Jokela et al. 1992). The chlorohumus lignins are relatively hydrophilic, with no known tendency to bioaccumulate. Nevertheless, they are absorbed by humans from consumed drinking water (Salkinoja-Salonen and Jokela 1991). The major portion of the resorbed organohalogens appear to be excreted in urine, and chlorine disinfection of sewage also adds to this load.

Soil residues

Chlorinated soil-bound residues have been observed. It is often thought that such residues may be of large molecular weight and poor solubility and therefore not accessible to bioaccumulation, but few studies support this. The origin of these residues has not been defined, but it may be both natural and anthropogenic. Bound residues formed in soil from chlorophenols used for wood, timber, and lumber treatment have been studied (Middeldorp et al. 1992; Laine et al. 1994). The soil-bound organic halogens, extractable into tetrahydrofuran or acetone, occur as halogenated molecules of 200 to 1000 and 200 to 10,000 g/mol, the larger sizes being found in soil with more than 5% organic matter.

Persistent and bioaccumulative organochlorine substances

The subset of persistent and bioaccumulating chlorinated substances have been identified as requiring special attention. These include PCBs, DDT, PCDD/PCDFs, dieldrin, aldrin, chlordane, endrin, heptachlor, hexachlorobenzene (HCB), mirex, and toxaphene (Ritter et al. 1995). Since 1970, the use of these compounds has been increasingly restricted and their production banned in many, but not all, nations. A brief summary of legislative and regulatory action for some of these compounds in Organization for Economic Cooperation and Development (OECD) countries and other industrialized nations is shown in Table 1-1. Table 1-1 should be judged as a representative summary compiled from several sources; however, the current regulatory status in all countries has not been verified. It should be further emphasized that regulatory controls are applied on an individual, national basis despite the transboundary movement of these compounds. Even at the national level, the presence of a regulation does not automatically lead to its efficient and thorough enforcement. In fact, there are cases in which regulations are routinely ignored. Further, while there have been recent efforts to limit or prohibit the export of hazardous wastes from industrial countries, there has been no agreement on how to assist non-OECD countries to replace those compounds that they presently produce, to recover their inventories of PCBs and other compounds, or to remediate contaminated sites in those nations. A United Nations Economic Commission for Europe protocol on persistent organic pollutants is under development (UN-ECE 1994).

The future emission of persistent and bioaccumulative OC compounds is related to sources that are less controllable. This lack of controllability can often be attributed to several factors:

1) the sources are located in countries that do not effectively regulate production;
2) existing stocks and uses (as transformers) of these substances will result in diffuse emissions long after production of the same substances has ceased; and
3) the origin of some chlorinated organics found in the environment is uncertain.

One aspect of the long-range transport of OCs deserves attention: the possible transport of persistent OC pesticides from tropical regions, where they are still widely used, to temperate and arctic regions. The southward shift in production to the tropics was evident in the mid-1970s (Goldberg 1975). The evidence that the Arctic, which has no direct industrial or agricultural sources or in situ contaminants, receives considerable inputs of OC and other chemicals is increasingly apparent (Barrie et al. 1992) and is discussed further in Chapter 2.

Table 1-1 Status of legislative and regulatory action for selected chlorinated pesticides and industrial chemicals

Country	PCBs	DDT	Hexachloro-cyclohexane (HCH)	Chlordane	Toxaphene	Dieldrin	Heptachlor	Endosulfan	Methoxychlor
USA	Manufacture ended 1977	Most agricultural uses prohibited 1972	All isomers except lindane prohibited since 1978[1]	Restricted use 1978; deregistered 1988	Prohibited 1983	Prohibited 1974; no longer produced	Limited use since 1978; deregistered 1988	Registered	Registered
Canada	Use of new product prohibited 1980	Deregistered 1986; most uses phased out 1970 to 1978	Mixed isomers prohibited 1971; use of lindane restricted	Use restricted since 1970; agricultural use eliminated 1985	Phased out in 1970s; deregistered 1982	Restricted 1977	Restricted 1970	Registered	Agricultural, other use approved; use decreased since 1970
Britain	Restricted 1980	Usage curtailed 1980s	Restricted	Restricted use 1981	Unknown	Restricted use	Registered use 1981	Registered	Unknown
Federal Rep. of Germany	Restricted 1978; banned 1989	Prohibited 1972	Banned 1977; lindane restricted	Deregistered 1986	Prohibited	Banned 1974	Prohibited 1981	Registered	Registered with restrictions
Former German Democratic Republic[2]	Importation and use prohibited 1984	Forestry application with permit; used on seed onions	Lindane permitted	Prohibited in agricultural chemicals	Registered with some restrictions	Prohibited in agricultural chemicals	Prohibited	Unknown	Unknown
Netherlands	Prohibited 1985	Prohibited 1981	Mixed isomers prohibited 1981	Prohibited 1981	Prohibited 1984	Prohibited 1981	Prohibited 1981	Restricted to apples 1987	Unknown
Sweden	Importation and handling prohibited 1973	All uses prohibited 1975	Lindane registered	Prohibited 1971	Classified hazardous product	Prohibited 1969	Agricultural product standards in effect	Registered with restrictions	Registered
Former USSR	Unknown use; standards exist	Prohibited 1970 agricultural use; production prohibited 1980s	Extensively used until 1986 when restricted	No use restrictions	Restricted except sugar beets, seed treatments 1986	Not restricted	Standards but no use restrictions	Registered	Registered with restrictions
India	Unknown	Registered	Registered	Registered for use and production	Unknown	Registered	Registered	Registered	Unknown

Table 1-1 *(continued)*

Country	PCBs	DDT	Hexachloro-cyclohexane (HCH)	Chlordane	Toxaphene	Dieldrin	Heptachlor	Endosulfan	Methoxychlor
Mexico	Unknown	Prohibited agricultural pesticide 1982	Restricted to certain agricultural uses 1988	Restricted for use on maize and sorghum since 1988	Restricted 1984 agri. use (cotton, maize, sorghum)	Prohibited agricultural pesticide 1982	Restricted to certain agricultural uses 1988	Unknown	Unknown

1 Lindane is available in the U.S. only on a permit system.

2 Prior to 1990; now under regulation of the Federal Republic of Germany.

This information has been taken from several sources, of which the most important are Barrie et al. 1992, International Registry of Potentially Toxic Chemicals (IRPTC) 1988, and Canadian Council of Resource and Environment Ministries (CCREM) 1987. It should be used only as a general indication of status in each country. In cases where environmental guidelines have been provided for a country (i.e., air, water, and/or food) and the product has not been specifically noted as prohibited, it has been assumed that the product is in use. The countries listed do not include all countries for which data are readily available, but they represent a geographic cross-section and are included because the available information was relatively complete.

Global usage of selected persistent organochlorines

Collecting information about historical and future global trends in use and emissions of persistent OCs is essential for interpreting present and predicting future environmental concentrations of these compounds in remote regions. This is a difficult task. The quality of the historical record of production, release, and destruction varies from nation to nation for each compound. In many cases, the absence of legislation that requires tracking chemical production and even manufacturing records results in the use of upper and lower bound estimates for production. A similar variation in data quality exists for current stocks of these compounds because of the poor records for both trade shipments and production records in non-OECD countries.

One example of the difficulty in establishing the status of global stocks has been noted for PCBs by Blectchy (1983) and again by De Voogt and Brinkman (1989). Similar difficulties exist for other compounds, especially those for which production may be continuing, such as for DDT. There is now a recognized project under the Global Emissions Inventory Activity of the International Geosphere Biosphere Program to gather global data on registration status, trends in regional use and emissions, and mode and time of application for major persistent and bioaccumulative OCs. Data on the historical production and use of persistent OC pesticides have been difficult to obtain and are of uncertain quality. There is no central global registry for this information. In many countries, records of production and use are unavailable, fail to distinguish between active ingredient and total formulation, and lack important details on the method of pesticide application. As a result, the actual use of a large fraction of the suspected production cannot be determined (Voldner and Li 1993, 1995). Current emission inventories comparing direct accounts of use and suspected total production have been compiled and include data from countries where continued use is suspected. One such inventory is shown in Table 1-2 (Voldner et al. 1994). International cooperation, changes in legislation and regulations, as well as improved record keeping by governments and other bodies are required to restrict usage as well as to improve estimates.

The examples that follow are not intended to offer a comprehensive list but to provide examples of compounds in addition to the recognized OC pesticides.

Polychlorinated biphenyls

There are 209 possible PCB congeners with 1 to 10 chlorines substituted on the aromatic rings, and many congeners are bioaccumulative to potentially toxic concentrations, especially in higher-trophic-level animals. PCBs have been produced deliberately in global quantities exceeding 1.5 million metric tons, with significant stocks still in existence (Blectchy 1983). As one of the most common, persistent, and bioaccumulative chlorinated substances, PCB has been banned globally and its history is widely known (Blectchy 1983).

Table 1-2 Pesticide inventories compiled by the Canadian Global Emissions Inventory Center[1]

Pesticide	Total use accounted for[2]	Suspected total production (1948 to 1993)[2]	Countries where continued use is suspected (1993)
Aldrin	505,400	ND[3]	Brazil, India, Thailand
Chlordane	7,800	ND	India, Mexico, Thailand
DDT	1,495,000	2,600,000	Brazil, India, Mexico, Thailand, Venezuela
Dieldrin	33,700	ND	None listed
Endrin	408,000	ND	None listed
HCH	550,000	ND	India, Mexico, Pakistan
Lindane	721,200	ND	Argentina, Canada, China, Colombia, Costa Rica, France, India, Spain, the U.S., Venezuela
Toxaphene	455,000	1,300,000	Colombia, Mexico, Nicaragua, the former USSR

[1] Source: Voldner et al. 1994.
[2] metric tons
[3] ND = not done for this inventory

Chlorinated benzenes, naphthalenes, and phenols

Polychlorinated benzenes (PCBzs) have been used as solvents or intermediates for synthesizing other chemicals, including insecticides. Polychlorinated naphthalenes (PCNs) were produced as commercial products (Halowaxes). Pentachlorophenol (PCP) was widely used as a wood preservative and for other pesticidal purposes, leading to a significant dispersion into the environment from direct production at a large number of processing sites, and to elution from wood products. Its use is now banned in several countries.

Chlorinated terphenyls and benzyltoluenes

Polychlorinated terphenyls (PCTs) were produced as replacements of PCBs and were given industrial names, such as Aroclor 5442, Aroclor 5460, or Leromoll 141. Polychlorinated benzyltoluenes (PCBTs) were predominantly used as hydraulic fluids in mines in Germany; their technical names are Ugilec 141, BP-Olex, SFD-0207 ZF, and Hydrocor HFD 25 A3. They consist of predominantly tetrachlorinated and some trichlorinated congeners (Ehman and Ballschmiter 1989).

Chlorinated paraffins

Chlorinated paraffins (CPs) are chlorinated *n*-alkanes with chain lengths of 10 to 30 carbon atoms, and their chlorine content varies between 40 and 70% by weight. CPs are currently used as solvents, plasticizers, lubricants, and flame retardants, in most cases in open systems. Estimates of produced and released amounts of CPs are scarce, but the estimated global production in 1988 was 340,000 metric tons (Svanberg 1983; Mukherjee 1990; Slooff et al. 1992). Little information is available on physicochemical aspects of these persistent hydrophobic chemicals, such as aqueous solubility, octanol/water partition coefficient (Sijm and Sinnige 1995), Henry's Law constant (HLC), and relative bioavailability, while more information is available on acute toxicity and bioaccumulation (UK Department of the Environment 1994). While there may be some environmental concern for short chain length CPs (C_{10-13}), a low level of environmental concern is expected for intermediate (C_{14-17}) and long chain length (C_{20-30}) CPs (UK Department of the Environment 1994). In some studies, CPs caused growth inhibition, reproductive, and carcinogenic effects (Sloof et al. 1992). CPs can be metabolized, but whether this is restricted to ß-oxidation of the primary carbon atom is subject to interpretation. Half-lives of radiolabeled CPs are long, indicating strong bioaccumulation of the parent compounds or their metabolites (Svanberg 1983; Mukherjee 1990; Slooff et al. 1992).

Industrial byproducts

A number of byproducts are mentioned to emphasize the need to better understand these reactions, to better control these reactions to minimize byproduct formation, and to better address release and emissions of byproducts.

Chlorinated benzenes, naphthalenes, and phenols

PCBzs and PCNs are formed and released to the environment as byproducts from various processes, such as chlorine production in the presence of graphite electrodes, municipal waste incineration, and copper roasting, and they are also present as contaminants in PCB formulations (Brinkman and de Kok 1980). Graphite electrodes for chlorine production may also have been a primary source for octachlorostyrene. Pentachlorophenol was often contaminated with PCDDs and polychlorinated phenoxyphenols (PCPPs) as byproducts. Pentachlorophenol was considered to be a significant source of PCDDs in the environment and may be a continuing source because of its release from preserved wood (Harrad and Jones 1992).

Trichlorophenol (TCP) deserves special mention, as it was used to produce the herbicide 2,4,5-T, a component of the Agent Orange used in Vietnam and the substance produced at the Seveso plant in Italy. Because of the alkaline conditions used, condensation of the TCP resulted in the production of 2,3,7,8-TCDD as a byproduct in up to ppm quantities. The explosive, runaway conditions of Seveso resulted in the production of between several hundred grams and 3 kilograms of 2,3,7,8-TCDD, based on conservative estimates from soil and vegetation analyses (Di Domenico et al. 1980). As a result, 2,4,5-T production may have been the single significant source of environmental release of TCDD, and its production is now banned.

Polychlorinated diphenyl ethers

Polychlorinated diphenyl ethers (PCDEs) compose a group of 209 congeners with physicochemical properties similar to those of the PCBs. The substitution pattern on the 2 phenyl rings is the same as that of the PCBs. PCDEs have never been produced in substantial amounts per se, but they are byproducts in the 1 to 10 mg/kg range in PCP. PCDEs are also formed in the chemistry of incomplete combustion, where they are precursors of the PCDFs. The degree of chlorination has a maximum, in this case at 2 to 3 chlorine per molecule. The similarity in the properties of these chemicals and the PCBs in terms of hydrophobicity and chemical stability could lead to similar regional and global distribution patterns for the PCDEs.

Waste incineration

Combustion or incineration processes have been associated with the production of organohalogen byproducts, including PCDD/PCDFs. A variety of unwanted and toxic halogenated products can be formed in poorly designed or poorly operating incinerators. Modern incinerators operating under optimal conditions have proven to reduce this formation drastically, and modern air pollution control devices have been shown to reduce such emissions.

*Polychlorinated dibenzo-*p*-dioxins and polychlorinated dibenzofurans*

There are 75 PCDD and 135 PCDF congeners with 1 to 8 chlorines on the aromatic rings. The PCDDs and PCDFs have not been deliberately produced except as analytical standards and for toxicology studies. As noted above, they are byproducts of several industrial reactions (Fiedler et al. 1990; Harrad and Jones 1992; USEPA 1994a), but the most significant sources of PCDDs and PCDFs appear to be their thermal formation during the incineration of municipal, industrial, and medical wastes (USEPA 1994a). Other potential sources include uncontrolled burning such as in household wood stoves, backyard burning, forest fires, and automobiles, although the use of unleaded gasoline has resulted in considerable reduction in releases. Ferrous and nonferrous metal industries may the major source of these substances at this time (USEPA 1994a).

The thermal formation of PCDDs, PCDFs, chlorobenzenes, and chlorophenols has been of special interest (Fiedler et al. 1990; Harrad and Jones 1992; USEPA 1994a). For PCDD/PCDFs, several countries have issued an emission guideline of 0.1 ng toxic equivalent (TEQ)/m^3, a reduction by a factor of 100 to 1000 compared to older incinerators. Many modern incinerators in Europe, the U.S., Japan, and Canada meet or exceed this guideline, but because of the cost of this improved incineration technology, many countries cannot afford its installation. For modern incinerators, the composition of waste is of little importance. However, many older incinerators are still operating, and these need to be modified or replaced.

Uncertainties

In several cases, there is a noteworthy discrepancy between observed concentrations and theoretically predicted concentrations of OC compounds. This can be illustrated with data for carbon tetrachloride: estimates of the atmospheric half-life and the global mean concentration in the troposphere predict that the annual input to the environment is 1 to 2 million metric tons, whereas the contribution from known anthropogenic sources is an order of magnitude less. Mono-, di-, and trichloroacetic acids provide other examples of low-molecular-weight OCs for which the major sources and pathways are not fully known. Analysis of rain and surface water samples has shown that these compounds can be present in μg/L (ppb) concentrations, i.e., in substantially greater concentrations than any other low-molecular-weight OCs (Laniewski et al. 1995). Theoretically, such concentrations may be caused by photochemically mediated transformation of tetrachloroethene (Frank 1991). However, the presence of chloroacetic acids in ice samples of preindustrial origin strongly indicates that natural sources also play an important role (Grimvall et al. 1994). Altogether, these examples show the need to elucidate not only sources but also environmental transformation of these compounds in order to fully understand and properly control their inputs.

Chapter 2

Distribution, Transport, and Transformation of Organochlorine Compounds

Organochlorine (OC) compounds have a wide range of physicochemical properties. Since environmental distribution and persistence is controlled by these properties, this range results in widely different distributions to different environmental compartments and phases and in widely different degrees of persistence in those compartments. Polychlorinated compounds, particularly polychlorinated aromatic compounds, can be persistent ($t_{1/2} > 6$ months) in the environment.

One group of OC compounds is of special concern: its members are of intermediate molecular weight ($200 < MW < 500$) and are semivolatile, lipophilic, and persistent in the environment. Included in this group are polychlorinated dibenzo-*p*-dioxins (PCDDs), polychlorodibenzofurans (PCDFs), polychlorinated biphenyls (PCBs), polychlorinated benzenes (PCBzs), and some older chlorinated pesticides like dichlorodiphenyltrichloroethane (DDT) and the chlordanes. The physicochemical properties of these compounds favor significant atmospheric concentrations that can result in their global redistribution by evaporation and atmospheric transport from tropical and temperate countries and by deposition in remote regions such as the Arctic and Antarctic. The significance of long-range transport and deposition is shown by well-documented contamination of biota in polar and other remote regions.

The persistence of OCs is determined by their degradation in air, water, and soil. The major pathway of transformation in air is reaction with hydroxyl radicals. In water, hydrolysis, microbial degradation, and photolysis have all been reported to occur. Likewise, in soils and sediments, biodegradation, hydrolysis, and photolysis have been reported. For the majority of OCs, environmental fate is not determined by chlorine substituents only, and a variety of degradation pathways are available.

Chapter Preview

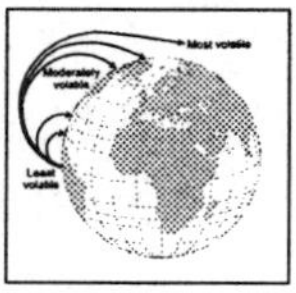

To comprehensively describe the myriad environmental processes that govern exposure of organisms to organochlorines (OCs) is not possible within the confines of one chapter. The number of compounds and processes to be considered is too great, and the available information is often incomplete. The past several years of research have resulted in substantial new knowledge about the environmental fate of different types of compounds. Simple physical and chemical characteristics of released compounds have proven useful in predicting their large-scale distribution among environmental compartments and phases: air, water, soil, sediments, and biota (Mackay et al. 1992a, 1992b, 1992c, 1993). Structure-activity relationship (SAR) models have been developed to predict such characteristics of compounds for which no empirical data are available (Meylan and Howard 1991, 1993, 1995; Meylan et al. 1992; Boethling et al. 1994). In addition, the amount of data regarding sources and ambient concentrations of chlororganics is rapidly increasing. When combined, these data and SAR models enable a systematic discussion of the environmental fate of organic compounds that have one or several chlorine atoms covalently bonded to a skeleton of carbon atoms.

Physicochemical Properties and Environmental Partitioning

Organochlorine compounds have a range of physicochemical properties. Since environmental distribution and persistence are controlled by these properties, this range results in widely different distributions to different environmental compartments and phases and in widely different degrees of persistence in those compartments. Polychlorinated compounds, particularly polychlorinated aromatic compounds, can be persistent ($t_{1/2}$ > 6 months) in the environment.

Organochlorine compounds exhibit a range of physicochemical properties that determine their transport pathways and their redistribution and partitioning in the environment. Properties of importance are vapor pressure (P, Pa), water solubility (S, mol/m^3), Henry's Law constant (HLC, Pa m^3/mol), octanol/water partition coefficient (K_{OW}), organic carbon/water partition coefficient (K_{OC}), and ionization constant (e.g., pK_a) for ionizable compounds (Mackay et al. 1992a, 1992b, 1992c). For compounds that are solids at ambient temperature, P and S of the subcooled liquid phase are usually used to remove the influence of crystalline forces. Some of these properties are related; e.g., HLC = P / S, and K_{OC} is correlated to K_{OW}. The environmental distribution of nonionic compounds can be estimated from P, S, HLC, $K_{OC,}$ and K_{OW}, which describe partitioning among air, water, soil, sediment, and organic phases. The latter include soil organic carbon, plant waxes, and the lipids of aquatic and terrestrial animals. Ionizable compounds represent a special class, since their phase distribution is strongly controlled by S, pKa, and the pH of water or soil. Some compounds such as pentachlorophenol (PCP) and chlorinated acetic acids are almost completely ionized at the pH of natural waters, and the lipophilicity and adsorption of the anion are low. In these cases, a distribution coefficient, D, is more appropriate than K_{OW}. The distribution coefficient at a specific pH is the product of the K_{OW} for the compound multiplied by the fraction of the compound protonated at that pH (Schellenberg et al. 1984). At greater ionic strengths, the formation of ion pairs that can partition into organic phases can also affect sorption behavior. Adsorption of chemicals can be reversible or nonreversible and, in such instances as adsorption to sediment followed by burial, can result in sinks for chemicals.

Table 2-1 shows a comparison of the predicted equilibrium distribution among air, water, soil, and sediments for compounds with differing values of HLC and K_{OW}. These estimates were made from a simple equilibrium partitioning model, assuming the relative phase volumes of sediment:soil:water:air = 1:90:2000:1,000,000 (Mackay et al. 1992a, 1992b, 1992c). The effect of adding chlorines to aromatic molecules can be seen by an increase in K_{OW}, resulting in greater partitioning into organic carbon phases of soil and sediments. Increasing K_{OW} also generally results in greater bioaccumulation.

Table 2-1 Environmental distribution of selected organic compounds

Organic compound	H Pa-m^3/mol	Log K_{OW}	Air %	Water %	Sediment %	Soil %
Low MW						
Chloroform	370	2.0	98	1	<1	<1
1,2-Dichloroethane	100	1.5	99	<1	<1	<1
1,2,4-Trichlorobenzene	140	4.0	82	2	<1	16
Chloroacetic acid	0.009	0.22	NA	NA	NA	NA
Intermediate MW						
Biphenyl	27	4.0	41	7	1	51
2,2'-Dichlorobiphenyl	22	5.0	14	1	2	83
2,2',4,4',5,5'-Hexachlorobiphenyl	13	7.0	<1	<1	2	98
Dibenzo-*p*-dioxin	12	4.4	11	5	2	82
2,3,7,8-Tetrachlorodibenzo-*p*-dioxin (2,3,7,8-TCDD)	3(?)	6.8	<1	<1	2	98
Hexachlorobenzene (HCB)	170	5.7	8	0.3	2	89
Lindane	0.2	3.7	NA	NA	NA	NA
PCP	0.002	5.1	NA	NA	NA	NA
Hexachloroethane	390	3.9	85	2	<1	13
High MW						
Chlorohumics	NA	–2	NA	NA	NA	NA
Chloroparaffins (C_{10}–C_{13})	~0.1	5.85–7.14	3	1	79	17

The accuracy required of physicochemical property estimates depends on the goals of the modeling exercise (see Chapter 4 for detailed discussion). For predicting the relative distribution of diverse compounds, such as those in Table 2-1, knowing HLC and K_{OW} within about an order of magnitude is adequate. Predictions of quantitative structure-property relationship models are within this range. The precision between laboratories in measuring physicochemical properties of greater-molecular-weight chlorinated organics is often no better. For example, reported HLCs differ by up to a factor of 10 for PCB congeners, and a similar range is reported for hexachlorobenzene (HCB) (Mackay et al. 1992a, 1992b, 1992c).

A greater degree of accuracy in predicting physicochemical properties is needed to understand the environmental transport and fate of persistent OC compounds (Hoff 1994). This is especially true for air–water gas exchange, where direction of the flux to or from the water is determined by fugacity differences of no more than a factor of 2. Unless HLC is known with greater accuracy than an order of magnitude, it cannot be used to confidently predict the direction and magnitude of gas exchange. Environmentally significant differences in the particle-gas distribution of PCBs in the atmosphere are found for congeners differing only 3-fold in vapor pressure (Falconer and Bidleman 1994). Although equilibrium models can give an indication of the net direction of the flux across the air–water interface, the rate of that transfer cannot be predicted without knowledge of mass transfer coefficients.

Some properties depend strongly on temperature: vapor pressure, water solubility, and Henry's Law constant. For many compounds, these have been measured only at 20 to 25 °C, and extrapolation is required to estimate their values at other temperatures. The effect of temperature on the partitioning of organic compounds in the environment is illustrated in the following examples:

1) Seasonal cycles of OC pesticides and PCBs in ambient air have been found in Southern Ontario. Concentrations were related to ambient temperature and were greatest in summer, intermediate in spring and fall, and lowest in winter (Hoff et al. 1992; Fiedler 1996).
2) The direction and magnitude of air–water gas exchange in the Great Lakes change seasonally with temperature for PCBs (Hornbuckle et al. 1994) and hexachlorocyclohexanes (HCHs) (McConnell et al. 1993). Greater volatilization occurs in summer, accompanying the warming of surface water and thermocline formation.
3) The net exchange direction for chemicals in the open ocean reflects differences in surface water temperature and atmospheric concentration, e.g., between the Bay of Bengal in the Indian Ocean and the atmosphere (Iwata et al. 1993; Streitmüller and Ballschmiter 1994; Jantunen and Bidleman 1995).
4) The aerosol/gas distribution of PCBs, OC pesticides, PCDDs and PCDFs, and other semivolatile organic compounds is inversely related to vapor pressure, and thus to temperature. Lower temperatures favor greater partitioning of these

compounds from the vapor phase to atmospheric particles and increase the likelihood of their removal by wet and dry deposition (Koester and Hites 1992a; Falconer and Bidleman 1994).

The properties of many semivolatile organic compounds favor their long-range atmospheric transport and accumulation in cold ecosystems such as the Arctic (United Nations Economic Commission for Europe [UN-ECE] 1994). Understanding these processes requires knowledge of physicochemical properties at polar temperatures. Recently, vapor pressures of several chlorinated and nonchlorinated compounds have been measured to –30 °C (Wania and Mackay 1994). Because of the fundamental importance of physicochemical properties in understanding the transport and fate of organic compounds, determining their variation as a function of temperature should receive top priority.

Atmospheric Deposition of Organochlorines

One group of OC compounds is of special concern: its members are of intermediate molecular weight (200 < MW < 500) and are semivolatile, lipophilic, and persistent in the environment. Included in this group are PCDD/PCDFs, PCBs, PCBzs, and some older chlorinated pesticides like DDT and the chlordanes. The physicochemical properties of these compounds favor significant atmospheric concentrations that can result in their global redistribution by evaporation and atmospheric transport from tropical and temperate countries and by deposition in remote regions such as the Arctic and Antarctic. The significance of long-range transport and deposition is shown by well-documented contamination of biota in polar and other remote regions.

The deposition by wet, dry, and fog/aerosol processes of chemicals transported through the atmosphere represents one of the most important fluxes of organohalogens to many parts of the environment. Wet deposition contains dissolved OCs with a variety of physicochemical properties, and dry deposition makes a substantial contribution of semivolatile, particle-associated compounds. The average adsorbable organic halogen (AOX) concentration in precipitation in Central and Northern Europe has been estimated at about 10 μg Cl/L (Enell and Wennberg 1991; Grimvall et al. 1994). A majority of the OCs detected with this group parameter are relatively hydrophilic substances, such as chloroacetic acids that are the most abundant of the low-molecular-weight compounds identified to date (Laniewski et al. 1993). However, a substantial fraction of the AOX content in precipitation is composed of relatively lipophilic, semivolatile, or nonvolatile neutral compounds (Laniewski et al. 1993).

An overview of processes controlling removal of OC compounds from the atmosphere is given here (for details, see Bidleman 1988). Precipitation of semivolatile organic compounds takes place by dissolution of gases in water droplets (gas-phase scavenging) and washout of chemicals bound to particles. Snow scavenging of particles also takes place, and sorption of gases by snow may be significant. There are few data to accurately describe the latter process, however. A simple equilibrium model is given in Equation 2-1 (Ligocki et al. 1985a, 1985b; Bidleman 1988):

$$W_t = W_p F + W_g(1 - \Phi) \qquad \text{(Equation 2-1)},$$

where W_t is the total washout ratio = C_{rain}/C_{air} (both in ng/L), W_p is the particle scavenging ratio (approximately 104 to 105), W_g is the gas scavenging ratio = RT/HLC at the rain temperature (where R is the natural gas constant and T is temperature), and Φ is the fraction of compound associated with particles in air. Φ can be estimated using the Junge-Pankow adsorption model (Pankow 1987; Falconer and Bidleman 1994) or from ambient air measurements that attempt to separate particulate and gaseous compounds with a filter-adsorbent collector (Pankow and Bidleman 1992; Falconer and Bidleman 1994). Some OCs for which gas-phase scavenging dominates are HCHs, dieldrin, and chlorophenols. Gaseous PCBs, DDT, and PCDD/PCDFs are poorly soluble, and their wet deposition is controlled by the washout of particles (Bidleman 1988; Falconer and Bidleman 1994). Because both vapor pressure and HLC are reduced at lower temperatures, rain scavenging of gaseous and particulate compounds is thermodynamically favored in cold seasons and climates. This may make ice crystals and snow important at higher latitudes. Increased wet deposition of PCDD and PCDFs has been found in winter, compared to summer (Koester and Hites 1992).

Dry particle fluxes of organic contaminants are related to the rate of deposition of the particles and the fraction of the compound on aerosols (Φ) (Bidleman 1988). Large particles have greater deposition velocities and dominate the flux, even though greater concentrations of organic compounds on a weight basis are usually found on submicrometer particles. In Chicago, dry deposition of PCBs was up to 3 orders of magnitude greater than in rural areas, indicating the likely contamination of southern Lake Michigan by urban plumes (Holsen et al. 1991; Holsen and Noll 1992).

Gas exchange of organic compounds across lake and ocean surfaces is a 2-way process in which invasion to the water body and evasion from the water body occur simultaneously. The net flux, which can be calculated from equations presented in Bidleman and McConnell (1995), is given by the difference between deposition and volatilization. Gas fluxes of PCBs and OC pesticides like DDT, toxaphene, and mirex account for a major share of the mass balance of these contaminants in the Great Lakes. For PCBs, volatilization exceeds deposition over most of the year (Hornbuckle et al. 1994). Since 1980, approximately 25,000 kg of PCBs have been lost from the water column of Lake Superior, apparently by volatilization (Jeremiason et al. 1994). Fluxes of HCHs in Lakes Michigan, Huron, and Erie change seasonally, from deposition over most of the year to volatilization in summer (McConnell et al. 1993). Calculations based on current levels of toxaphene in the Great Lakes airshed suggest that the gas-phase deposition to Lake Superior outweighs rain and dry deposition loadings (Hoff et al.

1993). Differences in the exchange direction and magnitude of the flux are found in different ocean regimes, depending on air and water concentrations and surface water temperatures (Iwata et al. 1993). Paired air and water measurements in the Bering and Chukchi Seas (Jantunen and Bidleman 1995) and the Arctic Ocean in the Canadian Archipelago suggest that the flux direction of α-HCH exchange in summer has reversed, from deposition between 1988 and 1990 to volatilization in 1992 and 1993. The oversaturation of α-HCH in surface waters apparently resulted from a decrease in atmospheric concentrations over the last decade, and especially within the last 3 years (Bidleman et al. 1995).

Several investigations that document the significance of atmospheric deposition of OCs and the relationship of deposition to accumulation in the food chain are cited in the USEPA Great Lakes Water Report (1993a). A mass balance study of the Great Lakes concluded that atmospheric deposition was a significant contributor of OC compounds. Wet and dry loading estimates to the Great Lakes for the mid-1980s were 1700 kg/y PCBs and 270 kg/y DDT (Strachan and Eisenreich 1988). Direct deposition was estimated to contribute 76 to 90% of total inputs to Lake Superior, 58 to 63% to Lakes Michigan and Huron, and 6 to 20% to Lakes Erie and Ontario (Strachan and Eisenreich 1988; USEPA 1993a). Atmospheric loading estimates were updated in 1992 with the availability of improved concentration data for air and precipitation. Comparison of the mid 1980s and 1990 to 1992 loadings indicates about a 3-fold decline in PCBs and DDT (Eisenreich and Strachan 1992). The clearest evidence of atmospheric deposition of OCs leading to contamination of biota was presented by Swackhamer and Hites (1988), who found that fish from Siskiwit Lake (Isle Royale, Lake Superior) contained residues of PCBs and several OC pesticides equal to or greater than those found in Lake Superior. Siskiwit Lake receives no water flow from Lake Superior, and its only route of contamination is through aerial processes.

Leister and Baker (1994) found that the average rainfall flux of PCBs in Chesapeake Bay compared closely to Great Lakes values. Atmospheric deposition is estimated to be the major source of PCBs and OC pesticides to the open ocean, greatly exceeding inputs from rivers. Gas exchange forms a large share of this budget (Bidleman et al. 1995).

Plants can accumulate OCs through direct wet and dry deposition and also through exchange with fog and air. Thus, plants can serve as sentinel monitors for OCs in air and also as the route by which these compounds are introduced to the terrestrial food chain. Pine needles have been used to follow the geographical and temporal trends of pesticides (Jensen et al. 1992; Strachan et al. 1993; Calamari et al. 1994) and PCBs (Kylin et al. 1994) across Europe. Accumulation of OCs has been measured in lichens from the upper Great Lakes region (Muir et al. 1993) and in lichens and mosses from arctic, temperate, tropical, and antarctic regions (Calamari et al. 1991). The quantity of OCs in the vegetation reservoir is unknown but is potentially large. Simonich and

Hites (1994) estimated that 40% of airborne polycyclic aromatic hydrocarbons (PAHs) that span the physicochemical range of the semivolatile OCs is deposited onto plant leaves, where they become transferred to the forest floor when leaves fall.

Distribution of persistent OCs in the tropical environment

Although many developed nations prohibited their production and usage more than a decade ago, some tropical countries still use persistent OCs such as chlordane and toxaphene in agriculture and public health (Mowbray 1988; Forget 1991; Ritter et al. 1995). According to a monitoring survey on the distribution of persistent OCs in riverine and estuarine waters from urban areas of Asia and Oceania (Iwata et al. 1994), greater concentrations were observed in tropical than in the temperate and colder regions. Prominent concentrations of OCs were also found in tropics air samples. Besides air and water, samples of human breast milk (Tanabe et al. 1990) and birds (Ramesh et al. 1992) revealed a similar pattern of contamination, with tropical regions registering greater levels. Contamination of foodstuffs by persistent OCs was also investigated in tropical Asia and Oceania, where the concentrations exceeded the acceptable daily dietary intake values (Tanabe, Kannan et al. 1991; Kannan, Tanabe, Quynh et al. 1992; Kannan, Tanabe, Ramesh et al. 1992; Kannan et al. 1994). These results strongly suggest the presence of major emission sources in tropical countries.

Countries located in the tropical belt often experience high temperatures and heavy rainfalls. Tropical agroecosystems may facilitate the rapid dissipation of toxic contaminants through air and water. In fact, a typical case was observed in the comparative study of OC distributions in tropical and temperate regions. For example, in Asia and Oceania, residue levels of OC contaminants in sediments (Iwata et al. 1994) and fish (Ramesh et al. 1992) varied less in tropical regions than in temperate, while in air and water, the opposite was observed (Iwata et al. 1994). In the Vellar River and its watershed in south India, the flux of HCH residues was modeled using the hydrokinetic parameters of the river and data on HCH concentrations in air and water (Takeoka et al. 1991). It was estimated that about 99.6% of the applied HCH in the paddy areas of this watershed was volatilized to the atmosphere and only 0.4% was transported by water to the estuary over a year. Moreover, about 75% of the waterborne flux to the estuary was also lost by volatilization to the air. Thus only 0.1% of the applied HCH was ultimately drained to the sea by way of the river. An HCH application test in a tropical paddy field also supported the rapid volatilization of residues into the air from water, rice plant, and soil (Tanabe, Ramesh et al. 1991). These observations illustrate that inputs of HCH (and possibly other contaminants) to tropical coastal water bodies through river discharge are less significant than inputs in temperate zones. The residence time of contaminants in the aquatic environment is quite short in tropical areas, whereas transfer of contaminants to the atmosphere is quite large.

The short residence time of toxic contaminants in the tropical water bodies might be favorable from the viewpoint of local environmental quality and animal health.

However, it may have wider implications for the global environment. Undoubtedly, the volatilized residues from the tropics disperse through the atmosphere on a global scale. In addition, because the open ocean serves as a sink for contaminants and because it contains vulnerable animals like cetaceans and pinnipeds, the possibility of open ocean effects cannot be ignored.

Transport and fate of organochlorines in the open ocean environment

The processes involved in the global transport and fate of organic compounds, including OCs like DDT and HCH, were reviewed by Ballschmiter (1992). These 2 OCs are particularly suitable models because they have both been subjected to widespread atmospheric transport. In a monitoring survey of air and surface seawater in various seas and oceans worldwide during the period of 1989 to 1990, Iwata et al. (1993) found HCH to be in greatest concentration among the OCs examined. Contamination was greatest in the northern hemisphere, with greater residue levels in tropical regions and cold waters near the Arctic. In contrast, concentrations of DDT were lower than those of HCH and displayed a different pattern of distribution; greater concentrations were found in the seas around tropical Asia. PCBs and chlordanes were more uniformly distributed.

The present distribution of these OCs in the world's oceans suggests a major change in the contamination pattern during the last decade. Until the early 1980s, there was greater contamination by OCs (HCHs, DDTs, and PCBs) in the mid-latitude oceans of the northern hemisphere, reflecting their use in developed countries such as Japan, Europe, and the U.S. (Tanabe et al. 1982, 1983; Tatsukawa et al. 1990). However, this pattern is not seen in the most recent samples. Goldberg (1975) predicted the transition of major DDT usage from developed to developing countries by the so-called "southward tilt." This prediction has proved correct, not only for DDT but also for HCH. PCBs and chlordanes are also known as worldwide contaminants, as indicated by their detection in arctic samples (Kawano et al. 1988; Muir et al. 1992; Thomas et al. 1992). The smaller geographical variations of PCB and chlordane concentrations in the open ocean suggests the expansion of their point source areas of contamination to tropical developing countries, but in the absence of evidence that PCB production in these areas has increased, the global distribution of "old" PCBs is likely a better explanation.

To elucidate the transport and fate of persistent OCs in global terms, an attempt was made to estimate the mass transfer of these compounds between air and water. HCH fluxes in the East and South China Seas and the Bay of Bengal indicated active transfer from air to surface seawater. These seas are a sink of HCHs because of their proximity to India and China, where HCH use is extensive. Interestingly, cold waters near the Arctic also showed the negative fluxes, indicating a significant sink for HCHs in that region (Takeoka et al. 1991). Similarly, DDT mass transfer also showed greater negative values in tropical waters, which are also a sink for DDTs (Iwata et al. 1993). In the case of PCBs and chlordanes, colder waters had greater negative values of flux, indicating a sink for these contaminants as well.

These findings support the assumption that the large quantities of persistent OCs used in the tropics are released into the atmosphere and dispersed through long-range global transport.

Transport and cold condensation in polar regions

The discovery over 2 decades ago of OC pesticides in biota from the Arctic (Bowes and Jonkel 1975) and Antarctica was an early warning of the global spread of these contaminants (Thomas and Hamilton 1988). Since then, atmospheric transport and accumulation of persistent OCs (PCBs, DDT and related compounds [DDTr], HCHs, and chlordanes) in the Arctic have been extensively documented (Cotham and Bidleman 1991; Barrie et al. 1992; Lockhart et al. 1992; Muir et al. 1992; Thomas et al. 1992; Iwata et al. 1993). Air samples from Antarctica also show continued transport of DDT, chlordanes, and HCHs (Bidleman et al. 1993; Larsson et al. 1992). Accumulation in polar regions is explained in part by "global fractionation" and "cold condensation" of compounds within the volatility range of PCBs and pesticides (Wania and Mackay 1993a; Mackay and Wania 1995). These authors make the following points concerning accumulation and potential impacts of OCs in the Arctic: As these contaminants travel from temperate and tropical regions to the poles, they are continually deposited and re-evaporated, and they fractionate according to their volatilities (Figure 2-1). The result is facile transport and condensation of compounds having intermediate volatility, such as HCB, and slower migration of a less volatile substance like DDT.

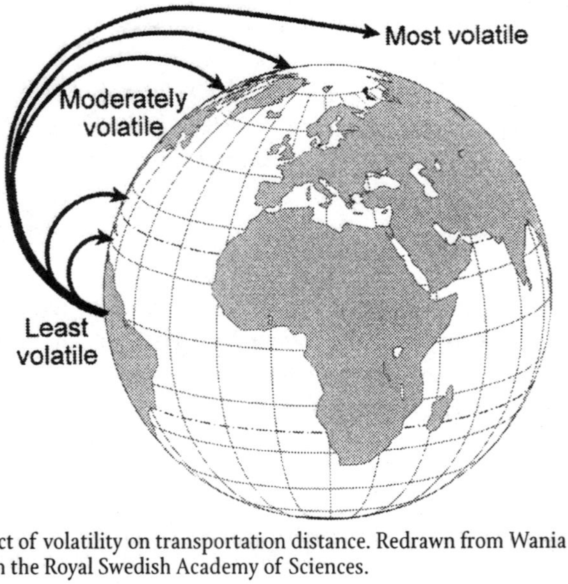

Figure 2-1 Effect of volatility on transportation distance. Redrawn from Wania and Mackay 1993a with permission from the Royal Swedish Academy of Sciences.

Furthermore, the characteristics of arctic ecosystems exacerbate the problems of OC contamination. Lower temperatures, limited biological activity, and relatively small incidence of sunlight should increase the persistence of OCs. Poor vertical mixing of the surface layer of the Arctic Ocean may increase the availability of OCs to the food chain. During snowmelt, the sudden release of a "shock wave" of stored-up contaminants may occur, coinciding with an especially sensitive phase of the aquatic life cycle. The considerable concentrations of toxic contaminants in arctic fish (Lockhart et al. 1992) and in terrestrial (Thomas et al. 1988) and aquatic mammals (Muir et al. 1992) document the significance of long-range atmospheric transport from lower latitudes.

Persistence of Organochlorines

The persistence of OCs is determined by their degradation in air, water, and soil. The major pathway of transformation in air is reaction with the hydroxyl radical. In water, hydrolysis, microbial degradation, and photolysis have all been reported to occur. Likewise, in soils and sediments, biodegradation, hydrolysis, and photolysis have also been reported. For the majority of OCs, environmental fate is not determined primarily by chlorine substituents and a variety of degradation pathways are available.

Persistence in the environment is an important property of chemicals, since long-range transport can extend the range of exposures of persistent contaminants far beyond the immediate area of release. Environmental transformations of contaminants can be subdivided into 3 sub-processes: 1) biotransformations, such as transformations by the microbial communities in the receiving environment; 2) chemical reactions, such as transformations like oxidation and hydrolysis by chemical reactivity of the contaminant; and 3) photolysis, such as transformation by interaction of the contaminant with sunlight or with photochemically produced reactive species.

The relative importance of these processes depends on the rates at which they occur under environmental conditions. The rates, in turn, depend on the chemical structure and properties of the contaminant, its distribution in the various compartments of the receiving environment, and the nature of that environment. Although useful information can be obtained from study of transformation in the laboratory, it is important to study transformations in the field because transformation rates are inevitably site-specific. Kollig (1988) described a useful set of criteria for evaluating the reliability of reported environmental rate and equilibrium constants.

Persistence of organochlorines in air

In the atmosphere, different transformations can occur for gaseous contaminants in the vapor phase, dissolved contaminants in the aqueous phases, and sorbed contaminants on the particulate phase. The most important vapor-phase degradation process is reaction with hydroxyl radicals. Other degradation processes are important only for selected chemical classes: direct photolysis and reactions with O_3 (alkenes and alkynes)

Table 2-2 Atmospheric degradation of selected organic compounds

Substance	Atmospheric $t_{1/2}$ (d)
Low MW	
$CHCl_3$	106
1,2-Dichloroethylene	320 (*trans*), 640 (*cis*)
1,2,4-Trichlorobenzene	37
Chloroacetic acid	17[a]
Intermediate MW	
Biphenyl	1.5
2,2'-Dichlorobenzene	8[a]
2,2',4,4',5,5'-Hexachlorobiphenyl	95[a]
Dibenzo-*p*-dioxin	0.7
2,3,7,8-TCDD	3.7[a]
PCDDs Chlorine atoms	
0	1
1	3
2	2 to 3
3	2 to 4
4	3 to 7
5	4 to 9
PCDFs Chlorine atoms	
0	4
1	3
2	4 to 6
3	5 to 10
4	8 to 18
5	15 to 29
HCB	627[a]
Lindane	7.9[a]
2,4-Dichlorophenol	3.5
PCP	23[a]
Hexachloroethane	Very long
High MW	
Chlorohumics	not volatile
Chloroparaffins (C_{10}–C_{13})	NA

[a] estimated

Sources: Atkinson 1985, 1989, 1995; Meylan and Howard 1993

and nitrate radicals (phenols) (Atkinson 1985, 1989). OC compounds are susceptible to all of these processes unless chlorine or other halogens replace all the hydrogens (chlorofluorocarbons [CFCs]) so that hydroxyl radicals have no hydrogen to abstract. Prediction of hydroxyl radical rate constants from chemical structure can be quite accurately accomplished (Atkinson 1989; Meylan and Howard 1993). This leads to relatively accurate predictions of the vapor-phase half-lives of these gaseous contaminants. Atmospheric half-lives have been calculated from either measured or estimated (E) hydroxyl radical rate constants for a variety of OC compounds and an assumed average hydroxyl radical concentration of 1.5×10^4 radicals per cm^3 (Table 2-2). With their extended lifetimes, the highly halogenated chemicals can be transported long distances, and the more volatile and stable chemicals are of concern as potential ozone depletors. For chemicals that have low vapor pressures ($< 10^{-5}$ Pa), most of the chemical will be partitioned onto airborne particulate matter. With few exceptions (PAHs), the degradation rates of chemicals in the adsorbed form are not well understood. In the gaseous or vapor phase, particularly in the upper atmosphere where shorter wavelengths are available, photolysis of chloroaromatics can be important (Bunce et al. 1991). Bunce et al. (1989) estimated that the atmospheric half-life of PCBs in sunlight should be several days but that chlorobenzenes would be less photolabile because of poorer spectral overlap with sunlight. These estimates are based on ideal assumptions, e.g., that the compounds are truly in the vapor phase, not adsorbed on particles.

Transformation reactions also take place in atmospheric water: cloud droplets, fog, and rain. Many oxidants are found in these droplets, including hydroxyl radicals, hydrogen peroxide and peroxyl radicals, singlet molecular oxygen, ozone, and nitrogen oxides. Many of these oxidants are coupled to the iron redox system (Faust 1994). Sedlak and Andren (1991) showed that hydroxyl radicals in aqueous solution quickly oxidized PCBs. Compared to gas-phase hydroxyl radical reactions (Table 2-2), aqueous rates were insensitive to the number of chlorines, decreasing by only a factor of 2 from dichloro- to pentachlorobiphenyls. Estimated half-lives of PCBs in cloud-water were < 10 d.

Photodegradation of OC compounds on particulate surfaces is highly variable, depending on the surface type and the wavelength and intensity of light. Koester and Hites (1992b) found large differences between the rates of photolysis of PCDDs and PCDFs sorbed on silica gel and fly ash. Half-lives of compounds on silica gel irradiated under laboratory conditions ranged from 3 to 14 h for PCDFs and 88,270 h for PCDDs, whereas negligible loss of PCDDs and PCDFs was found on fly ash after 200 h. By comparison, the half-life of 2,3,7,8-tetrachlorodibenzo-*p*-dioxin (2,3,7,8-TCDD) adsorbed to the surface of vegetation was 44 h in natural sunlight (McCrady and Maggard 1993).

Persistence of organochlorines in water

Hydrolysis of organochlorines in water

"Hydrolysis" is used broadly to refer to all thermal reactions of contaminants with water, including both true hydrolysis, i.e., substitution involving water or hydroxide, and elimination or oxidation reactions. This is because most protocols follow disappearance of the parent material so that these different mechanisms are rarely separated, although they can result in the formation of different products. The products of dehydrohalogenation reactions of halogenated compounds, such as the production of dichlorodiphenyldichloroethylene (DDE) from dehydrohalogenation of DDT, may be more persistent or toxic than the parent compound. Hydrolysis and elimination reactions such as dehydrohalogenations may also be catalyzed by an acid or a base. In recent years, systematic work on hydrolysis and other reactions of chlorinated pollutants under conditions that could be extended to the environment has been reported, and a number of compilations of hydrolysis rate constants are available (Mabey and Mill 1978; Mill and Mabey 1988). Jeffers et al. (1989) determined hydrolysis rate constants for a series of 18 chlorinated methanes, ethanes, ethenes, and propanes. Reactivity in basic solutions was found to depend on the acidity of the most

Table 2-3 Comparison of hydrolysis half-lives at pH7 and 25 °C for selected OC compounds

Substance	Half-life
Chloromethyl methyl ether	0.03 h
Chlorambucil	1.7 h
2,2-Dichloropropane	15 h
Dichloroethyl ether	22 h
Heptachlor	1 d
1,4-Dichloro-2-butene	3.2 d
Pentachloroethane	3.6 d
Chloronaphazine	9 d
1,3-Dichloro-2-propanol	9.3 d
1,1,2,2-tetrachloroethane	130 d
Lindane	206 d
Methyl chloroform	1 y
1,1,1-Trichloroethane	1.1 y
2,3-Dichloro-1-propanol	1.4 y
1,2,4-Trichlorobenzene	3.4 y
1,2,3-Trichlorobenzene	4.9 y
ß-Chloronaphthalene	8.3 y
Toxaphene	10 y
2-Chloroethanol	10 y
1,2-Dichloropropane	15.8 y
Carbon tetrachloride	41 y
1,2,3-Trichloropropane	43.9 y
1,1,1,2-Tetrachloroethane	46 y
1,1-Dichloroethane	61 y
1,2-Dichloroethane	71 y
1,1,2-Trichloroethane	139 y
Chloroform	1,800 to 3,000 y
cis-Chlordane	>197,000 y
HCB	zero hydrolysis
Hexachloroethane	zero hydrolysis

Sources: Ellington et al. 1987a, 1987b, 1988; Jeffers et al. 1989; Jeffers and Wolfe 1989)

reactive hydrogen atom in the molecule. Neutral hydrolysis depended on the strength of the C–Cl bond and the degree of steric hindrance. Aqueous half-lives under environmental conditions ranged from 36 h to 106 y. Roberts et al. (1993) used rate constants for 28 polychlorinated and polybrominated alkanes to develop structure-activity relationships for dehalogenation reactions. Rates varied by more than 8 orders of magnitude and exhibited no simple relationship to the number of chlorine substituents. In fact, addition of chlorine does not necessarily increase resistance of a chemical to hydrolysis. For example, the phenyl ester of dichloroacetic acid has an estimated hydrolysis half-life of 3.7 minutes under environmental conditions, compared to 38 d for phenyl acetate (Mabey and Mill 1978). The range of hydrolysis half-lives of more than 5 orders of magnitude and the inconsistent effect of chlorine addition on half-life do not support the suggestion that OCs can be treated as a "class" of compounds.

Hydrolysis half-lives for a variety of chlorinated organic compounds (Table 2-3) indicate that, for some contaminants, hydrolysis can be an important process, but its importance is dependent on the structure of the contaminant. Small changes in structure can result in large changes in rate constant because of changes in reaction mechanism. This is illustrated by the dichloropropanes: half-lives for hydrolysis at 25 °C and pH 7 range from 35 h for 2,2-dichloropropane to 15.8 y for 1,2-dichloropropane. Molecules that seem structurally similar can react by different mechanisms to give different products. For example, DDT reacts by simple dehalogenation to produce an ethene (DDE) as the major product, whereas for the chemically similar methoxychlor, the ethene is a minor product and the major product results from a 1,2-(p-methoxyphenyl) shift.

A number of environmental factors can potentially alter hydrolysis rate constants. These include temperature, pH, ionic strength, metal ion catalysis, and sediments. Therefore, hydrolysis rate constants determined in the laboratory need to be verified in the field. Temperature exhibits a strong effect. Jeffers and Wolfe (1989) calculated that the hydrolysis half-life of CCl_4 varied from 40 to 2790 y over a temperature range of 25 to 0 °C; a similar calculation for CH_3CCl_3 over the same temperature range yielded half-lives from 1 to 74 y. Another important variable could be pH. The values in Table 2-3 represent the sums of both neutral and catalyzed processes. If the rate-controlling processes are catalyzed by acid or base, pH will have a strong effect on half-life. For example, the rate-controlling process for many compounds is base-catalyzed dehydrohalogenation, and the pH of many natural waters is greater than 7. In such cases, a change in 1 pH unit reduces the half-life by up to an order of magnitude. Dissolved components like humic substances and metal ions can potentially influence rate constants (Perdue and Wolfe 1983; Macalady et al. 1989), probably through processes known as general acid and base catalysis or, more generally, buffer catalysis. Several authors have concluded that the concentrations of these substances in natural waters are too low for them to have a measurable catalytic effect (Mabey and Mill 1978; Perdue and Wolfe 1983; Macalady et al. 1989). These conclusions are supported by hydrolysis studies of malathion and hexachlorocyclopentadiene, where hydrolysis rates in river water and distilled water were comparable (Wolfe et al. 1977, 1982). However, recent studies on the environmental chemistry of substituted benzothiazoles indicate that, for at least one compound, buffer catalysis may be important (Brownlee et al. 1992). The C–Cl bond in chlorinated aromatics is not amenable to hydrolytic cleavage.

Microbial degradation of organochlorines in water

Rates of microbial degradation are believed to be second order and thus are proportional to both the concentration of the organic chemical, or more correctly the bioavailable concentration, and the number of organisms present with the enzymatic capability to degrade it (Boyle 1989). For some chlorinated compounds that are similar to naturally occurring compounds, it is likely that the enzymes necessary for metabolism are already present in the microbial population of the receiving water. In the case of compounds degraded by enzymes that need to be induced, these enzymes may already be present if there has been a past exposure or if the rate of discharge is relatively constant. For episodic discharges, there may be a lag period involved before the compound starts to degrade. If co-metabolism, such as a transformation occurring only during metabolism of other substrates, is an important pathway for a chemical, the available concentration of the co-metabolites is also of importance.

Laboratory studies of biodegradation have provided information about mechanisms, steric pathways, and relative degradability of contaminants. Rochkind et al. (1986) reviewed data on the microbial decomposition of many of the common families of chlorinated aromatic pollutants, including chlorobenzenes, chlorophenols, chlorophenoxy herbicides, chlorinated anilines, PCBs, DDT, and chlorinated dioxins and furans. They identified several general biochemical pathways for the potential decomposition of chloroaromatics. The chlorophenoxy herbicides and the chlorobenzenes can be metabolized to chlorophenols. Similarly, a variety of phenylamide herbicides and other aromatics with nitrogen-containing substrates are metabolized to chloroanilines. The most common route of biodegradation of chlorophenols was to chlorocatechols (chlorocatechols are also among the degradation products of chloroanilines). Several degradation pathways of chlorobenzoic acids were also described. In general, addition of chlorine reduces the biodegradability under aerobic conditions, with highly chlorinated chemicals being extremely resistant to biodegradation and lesser chlorinated chemicals exhibiting fairly rapid biodegradation (Boethling et al. 1994). More highly chlorinated aromatic compounds, like some PCBs, dioxins, and furans, are resistant to aerobic microbial degradation.

After reviewing previous attempts by several investigators to use qualitative degradation data to develop models that classify chemicals as biodegradable or nonbiodegradable, Howard et al. (1992) developed a structure–biodegradation model that was later updated by Boethling et al. (1994). Well-documented and widely available biodegradation data and a set of structural fragments that had been selected based on conventional knowledge of the effects of chemical structure on biodegradability were used to develop a model for predicting aerobic biodegradability from chemical substructures. A training set of 264 chemicals was used to calculate coefficients that predict the effect of a variety of substructures, including aliphatic and aromatic chlorine, on biodegradation rate. For both aliphatic and aromatic chlorine, the coefficients were negative for both linear and nonlinear models, confirming that these substructures decreased biodegradability. However, the magnitudes of these coefficients were less than those of a variety of other substructures that did not contain chlorine, such as PAHs, aromatic nitro, N-nitroso, aliphatic ether, and tertiary alcohol. These results suggest that, although the addition of a chlorine will decrease biodegradability of organic structures, the degradability generally remains within the range of degradability of other nonchlorinated structures. Thus, there is little justification for considering substances with few chlorines as unusually persistent, requiring different treatment than other organic substances. More highly chlorinated substances can be more persistent and likely deserve special attention when they are released to the environment.

Although many studies show that general microbial degradation pathways for most of the common chlorinated aromatic compounds exist, whether these pathways are important under environmental conditions depends on a number of site-specific factors, such as the type of microbial community, nutrient availability, and the availability of more easily metabolized compounds (Swindoll, Aelion, Pfaender 1988). Since we still do not understand fully all of the factors that determine biodegradation rates in the field, it is usually difficult to extrapolate quantitatively from laboratory data to field conditions.

Battersby (1990) and Banerjee et al. (1984) have reviewed the controlling factors and kinetic expressions that describe biodegradation rates. The most common rate expression is derived from Monod's equation for bacterial growth. This expression is most useful when the chemical under study is the sole carbon source. However, at low concentrations, transformation of a substrate may be used for maintenance of energy requirements rather than for growth (Larson 1980). In cases where contaminant transformation does not lead to growth, Monod's equation is not useful.

In the environment, where growth often is due to a variety of substrates and the one studied may not be growth limiting, first- and second-order kinetic expressions may be more appropriate. Where the contaminant concentration is low compared to other potential growth substrates, the dependence of biodegradation rate on substrate concentration and biomass leads to second-order kinetics (Baughman et al. 1980; Paris et al. 1981). Castro (1993) examined the applicability of the second-order rate law to the biodehalogenation kinetics of alkyl halides. At low cell densities and low substrate concentrations, the second-order rate law was valid in all cases, but toxic effects and competing reactions made the kinetics more complex at greater substrate concentrations and cell densities. Factors influencing biomass have a corresponding effect on biodegradation rates. For example, the effect of nutrient additions is to increase degradation rates and pseudo-first-order rate constants because of an increase in biomass with no change in second-order rate constants (Paris and Rogers 1986). Second-order rate constants offer the promise of predictability if reliable techniques can be developed to estimate the biomass actually transforming the contaminant. The techniques currently used to estimate this biomass are summarized by Klecka (1985), Battersby (1990), and Kuenemann and Battersby (1992).

The type of bacterial community is an important factor in determining the significance of biodegradation. Aquatic microbial populations can be subdivided into at least 4 distinct communities: free floating, attached to suspended particles, attached to solid substrates, and sedimentary. In shallow flowing water, attached communities can dominate (Gantzer et al. 1988; Lewis and Gattie 1988) so that, in many rivers, the presence of solid substrates supporting these communities is of primary importance. In these cases, surface area rather than biomass becomes the important normalizing factor (Lewis and Holm 1981; Lewis and Gattie 1988).

Photolysis of organochlorines in water

Environmental transformations of contaminants can be initiated by direct and indirect photochemical processes. Direct photolysis of an OC compound in the environment occurs when the contaminant itself absorbs sunlight, causing an alteration of the electron distribution of the contaminant that ultimately leads to the formation of products. Indirect photolysis occurs when a photoreaction of a substrate is initiated by absorption of light by species other than the substrate. Potential mechanisms of indirect photolysis are available, including energy transfer, electron transfer (Skurlatov et al. 1983), generation of reactive intermediates, surface catalysis on sediments (Langford and Carey 1987), and algal enhanced photolysis (Zepp and Schlotzhauer 1983). The most important is likely generation of reactive species (Cooper, Zika et al. 1989; Hoigne et al. 1989). Irradiation of humic waters by sunlight has been shown to lead to the formation of such reactive species as singlet oxygen (Zepp et al. 1977; Baxter and Carey 1982), superoxide ion (Baxter and Carey 1983; Draper and Crosby 1983b), and hydrogen peroxide (Draper and Crosby 1981, 1983a; Cooper et al. 1988; Cooper, Lean, and Carey 1989).

The rate of the direct photolysis of a contaminant in the environment will depend on the intensity of the light available, the amount of that light absorbed by the contaminant, and the relative importance of the competing excited state processes described earlier. In addition to depending on properties of the compound like absorption spectrum and quantum yield, rates of environmental photolysis also depend on the fraction of light that is absorbed by the contaminant in competition with the light-absorbing properties of the natural water body itself. Photochemical half-life in a water body varies with rate of the photolysis but also depends on the depth of the water. In systems where the overall depth is greater than the depth of light penetration of the active wavelengths, a reservoir of unirradiated substrate exists. This dependence on parameters such as depth and water absorption means that environmental photolysis rates are site-specific (Zika and Cooper 1987).

The absorption spectra of the majority of organic compounds usually exhibit maxima in the ultraviolet region well below the wavelengths available in sunlight at the earth's surface. These wavelengths begin at approximately 280 nm. Unless they possess a substituent group or substructure that absorbs sunlight, chlorinated aliphatics are such weak absorbers over the wavelength range available at the earth's surface that direct sunlight photolysis in water is usually unimportant. For chlorinated aromatics, $\pi \rightarrow \pi^*$ and $n \rightarrow \pi^*$ transitions result in a weak, but not negligible, overlap with the sunlight spectrum. Chlorinated aromatics, particularly those with substituents that shift their absorption maxima toward lower energies, may absorb enough light for direct photolysis to be an important pathway. An example is the chlorophenol family, whose phenolate ions are significant absorbers of the wavelengths available in sunlight (Zika and Cooper 1987).

Photoreactions of chloroaromatics in nonpolar solvents are dominated by homolytic cleavage of the C–Cl bond in the $\pi \rightarrow \pi^*$ excited states to produce organic radicals and chlorine atoms. The organic radicals usually react with the H-atoms in the solvent. Water is not a good H-atom donor, nor does it stabilize $\pi \rightarrow \pi^*$ excited states (Zika and Cooper 1987). Thus, many of the literature studies on photoreactivity of chlorinated compounds in nonaqueous solvents are of limited use in predicting the importance of photodecomposition in the environment, except for cases where the irradiated compartment is organic rich and can function as an H-atom donor (leaf waxes). A good example of the differences between the organic phase and water is the chlorobenzene family (Boule et al. 1987). In organic solvents, chlorobenzenes react through photosubstitutions and photoreductions, and rates increase with increasing chlorine content. In water, photohydrolysis is the predominant reaction pathway, and rates decrease with increasing chlorine content.

Under appropriate conditions, photodegradation can be the main pathway of degradation. The rate expressions for direct photolysis have been developed (Zepp and Cline 1977). For example, many studies have shown that photodegradation is the main environmental degradation pathway for chlorinated phenols. The products depend on the Cl substituent. For *ortho*-Cl substituted phenols, photolysis results in ring contraction to form cyclopentadienoic acids, while for *meta*-Cl and *para*-Cl substituted phenols, hydroxylation to resorcinol and hydroquinones is important (Boule et al. 1987). The lack of a pH dependence in these photohydrolysis reactions indicates that they result from reaction of the excited state with water, not hydroxide. In the case of polychlorinated phenols, the initial chlorinated hydroquinone and resorcinol photoproducts can then undergo oxidation reactions to quinones and ring cleavage reactions to chlorinated di-acids (Wong and Crosby 1978). Photochemical half-lives for chlorophenols range from 2 to 3 h in shallow systems to months in deeper systems (Lee et al. 1982; Pignatello et al. 1983; Crossland and Wolff 1985; Hwang et al. 1987; Svenson and Kaj 1989).

For chlorinated organic compounds whose spectra overlap poorly with sunlight wavelengths, direct photolysis is usually far less important. An example is the PCB family. PCBs absorb weakly in the 280 to 300 nm range. Using a simple analysis, Bunce et al. (1978) calculated that, in shallow waters and assuming ideal light absorption characteristics (no sediment), an average PCB might lose one chlorine per year.

Photolysis in lake and river water is often faster than in distilled water. These differences are related to the presence of dissolved organic compounds, such as humic acids or proteins, and the process is known as "indirect photolysis." Indirect photoreactions can be significant for some substrates. For example, the photolysis half-life of methoxychlor in filter-sterilized natural waters exposed to sunlight was 2 to 5 h compared to 4 to 5 months for the direct photolysis (Zepp et al. 1976). An example of this is the enhanced transformation of the pesticide mirex to photomirex, which occurs upon the addition of humic acid in laboratory experiments (Mudambi and Hassett 1988). Subsequent field work showed that photomirex is produced in the water column of Lake Ontario (Mudambi et al. 1992). Half-lives of 1,2,7,8-tetrachlorodibenzofuran and 2,3,7,8-tetrachlorodibenzofuran in river and lake water under natural sunlight were very short (4 to 6 h), and photodegradation in these waters proceeded about twice as rapidly as in distilled or high-performance liquid chromatography (HPLC)-grade water (Dung and O'Keefe 1994). Under midsummer sunlight, Friesen et al. (1990) found that

5- and 7-chlorinated PCDDs photolyzed 12 to 15 times faster in lake water than in distilled water or acetonitrile.

Effluents from pulp mills employing chlorine bleaching are a major source of chlorinated organics to surface waters. The chlorine in these macromolecules is often measured using the sum parameter AOX. From 25 to 70 % of this AOX is degradable in pulp mill effluent treatment systems (Gergov et al. 1988; Jokela and Salkinoja-Salonen 1992). Laboratory studies have shown that pulp mill AOX is subject to photodegradation in the environment (Caron and Reeve 1991; Millar 1993). The dechlorination half-life of pulp mill AOX in full sunlight in a roof-top experiment was on the order of 2 weeks (Millar 1993). Based on AOX disappearance rate in a river downstream from a pulp mill employing primary treatment, Carey (1994) estimated the downstream distance required for a 50% reduction in AOX concentration at 188 km, corresponding to about 5 d of irradiation based on the flow times in the river and assuming plug flow. The AOX remaining after secondary treatment is more persistent. Saski et al. (1991) estimated a half-life of > 1 y in an aquatic ecosystem. A long-term study in Lake Vattern, Sweden revealed that a major fraction of AOX discharged by a bleached kraft mill into the lake was removed from the water column by degradation (Grimvall et al. 1991). The 6-month half-life estimated in this study was considered an overestimation by its authors, who were unable to correct for significant natural inputs of AOX to Lake Vattern.

Persistence of organochlorines in soils and sediment

Biodegradation of organochlorines in soils and sediment

Aerobic biodegradation in soil has been well studied for a number of OC chemicals, especially the pesticide and pesticide metabolites. Some of the older, highly chlorinated pesticides such as chlordane and mirex were extremely resistant to biodegradation (Howard, Boethling et al. 1991; Howard, Michalenko et al. 1991), with soil half-lives of years. Newer pesticides are less chlorinated and are more biodegradable in soil. Chlorophenols are particularly well studied because they are herbicide degradation products (2,4-dichlorophenol and 2,4,5-trichlorophenol) or are used directly for wood preservation (PCP). The lower-chlorinated phenoxy herbicides such as 2,4-D are generally rapidly degraded, while the higher-chlorinated 2,4,5-T is more persistent.

In most soils and sediments with sufficient nutrient inputs to have significant microbial activity, biotransformation by natural microbial populations is the most significant pathway controlling persistence. Natural populations readily degrade a variety of organic compounds. For example, Swindoll, Aelion, Dobbins et al. (1988) studied the degradative abilities of microbial communities in pristine aquifers by measuring incorporation of radiolabeled substrate into cell biomass or metabolism to $^{14}CO_2$. The natural microbial community was able to degrade a variety of natural substances and xenobiotics, including chlorinated aromatics. Although microbially mediated transformations have been observed in such systems, contaminant degradation often is not complete. Examples of transformations of contaminants known to occur in soil and sediments include reductive dehalogenation of OCs, reduction of aromatic nitro-compounds, and formation of organosulfur compounds.

Microbial reductive dehalogenation, which is defined as removal of a halogen and addition of electrons, has recently been comprehensively reviewed (Kuhn and Suflita 1989a; Mohn and Tiedje 1992). Two processes were identified: 1) hydrogenolysis, in which the halogen is replaced by hydrogen, and 2) vicinal reduction, in which halogens bonded to neighboring carbons are replaced by a double bond. Reductive dehalogenations usually have been studied in the laboratory with enrichment cultures and sewage sludges under anoxic conditions. However, recent evidence suggests that these reactions may also occur with aerobic organisms under aerobic conditions. In a study of the biodehalogenation kinetics of alkyl halides, "oxidizer" organisms reductively dehalogenated alkyl halides under aerobic conditions faster than the anaerobic organisms did it anaerobically (Castro 1993). There is a relatively small dataset documenting reductively dehalogenated alkyl halides actual environmental occurrence and a still smaller dataset documenting rates under environmental conditions. A number of authors have observed the dehalogenation of simple chloroaliphatics such as chloroethanes, chloroethylenes, and chlorocyclohexanes in soils and groundwater (Kleopfer et al. 1985; Barrio-Lage et al. 1986; Bachmann et al. 1988).

Reductive dehalogenation of chloroaromatic compounds has also been reported. Chloroaromatics that have been observed to undergo reductive dehalogenations in soils and sediments include PCBs (Brown et al. 1984; Brown, Bedard et al. 1987; Brown, Wagner et al. 1987; Quensen et al. 1988; Beurskens, Dekker et al. 1993), methoxychlor (Muir and Yarechewski 1984), halobenzoic acids (Suflita et al. 1982; Horowitz et al. 1983), chlorophenoxyacetic acids (Suflita et al. 1984; Mikesell and Boyd 1985), toxaphene (Lee et al. 1977; Williams and Bidleman 1978; Harder et al. 1983), chloroanilines (Kuhn and Suflita 1989b), DDT (Guenzi and Beard 1967), chlorobenzenes (Beurskens, Mol et al. 1993; Beurskens, Dekker et al. 1994), and chlorophenols (Murphy et al. 1979; Mikesell and Boyd 1986). Recently, it has been suggested that several chlorinated dioxin and furan congeners also undergo microbial dehalogenation reactions in sediments (Beurskens, Dekker et al. 1993). The dechlorination of chloroaromatics appears to involve enzymes that are different from those involved in the chlorination of aliphatics. Peijnenburg et al. (1992) reported quantitative structure-activity relationships (QSARs) for reductive transformations of 45 halogenated aromatics in anoxic sediment. Disappearance rate constants were related to 4 molecular descriptors: carbon–halogen bond strength, summed Hammett sigma constants of other substituents, inductive constants of these substituents, and steric factors of the substituents. These results indicate that the overall structure of the compound, not just the presence of a C–Cl bond, is the determinant of disappearance rates.

There is little information concerning the rate at which these transformations occur under field conditions. However, Beurskens, Mol et al. (1993) compared PCB concentrations in a dated slice of a sediment core with those from an archived surface-sediment sample of the same age. Based on the assumption that the kinetics are first order, they estimated a dechlorination half-life of 9 y for PCBs in the anaerobic conditions at the bottom of Lake Ketelmeer in the Netherlands.

For phenolic substances, both microbiological methylation and demethylation can occur in soils and sediments. For example, in the Baltic, a methylation/demethylation cycle for chlorophenolics, particularly chloroguaiacols, occurs in sediment to produce a mixture of chloroanisoles and chloroveratroles (Neilson et al. 1983). In these studies, the parent chlorophenolics originated in pulp mills. The microbial degradation of greater-molecular-weight chlorinated organics from these pulp mills produced the same mix of chloroanisoles and chloroveratroles. Since these chloroanisoles and veratroles are neutral compounds, they should be more lipophilic than the parent chlorophenols and should exhibit a greater tendency to bioaccumulate. However, transformation of chloroveratroles themselves in sediments also occurs, and the products of these transformations may be strongly bound to the sediment (Remberger et al. 1986). Therefore, accumulation of chloroveratroles in sediments and subsequent transfer to organisms may be unlikely (Neilson et al. 1990). Since the chloroveratroles and anisoles have low taste and odor thresholds, these methylation reactions could be viewed as sources of potential taste and odor problems.

Hydrolysis of organochlorines in soils and sediment

The production of organosulfur compounds from halogenated alkanes in groundwater was reported by Schwarzenbach et al. (1985). The reaction between the halogenated alkane and the sulfur-containing derivative was apparently abiotic. However, the production of hydrogen sulfide (and other sulfides) from decaying organic matter in anaerobic soils is known to be microbial. Thus, the organosulfur production appears to be initiated by the microbiological formation of sulfur-containing species that can act as nucleophiles and attack suitable functional groups. This type of reaction may explain the observations of Buser and Muller (1986), who reported the occurrence of methylthio-derivatives of PCBs in lake sediments. In this case, the route of formation could be either nucleophilic displacement of chloride by hydrogen sulfide, followed by microbiological methylation, or direct nucleophilic attack by methyl sulfide. Since these reactions appear to take place under conditions of permanent burial in aquatic ecosystems, they likely deserve more study.

Photolysis of organochlorines in soils and sediment

In some cases, the same photochemical reactions observed in homogeneous solution could be expected to occur for molecules sorbed on soils and for molecules on soil or sediment surfaces. However, there are a several ways in which photochemistry at the interface can be different from that in solution. These include changes in reactivity due to physical effects, quenching or sensitization reactions, reactive species (Gohre and Miller 1983), stabilization of intermediates (Gohre et al. 1986), and photoredox chemistry involving the substrate (Cunningham et al. 1988). In some cases, photolysis on soils leads to products different from those produced in solution. For example, in the photolysis of octachlorodioxin in hydrocarbon solvents, 1,4,6,9-tetrachlorodioxin is the major tetrachlorinated photoproduct (Buser et al. 1978a, 1978b). On soils, the 2,3,7,8-congener is the major tetrachlorodioxin produced (Miller et al. 1989). The photoreactivity of sediment-sorbed compounds can change with age, apparently as a result of diffusion of the compound from the surface, where it is photoreactive, into the interior of the particle, where it is not (Miller and Zepp 1979). The efficiencies of direct photolysis of OCs on soils and sediments are generally low and are inverse functions of organic content. This suggests that quenching reactions involving energy transfer from the excited state of the contaminant to the organics in the matrix are predominant pathways. This quenching results in low efficiencies for direct photolysis of contaminants sorbed on soils and sediments. Hebert and Miller (1990) found that direct photolysis was restricted to the photic depth of soil (0.2 to 0.4 mm), while indirect photolysis on soils was slightly deeper.

In theory, adsorption of compounds to surfaces may lengthen the maximum absorption wavelength of the compound. Studies of the adsorption spectra of organics capable of hydrogen bonding indicate that substantial shifts in spectra occur when these molecules are adsorbed to silica gel and clays. For compounds that have photoactive bands just outside the wavelength range available in sunlight, such changes in spectra could result in greater light absorption and enhanced reactivity.

The relatively small quantity of soil and sediment surfaces irradiated (compared to the unirradiated reservoir) and the low efficiency of most of these reactions under environmental conditions suggest that photolysis of sorbed contaminants is unlikely to be an important pathway for degradation.

CHAPTER 3

Bioaccumulation and Metabolism of Organochlorine Compounds

Bioaccumulation is the process whereby chemicals accumulate in the tissues of organisms in concentrations that exceed those in the ambient environment (air, water, sediments, and soil). Bioaccumulation occurs through chemical uptake by biota and transfer through food chains. The magnitude of bioaccumulation is controlled by the properties of the chemical, the organism and its food web, and the ecosystem.

Bioaccumulation enhances exposure of organisms to chlorinated organic compounds. Since risks of toxicity increase with a chemical's increased concentration in susceptible tissues, bioaccumulation itself is a negative property.

The most important chemical properties known to determine the extent of organochlorine (OC) bioaccumulation are 1) the equilibrium partition coefficients of the chemical between the lipids of the organism and the external medium, such as water, air, and sediments, and 2) the structural characteristics of the chemical, including polarity, that control the chemical's persistence and rate of metabolic transformation.

Bioavailability is the fraction of total concentration of a chemical in an environmental medium that is available for uptake by an organism through a particular exposure route. Bioavailability is controlled by a combination of chemical properties; morphological, biochemical, and physiological attributes of the organism; and exposure conditions. Exposure concentrations, when expressed as the bioavailable fraction of total chemical in the medium, indicate the inherent bioaccumulation potential of a chemical.

Bioavailability is greatly influenced by hydrophobicity and other physicochemical properties. In general, more chemical is present in a bound form in air, water, soil, and sediment as hydrophobicity (and consequently, the tendency to sorb to carbon-containing particles) increases. This results in decreased availability of the chemical for direct uptake from these media. However, for some species such as filter-feeding organisms or clams that digest these particles, sorption of the chemical to particles enhances bioavailability.

CHAPTER PREVIEW

Hydrophobicity of chemicals that contain one or more aromatic rings increases with chlorine number. The presence of chlorine in an aromatic molecule does not necessarily make a compound a bioaccumulative substance, but there is a large body of scientific evidence that increased chlorination does, in general, give the compound a greater tendency to bioaccumulate in organisms. This general tendency for bioaccumulation of chlorinated aromatic chemicals does not always extend to chlorinated aliphatics, pesticides, and polymers.

Polarity of chemicals facilitates substitution reactions that are catalyzed by enzymes in organisms at all trophic levels. Generally, the higher the organism is in the food chain, the faster its metabolism. The most important exceptions to this rule occur when microorganisms significantly metabolize specific compounds.

Within a chemical class, metabolism sometimes varies enormously. For example, only a few polychlorinated dibenzo-*p*-dioxins (PCDDs), polychlorinated dibenzofurans (PCDFs), polychlorinated biphenyls (PCBs), and toxaphenes bioaccumulate through the food chain to birds and mammals, probably because these specific bioaccumulative congeners are very resistant to metabolism, whereas other congeners with similar hydrophobicities are much more readily metabolized. Normally, less persistent chemicals are formed, which leads to decreased bioaccumulation potential of these metabolites. However, other metabolites that are more persistent, such as methylated chlorophenols and methylsulfone-PCBs, may be formed.

Covalently bound chlorine atoms decrease microbiological degradation, sometimes leading to formation of toxic intermediates and hence impairing or slowing down the degradation that occurs in the aqueous environment and soil. Because of the potential for persistence and bioaccumulation, chemicals containing more than 2 chlorine atoms per molecule should be subjected to more rigorous risk assessment.

Direct exposure of plants and animals to halogenated hydrocarbons can result in adverse effects to organisms higher in food chains if these chemicals are bioaccumulated. Knowledge of bioaccumulation potential is therefore crucial to assessing exposure of organisms in food chains. Some halogenated compounds, e.g., modern pesticides, are designed to degrade rapidly by physical and microbiological processes. Other compounds may be persistent and potentially toxic, but they do not exert a toxic effect because their concentrations are low. However, even at low ambient concentrations, some halogenated compounds may be toxic at higher trophic levels because of bioaccumulation.

Processes Affecting Bioaccumulation

Bioaccumulation is the process by which a chemical is taken up from the exposure medium (air, water, sediment, soil, or diet) into an organism. This requires dissolution in surface lipids or waxes in the case of plants, microorganisms, and plankton. Bioaccumulation from water is an important phenomenon because of the large fugacity gradients that exist between water and aquatic biota for hydrophobic organic chemicals. Uptake by respiratory organs (gills) exposed to water can be an important route into aquatic animals. Uptake by fish through ingestion of food becomes the major route of uptake for very hydrophobic organic chemicals (log $K_{OW} > 5$). Absorption of contaminants dissolved in dietary lipids is the only important route for air-breathing animals. Table 3-1 lists some of the important processes affecting bioaccumulation.

Table 3-1 Important processes affecting bioaccumulation

Category	Process
Bioavailability and uptake	
	Feeding ecology
	Trophic status, whether aquatic or marine, including mammals and birds
	Temperature, for terrestrial or airborne exposures
	Partitioning between particulate matter, water, and air
	Speciation and binding to dissolved organic carbon
	Dissociation in response to environmental pH
Absorption	
	Respiratory uptake
	Surface partitioning
	Gastrointestinal tract uptake
	Maternal transfer in utero or in ovo
Distribution	
	Tissue distribution kinetics
	Bioenergetics in relation to growth and temperature
	Body and tissue composition, such as lipid concentrations and presence of binding proteins
	Speciation in relation to pH in the body
Metabolism	
	Presence of enzymes and their activity
	Induction of metabolic activity
	Availability of co-factors
	Metabolite pathways in the organism
Excretion	
	Through the respiratory surface
	Through the fecal route
	Through the urinary route
	Through the dermal route
	Through the reproductive route by egg gonadal production, milk, and shedding the placenta

Bioavailability and uptake

Concentrations of chemicals based on total mass in air and water, including particulates, may be poor measures of exposure, as environmental factors can modify bioavailability of chlorinated organics. If contaminants are adsorbed mainly on particles, they often (but not always) are less bioavailable than in the unbound phase. For example, bioaccumulation of 2,3,7,8-tetrachlorodibenzo-*p*-dioxin (2,3,7,8-TCDD) was enhanced in mountain whitefish downstream from a pulp mill because the fish were consuming filter-feeding caddisfly larvae. In contrast, benthic-feeding longnose suckers associated with relatively uncontaminated sediments did not accumulate significant amounts (Owens et al. 1994). The presence of dissolved humic material and colloids is also known to affect the bioaccumulation of hydrophobic organic substances (Leversee et al. 1983; Carlberg et al. 1986; Black and McCarthy 1988; Servos et al. 1989). Freely dissolved concentrations in water, which are most relevant for assessment of bioaccumulation potential, are very difficult to determine in practice. In sediments, porewater concentration is the determinant in bioconcentration by benthic invertebrates. In air, the accumulation of gas-phase concentrations of organochlorines (OCs) by plant surfaces is affected by temperature, which influences gas–particle partitioning and air–leaf exchange (Paterson et al. 1991, 1994).

Stability and bioavailability in the environment determine environmental concentrations and therefore exposure. The chemical's properties and the organism's ability to excrete the compound (with or without metabolism) determine the degree of bioaccumulation. The most important physicochemical property is hydrophobicity, which is related to solubility and vapor pressure. The ability to dissolve in lipids or waxes is also important in bioaccumulation of organic compounds but is less sensitive to structural features, such as number of chlorines, than is hydrophobicity. The tendency of organic chemicals to move from the medium to organism lipids is frequently modeled by the octanol/water partition coefficient, K_{OW}, in the case of water. Compounds with log $K_{OW} > 3$ will tend to bioconcentrate in aqueous environments (Figure 3-1).

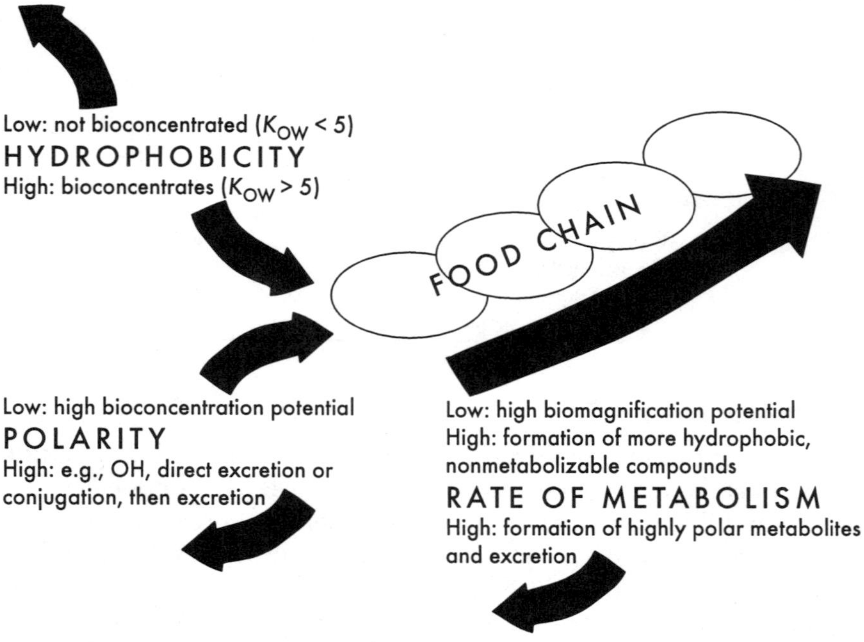

Figure 3-1 Chemical properties that strongly influence bioaccumulation

Many organisms do not have efficient mechanisms for excreting nonpolar, hydrophobic compounds but are able to excrete water soluble compounds or form more polar compounds through enzyme-mediated conjugation with glucuronic acid, sulfate, and glutathione. Ionizable compounds such as chlorophenols have a greater K_{OW} at low pH and are known to bioconcentrate more at these pHs in aquatic organisms (Howe, Marking, Bills, Boogard, and Mayer 1994; Howe, Marking, Bills, Rach, and Mayer 1994). However, the phenolic –OH renders chlorophenols easily conjugated and excreted. Thus, highly polar, basic and acidic compounds have low bioaccumulation potentials.

For species higher in the food chain, exposure to OC compounds occurs mainly through food. As a result, uptake from the gastrointestinal tract is the primary process influencing an organism's actual exposure. Net transport of OC compounds from the gastrointestinal tract to tissues is influenced by hydrophobicity, which is a function of the number of chlorine atoms per molecule. Hydrophobicity plays a highly significant role in net accumulation from the gastrointestinal tract for such polychlorinated polycyclic aromatics as polychlorinated biphenyls (PCBs), polychlorinated dibenzo-*p*-dioxins (PCDDs), and polychlorinated dibenzofurans (PCDFs). Eight or more chlorine atoms in these molecules may significantly reduce bioavailability after oral uptake (Opperhuizen et al. 1985; Niimi and Oliver 1988).

Metabolic rate and production, including growth, lipid deposition, and reproduction, control the food requirements of an organism and, thus, the rate of contaminant intake. The relationship between metabolic rate and contaminant intake is an allometric function of body weight: $W^{0.6}$ to $W^{0.8}$, so that food requirements per unit body weight decrease with increasing size, and contaminant uptake is also reduced. It is important to distinguish between warm-blooded (homeotherm) and cold-blooded (poikilotherm) animals. Homeotherms require significantly greater energy for the same body weight in order to maintain body temperature. Poikilotherms do not have this requirement, and their metabolic rate is therefore a function of ambient temperature. It has been estimated that a 1-kg gull in Lake Ontario would consume approximately 20 times more forage fish per year than a salmon or trout feeding on the same fish (Braune and Norstrom 1989).

Growth dilution may modify the concentration of chlorinated compounds in an organism. There is a significant difference in growth rate between homeotherms and poikilotherms. For example, fish usually have a rapid growth rate as juveniles, after which growth rate slows but continues over their lifetime. In contrast, birds and mammals attain their adult weight early in their lifetimes.

Water in aquatic ecosystems interfaces with microbial biofilms that actively sorb organic chemicals. This can lead to a large biomagnification of many OC compounds, such as PCBs. Microbes in the biofilms may biotransform some chemicals in such a way that more bioaccumulative chemicals are accumulated in food chains. The best known biochemical examples of this are biomethylation of phenolic hydroxyl groups, decarboxylation of organic acids, and conversion of organic sulfates and sulfones into sulfydryl compounds. These biotransformation processes may increase the K_{OW} by 10-fold or more (Neilson et al. 1990).

Distribution and metabolism

Once absorbed by the gastrointestinal tract, OC compounds are transported, mainly by binding to lipoproteins in the blood, to different tissues and organs in the body. Thus, bioaccumulation of halogenated hydrocarbons in mammals is observed not only in adipose tissue or blubber but also in the liver, bone marrow, and brain tissue. Even in brain tissues, concentrations in the mg/kg range have been observed for *p,p'*-dichlorodiphenyldichloroethylene (DDE), several PCB congeners (138 and 153), and components of toxaphene and chlordane (Tanabe et al. 1981). Hexachlorocyclohexane (HCH) enantiomer (+) –HCH accumulates preferentially in the brain (Mössner et al. 1992).

In all species, chlorinated aromatic compounds accumulate in tissues in proportion to the percentage of lipid. Exceptions are the 2,3,7,8-substituted PCDDs and PCDFs, which also strongly accumulate in the livers of mammals and birds. This greater liver retention has, in part, been attributed to the presence of inducible binding proteins for these compounds, such as cytochrome P4501A2 (Poland et al. 1989).

Many animals can enzymatically transform nonpolar xenobiotic compounds to more polar compounds that may be excreted (Figure 3-1). For alkyl-halogenated compounds, an important mechanism is enzymatic displacement of halogen with glutathione. Aromatic compounds are transformed mainly by cytochrome P450 mixed-function oxidases (MFOs) that add oxygen. If a moderately stable epoxide is formed, the molecule may be carcinogenic or may be transformed into nonmetabolizable aryl-methylsulfones. In other cases, very stable epoxides are formed that are not subject to attack by epoxide hydrolases or glutathione-S-transferases. Thus, heptachlor is metabolized to heptachlor epoxide, and several chlordane-related compounds are metabolized to oxychlordanes, which have a high bioaccumulation potential because they are not readily metabolized. The primary metabolites of chlorinated aromatics are phenols, which are conjugated and excreted in bile or urine. In some cases, dehydrohalogenation occurs, leading to a metabolically stable, hydrophilic metabolite. Such is the case for *p,p'*-dichlorodiphenyltrichloroethane (DDT), which is metabolized to *p,p'*-DDE by many organisms (Tanabe et al. 1988; Sijm and Opperhuizen 1989; Opperhuizen and Sijm 1990).

In vertebrates, the liver is the primary site for metabolism of chlorinated aromatic compounds. Metabolism of OC compounds is strongly governed by the presence of adjacent hydrogen atoms in the molecule, which facilitate the introduction of oxygen by the cytochrome P450 system. The number and position of the chlorine atoms can reduce the presence of adjacent hydrogens in the molecule. As a result, metabolic conversion to more hydroxylated or sulfonated polar metabolites is strongly hindered by the presence of chlorine atoms at these critical positions. Subsequent glucuronidation and elimination of the chlorinated compound from the body is significantly reduced.

Metabolism of many chlorinated compounds in the liver is highly stereospecific. Thus, the presence of chlorine atoms in critical molecular positions has a strong impact on the metabolic detoxification process because the parent compound is often more toxic than the metabolite. Metabolism of PCDDs and PCDFs is inhibited when chlorine atoms are present on all four 2, 3, 7, and 8 positions (Sijm, Wever, and Opperhuizen 1993). The resulting extended body retention of these PCDDs and PCDFs in combination with their strong enzyme-inducing properties make these compounds the most powerful known inducers of cytochrome P450.

Excretion

As discussed earlier, highly polar compounds may be taken up, but they are easily conjugated, making them water soluble and excretable in feces, through bile, or in urine. A less efficient means of excretion, but probably an important one for nonmetabolized chlorinated compounds such as PCB-153, is partitioning into gut contents, as occurs in the lower gut for hexachlorobenzene (HCB) in rats (Rozman et al. 1981); this process has also been inferred for excretion of mirex and *p,p′*-DDE in herring gulls (Clark et al. 1987).

Due to the great affinity of chlorinated aromatic compounds for lipid compartments, maternal transfer through lactation, especially for marine mammals, is a primary route for chemical elimination of chemicals from females and for exposure of the suckling young (Addison and Brodie 1987; Tanabe et al. 1988). Depending on the lipid content of the milk, the transfer becomes quantitatively more important. In all mammalian species studied so far, lactational exposure is greater than in-utero exposure. From a toxicological point, however, the smaller amounts transported across the placenta may be more important because of the greater sensitivity of the developing fetus. In birds and fish, chemical transfer from yolk lipids to the embryo, as the yolk is used up during early development and growth, is an important exposure pathway (Miller 1993; Walker et al. 1994).

Biomagnification

Biomagnification occurs when lipid-normalized concentrations of a chemical increase with position in a food chain. A large number of processes may affect bioaccumulation (Table 3-1). The rate of uptake of the chemical from food into the organism is increased through changes in chemical activity in the food during the digestive process (Connolly and Pederson 1988; Gobas 1993). Although steady state between these uptake and excretion rates may be established, the concentration in the organism will be greater than that in the diet. If substances have low or zero rates of metabolism and growth dilution, the excretion rate may be so slow that steady state is not reached and the concentration builds up with age, as has been shown for marine mammals, especially males (Addison and Smith 1974), and fish (Sijm et al. 1992). Female marine mammals have lactation as an additional route of excretion and are more likely to come to steady state with their diet. Biomagnification factors (BMFs) between trophic levels are in the order of 3 to 20 on a lipid-weight basis (Thomann 1981; Braune and Norstrom 1989; Muir et al. 1992; Suedel et al. 1994). Thus, even in relatively uncontaminated environments like the Arctic, marine mammals and polar bears attain mg/kg levels of contaminants such as PCBs and chlordane in fat (Norstrom and Muir 1994).

Within chlorinated and nonchlorinated compounds, there are ranges of properties and bioaccumulation potentials in different groups of organisms (summarized in Table 3-2).

Tools for the Assessment of Exposure and Bioaccumulation of Organic Chemicals

Halogen-specific analysis can be used as a first tool to assess the range of concentrations or organohalogens that may be present in organisms or their environmental matrix. These include tests for adsorbable organic halogen (AOX), solvent-extractable organic halogen (EOX), and purgeable organic halogen (POX). For these analyses, the detection range varies from ng/L to µg/L. The assays are not specific to halogen type (Cl, Br, I, or [less sensitive] F) nor to the chemical class of compounds; therefore, they are best used to select samples and environments for component-specific analysis rather than for direct quantification of exposure.

Currently, available tools for the assessment of exposure and bioaccumulation of organic chemicals can be grouped in 2 broad classes. First, measurements of concentrations of a particular chemical in the environment can be used to assess the extent to which the substance bioaccumulates. An example is the biota–sediment accumulation factor (BSAF) method, where observed concentrations of the chemical in organisms and sediment are used to construct an organism–sediment concentration ratio that expresses the extent of chemical bioaccumulation. BSAFs have recently allowed estimation of bioaccumulation factors in fish and herring gull eggs for chemicals whose concentrations in water are below analytical limits of detection (USEPA 1995). BSAFs can be used in retrospective assessments of chemicals, giving a historical perspective to evaluations of present or future exposure conditions. BSAFs also are useful for identifying organic chemicals that have reduced bioaccumulation potential due to biotransformation. This method is inherently reactive, as bioaccumulation of organic chemicals in the environment must occur before it can be recorded. However, in combination with proactive methods, BSAF can play an important role in adaptive environmental management strategies for organic chemicals.

Table 3-2 Bioaccumulation as influenced by chemical properties and trophic levels

Chemical class	Log K_{OW}[1]	Polarity[2]	Metabolism[3]	Bioaccumulation factor[4]
PCBs	3 to 8	–	MO: – –	MO: ++
			F: –/+	F: +/–
			B: +	B: +
			M: ++	M: –/+
PCDDs/PCDFs	3 to 8	+	MO: – –	MO: ++
			F: +/–	F: –/+
			B: +	B: –/+
			M: ++	M: –/+
Polychlorinated benzenes (PCBzs)	3 to 6	–	MO: – –	MO: ++
			F: +/–	F: ++
			B: +	B: +/–
			M: ++	M: +/–
Polychlorinated naphthalenes (PCNs)	3 to ?	–	MO: ?	MO: ++
			F: +/–	F: +/–
			B: +	B: +/–
			M: +	M: +/–
Polychlorinated terphenyls (PCTs)	high	–	MO: ?	MO: ++
			F: ?	F: ++
			B: ?	B: ?
			M: ?	M: ?
Polychlorinated benzyltoluenes (PCBTs)	high	–	MO: ?	MO: ++
			F: ?	F: ++
			B: ?	B: ?
			M: ?	M: ?
Polychlorinated diphenyl ethers (PCDEs)	4 to 8	+	MO: ?	MO: ?
			F: –/+	F: +
			B: ?	B: ?
			M: ?	M: ?
Polychlorinated phenols (PCPs)	2 to 5	++	MO: +/–	MO: –
			F: +	F: +/–
			B: ++	B: –
			M: ++	M: –
Polychlorinated anilines (PCAs)	3 to 6	++	MO: ?	MO: ?
			F: +/–	F: +/–
			B: +	B: ?
			M: ++	M: ?
Polychlorinated paraffins (CPs)	5 to 8	++	MO: ?	MO: ++
			F: –/+	F: ++
			B: ?	B: ?
			M: ?	M: ?
Chlorinated aliphatics (C_1–C_3)	< 3	–	MO: +/–	MO: ?
			F: +	F: ?
			B: ++	B: ?
			M: ++	M: ?
Chlorinated pesticides (old)	(all) 3 to 8		(all) MO: ?	(all) MO: +
Dieldrin		+	(all) F: ?	(all) F: +
DDT		–	(all) B: ?	(all) B: +
Mirex		–	(all) M: ?	(all) M: +
Chlordane		–		
HCH		–		
Chlorinated pesticides# (modern)			(all) MO: ++	(all) MO: –
Sulfonylurea (herbicides)	< 1	+	(all) F: ++	(all) F: –
Triazine (herbicides)	< 3	+	(all) B: ++	(all) B: –
Chloroacetamides (herbicides)	2 to 4	+	(all) M: ++	(all) M: –
Acetanalide (fungicides)	< 2	+		
Triazole (fungicides)	< 3	+		
Organophosphorous compounds (insecticides)	2 to 4	+		
Chloropolymers (lignins, humic, soil)	low to moderate	+(+)	MO: ?	MO: ?
			F: ?	F: ?
			B: ?	B: ?
			M: ?	M: ?

[1] octanol/water partition coefficient
[2] containing apolar group, such as –OH, –NH2, –O–
[3] + some/all congeners metabolized, ++ congeners rapidly metabolized, – no congeners metabolized
[4] + some/all congeners bioaccumulated, – no congeners bioaccumulated
?=unknown, MO=microorganisms, invertebrates; F=fish; B=birds; M=mammals; #=some contain chlorine atoms

The second class of tools for estimating bioaccumulation of organic chemicals is based on the application of knowledge of the mechanism of bioaccumulation of organic chemicals in organisms. This can be done before the chemical is introduced in the environment and, as long as some basic chemical properties can be assessed, even before the substance is manufactured.

Different types of predictive modeling tools are available for assessing exposure and bioaccumulation of organic chemicals in biological organisms. Mechanistic, process-oriented kinetic models are discussed in Chapter 4. Empirical and equilibrium partitioning models are summarized below.

Empirical partitioning models

Bioaccumulation factors (BAFs) can be linked to the output of chemical fate models in order to predict concentrations of OCs in biota from predicted concentrations in air, surface waters, suspended solids, sediment, and soil. In the case of aquatic biota, dissolved concentrations in water and concentrations on suspended solids and surface sediments are required. The simplest modeling approach for predicting concentrations in fish is to use the bioconcentration factor (BCF), the ratio of the concentration of chemical in an organism or in a specific tissue of an organism to the dissolved concentration in water. BCFs are available for a large number of chemicals and equations that relate log K_{OW} to log BCF (Veith et al. 1979; Connell 1988; Connell and Schüürmann 1988; Mackay et al. 1992a, 1992b, 1992c). This approach is most suitable for chemicals with log $K_{OW} < 4$, i.e., those that would be found mainly in the dissolved phase in natural waters. Porewater-to-biota bioconcentration can also be modeled for moderately hydrophobic compounds in sediment, using BCFs (Landrum 1989; Lake et al. 1990), provided that the concentration of the dissolved phase can be estimated. The BCF approach assumes that equilibrium is reached between the chemical in fish and the chemical in water.

For more hydrophobic OCs (log $K_{OW} > 5$), which are associated mainly with sediments and suspended particles in natural waters, BCFs tend to underestimate concentrations in fish, especially in top predators, because of contributions from food-chain transfer (Connolly and Pederson 1988; Thomann 1989). BAFs, whether measured or predicted with models, are used in many aquatic risk assessments because they are the most direct expression of bioaccumulation potential, especially for pelagic species. Bioaccumulation factors must be clearly defined and applied in order to avoid large errors associated with variation in chemical bioavailability and organism bioenergetic or pharmacokinetic factors (Cook et al. 1994). For example, the most generically applicable BAF for a fish species is based on the lipid-normalized concentration of chemical in tissues (C_ℓ) and the concentration of freely dissolved chemical in water (C_w^{fd}):

$$BAF^{fd}_\ell = C_\ell / C_w^{fd} \quad \text{(Equation 3-1).}$$

Hydrophobic OCs with log $K_{OW} > 6.5$ can rarely be measured in the freely dissolved phase (especially chlorinated dioxins and furans). The BSAF, calculated from the concentration of the substance in the organism lipid divided by the concentration in sediment organic carbon (Ankley et al. 1992; Cook 1994; Parkerton et al. 1993), can be used to estimate the concentration in benthic and pelagic organisms. This approach has the advantage that sediment concentrations are generally known and the uptake pathway, e.g., from pore water and/or particle ingestion or assimilation for benthic invertebrates, need not be defined. However, data on BSAFs are available for relatively few compounds, and results are mainly from field measurements in which the state of sediment–water equilibrium has not always been well characterized.

To evaluate the state of disequilibria, BSAFs can be written as

$$\mathrm{BSAF} = \mathrm{R_{BS}} \bullet K_{OW} / K_{OC} = \mathrm{R_{BW}} \bullet \mathrm{R_{WS}} \bullet K_{OW} / K_{OC} \qquad \text{(Equation 3-2)},$$

where octanol is assumed to be a surrogate for lipid (USEPA 1993a). At equilibrium, the disequilibrium factors for organism/water (R_{BW}), water/sediment (R_{WS}) or organism/sediment pore water (R_{BS}) = 1.

Disequilibria may exist between sediment pore water (C_S^D) and overlying water (C_W^D): $R_{WS} = C_W^D / C_S^D > 1$. Filter-feeding organisms may selectively ingest contaminated particles that are not equilibrated with the water or bottom sediment. BSAF values for PCDDs and other OCs have been shown to have a parabolic relationship with log K_{OW} (Parkerton et al. 1993; Endicott and Cook 1994), with the greatest BSAFs observed at log $K_{OW} = 5.5$ to 6.0. This may be due to decreased membrane permeability of superlipophilic organics, reduced accumulation through water, and lower assimilation efficiency from food (Opperhuizen et al. 1985; Ankley et al. 1992; Parkerton et al. 1993). Data requirements for calculation of concentrations in fish or invertebrates using the BCF are relatively straightforward: concentrations of the dissolved phase in water and the BCF or log K_{OW} value are required. Using the BSAF concentrations of the OC in sediment requires knowledge of sediment organic carbon and lipid percent in fish, along with knowledge of sediment–water disequilibria.

Equilibrium partitioning models

Equilibrium partitioning assumes that chemical bioaccumulation is due to thermodynamically controlled partitioning of the chemical between the tissues of the organism and such external media as water for fish and aquatic macrophytes, sediment and pore water for benthic invertebrates, or air for terrestrial plants. This partition coefficient reflects the affinity of the chemical or its solubility in the tissues of the organisms and those in the external media. For hydrophobic organic chemicals, lipids are often considered the most important compartment of the organisms into which this partitioning occurs. Hence, the bioaccumulation of hydrophobic organic chemicals is often viewed as a partitioning into lipids. In the aquatic environment, this form of bioaccumulation is often referred to as "bioconcentration," and it can be satisfactorily mimicked and represented by the octanol/water partition coefficient K_{OW}. The model that describes this for fish, phytoplankton, and aquatic macrophytes is

$$BCF = C_b / C_w^{fd} = f_\ell \bullet K_{OW} \quad \text{(Equation 3-3)},$$

where BCF is the bioconcentration factor (L/kg), C_w^{fd} is the bioavailable or freely dissolved concentration of the chemical in the water, and f_ℓ is the lipid content in the organism.

For benthic organisms, equilibrium bioconcentration has alternatively been expressed as

$$C_B/C_S = d_L/L_B = C_S \bullet d_{OC}/OC = K_{L\,W} \bullet C_P \quad \text{(Equation 3-4)},$$

where C_B is the concentration of chemical in the benthic invertebrate (µg/kg wet weight), C_S is the concentration in the sediments (µg/kg dry weight), C_P is the truly dissolved concentration of chemical in the pore water (µg/L water), L_B is the lipid fraction of the benthos (kg lipid/kg organism), d_L is the density of the lipids of the benthos (kg/L), OC is the organic carbon fraction of the sediments (kg organic carbon/kg organism), d_{OC} is the density of the organic carbon fraction of the sediments (kg/L), and K_{LW} is the dimensionless lipid/water partition coefficient. An interesting aspect of this model is that C_B / C_S should be approximately similar for organic chemicals, namely $L_B \bullet d_{OC} / OC \bullet d_L$, or simply L_B / OC, since d_L and d_{OC} are approximately the same. Thus, e.g., if L_B is 6% and OC is 2%, then C_B is approximately 3 times greater than C_S. Although this model is simple, and physiologically more detailed models can be derived, this equation has been shown to be in good agreement with field data (Gobas et al. 1989).

For terrestrial plants, the model is

$$BCFV = CBV / CA \quad \text{(Equation 3-5)},$$

where BCF_V is the plant–air volumetric BCF, C_{BV} is the concentration in the plant on a volume basis (mol/m^3), and C_A is the concentration in air (mol/m^3) (Paterson et al. 1991). Paterson et al. (1991) showed that BCF_V is correlated with water/air (K_{WA}) and octanol/air (K_{OA}) partition coefficients:

$$BCFV = 0.2 + 0.7K_{WA} + 0.05\,K_{OA} \quad \text{(Equation 3-6)}.$$

K_{WA} is a dimensionless Henry's Law constant (HLC). K_{OA} is calculated from the ratio of K_{OW} / K_{WA} or, more correctly, is obtained by direct measurements (Harner and Mackay 1995).

These simple equilibrium partitioning models have been effective tools for the estimation of the bioaccumulation potential of hydrophobic organic chemicals with log K_{OW}s between 2 and 5 in phytoplankton (Geyer et al. 1984, 1991; Mailhot 1987; Gobas et al. 1991), zooplankton (Hawker and Connell 1989), aquatic macrophytes (Gobas et al. 1991), fish (Veith et al. 1979), benthic invertebrates (Shea 1988; Gobas et al. 1989; DiToro et al. 1991; Geyer et al. 1991), and terrestrial plant species (Bacci et al. 1990; Paterson et al. 1991). The simple partitioning models have not been applied to mammals and birds. These models ignore the importance of several biological factors and environmental conditions, including metabolic transformation, routes of chemical uptake (e.g., skin absorption, gill uptake, and dietary uptake), routes of elimination (e.g., gill elimination and reproductive, fecal, and urinary excretion), growth, biomagnification, and temperature (Sijm et al. 1992; Sijm, Part, and Opperhuizen 1993). By providing the BAF at equilibrium (the BAF that will eventually be achieved), the models ignore the concentration–time curve leading up to equilibrium. The reason for the success of the models is that many of these factors can indeed often be ignored under environmental conditions. For example, for many chemicals with a log K_{OW} between 2 and 5, the time to equilibrium ranges from hours in small organisms such as phytoplankton to several days in large organisms such as rainbow trout. This means that many of these chemicals can indeed be found at equilibrium under actual environmental conditions. It also means that, when ambient concentrations change, the chemical adopts a new equilibrium relatively quickly.

The partitioning models have been particularly useful for estimating the bioaccumulation of OCs. The reason is that chlorine substitution often, but not always, gives the chemical a resistance to metabolic transformation, which is the most important factor causing the partitioning model to fail. It is fortunate, however, that this failure always results in an underestimation of the BAF. Hence, the simple partitioning models are still useful in estimating the maximum extent of chemical bioaccumulation. Bioaccumulation in some organisms, in particular birds and mammals, cannot be described with partitioning models.

Models for assessing bioavailability

Processes that affect the amount of chemical available for bioaccumulation are often broadly classified under "bioavailability." These processes often affect the apparent, rather than the actual, ability of chemicals to bioaccumulate. The most important known processes are

1) sorption of chemical to organic carbon in the water phase;
2) dissociation of chemical in the water phase;
3) biotransformation by the organism during exposure;
4) diffusion of chemical out of ingested matrices, such as sediments in benthic invertebrates and particles falling on plants;
5) transfer of the chemical from one compartment (water) into another (sediment, sludge), leading to exposure through multiple routes; and
6) microbial modification of a chemical in the environment or the gastrointestinal tract in such a way that its bioaccumulation potential changes.

Bioavailability is controlled by a combination of environmental factors (e.g., pH, concentration of particulate organic carbon, and redox potential) and chemical properties (Landrum 1989; Ankley et al. 1992). The most important chemical factors that may limit chemical diffusion rates in the gastrointestinal tract are K_{OW}, the dissociation constant, and molecular size (Gobas et al. 1994; Opperhuizen et al. 1995).

The process of dissociation of chemicals is very well defined, and the models for estimating the extent of dissociation are very good. In essence, it is assumed that organic chemicals can be absorbed when they occur in the nonionic or nondissociated form. Simple acid base chemistry models can be used to assess what fraction of the chemical is in this nonionic form, given the pH of the environment and the dissociation constant of the chemical.

The models that are used for estimating bioavailability of organic chemicals due to sorption of chemicals to particulate and dissolved organic carbon in the water phase are also well defined. However, greater uncertainty should be anticipated because of site-specific interactions between the chemical and the sorbent that are currently not well understood, such as the effect of sorbent or particle size and chemical composition of the sorbent. The model is usually expressed as the fraction of chemical in water that is freely dissolved (f_{fd}):

$$f_{fd} = 1 / [1 + (DOC)(K_{doc}) + (POC)(K_{poc})] = 1 / [1 + (DOC / 10 + POC)(K_{OW})] \quad \text{(Equation 3-7)},$$

where DOC and POC are the concentrations (kg/L) of dissolved and particulate organic carbon, respectively, in the water. Under most natural conditions, this equation may give a reasonable estimate of the fraction of chemical in the water that is available for uptake by aquatic organisms. The greatest source of uncertainty is associated with the approximation of the DOC partition coefficient, (K_{doc}), as $K_{OW} / 10$ (USEPA 1995).

Diffusion rates of chemicals from certain matrices are controlled by K_{OW} (in aqueous media) and molecular size. Very little is known about this phenomenon. This form of bioavailability of chemicals in "food" matrices (benthos) or particles (plants) is likely to be important for chemicals that are large molecules and that have a high K_{OW}, and it is more important for benthic invertebrates and plants than for fish, mammals, and birds.

Organic chemicals may become nonavailable if the molecular diameter exceeds 0.96 nm (Opperhuizen et al. 1985) so that the molecules may be unable to diffuse through membranes. There is considerable uncertainty on this issue, as molecules with the same and larger dimensions have been observed to be absorbed and bioaccumulated (Muir et al. 1986; Gobas et al. 1989; Sijm, Part, and Opperhuizen 1993). Thus, at this stage, the threshold levels for membrane diffusion limitations to uptake of chemicals have not been well defined.

In summary, there are various methods or tools to assess bioaccumulation and biomagnification potential of organic chemicals in the environment. These tools vary from empirically derived BAFs, such as BAF and BSAF, to detailed mechanistic descriptions of bioaccumulation and food-chain transfer. The models are not specific to OCs but have been developed and tested using OC substances (Sijm, Wever, and Opperhuizen 1993). The choice of bioaccumulation measurement and prediction tools is largely dependent on the goal of the investigation. In the absence of metabolic transformation, most models show that the extent of bioaccumulation is closely related to the K_{OW} of the chemical. Hence, if K_{OW} increases, exposure to the chemical and risk of effects become more important. However, when metabolic transformation occurs, the bioaccumulation potential can be much smaller than estimated from K_{OW} and may be virtually 0. Methods for estimating metabolic transformation rates based on chemical properties are very poorly developed. Therefore, it is difficult to estimate the bioaccumulation potential of metabolizable chemicals, unless the rates have actually been measured for the organism or systematic analysis of measured BAFs such as the BSAF is accomplished for the chemicals of concern.

Specific Chemical Classes

Polychlorinated biphenyls, polychlorinated dibenzo-*p*-dioxins, and polychlorodibenzofurans

PCBs, PCDDs, and PCDFs are the classes of chlorinated aromatic chemicals that have attracted the most attention because many members, or congeners, bioaccumulate to potentially toxic concentrations, especially in high trophic-level animals (Olsson et al. 1985; Safe 1989, 1990). Of the nonpolymeric chlorinated compounds with potential for bioaccumulation, PCBs also had the greatest global production (Blectchy 1983). There are 209 possible PCB congeners with 1 to 10 chlorines substituted on the aromatic rings, and 75 PCDD and 135 PCDF congeners with 1 to 8 chlorines on the aromatic rings. In practice, as well as in theory, a large number of these congeners are emitted to the environment.

Because of the 2- to 3-ring aromatic skeleton of these compounds, the K_{OW} of even the monochloro congeners is > 3. The majority of the congeners of concern are tetrachloro to octachloro congeners, which have K_{OW}s from 5 to 8 (Mackay et al. 1992c). PCBs, PCDDs, and PCDFs also are not easily degraded by microbial action or physical processes and therefore are highly persistent in all media. These properties satisfy the criteria for high bioconcentration in aquatic or marine environments, and BCFs from water to plankton are in the order of 10^5 to 10^7 (Swackhamer and Skoglund 1992). The most highly chlorinated members of these classes (octachloro and decachloro) have K_{OW}s > 8. The bioavailability of these compounds is reduced in aquatic or marine ecosystems by partitioning to particulate matter, and bioconcentration potential may be significantly reduced in habitats where there are greater concentrations of particulate organic matter.

PCBs, PCDDs, and PCDFs are nonpolar compounds that cannot be excreted or transformed to polar excretable compounds without the introduction of a polar functional group through metabolism. None of these compounds are metabolized at a significant rate by invertebrates, and therefore they biomagnify through the diet in invertebrate food chains (Thomann 1989; Gobas 1993).

PCDDs and PCDFs that do not possess chlorines at all four 2, 3, 7, and 8 positions do not bioaccumulate in vertebrates (except cetaceans). It is not clear which enzyme system or excretory mechanism is responsible for the clearance of the non-2,3,7,8-substituted PCDDs and PCDFs. Cetaceans, on the other hand, appear able to metabolize (or excrete) 2,3,7,8-substituted PCDDs and PCDFs and accumulate only the non-2,3,7,8-substituted congeners (Norstrom et al. 1990). Clearance rates of 2,3,7,8-substituted PCDDs and PCDFs are low enough (except in cetaceans) that biomagnification occurs in food chains, in the order of 3- to 10-fold per trophic level (Endicott and Cook 1994).

PCBs bioaccumulate in invertebrates and most fish with little metabolic alteration. Thus, excretion of less chlorinated compounds may occur across the gills in fish, but generally PCBs are bioaccumulated by fish with little change in patterns (Norstrom 1988). Vertebrates higher in order than fish, including reptiles, birds, and mammals (except marine mammals), possess hepatic cytochrome P450 mono-oxygenase enzyme systems that are efficient at degrading certain PCB congeners. C4502B enzymes are responsible for metabolism of PCB congeners that are not substituted by chlorine at one or more *meta-* and *para-*positions. Thus, PCB congeners with at least one unsubstituted *meta-* and *para-*position are not biomagnified in higher animals. As for PCDDs and PCDFs, cetaceans have a metabolic capability different from other mammals (Tanabe et al. 1988; Norstrom et al. 1992). They do not degrade *meta-* and *para-*unsubstituted PCB congeners to any extent. Seals may be intermediate to cetaceans and land mammals in this regard (Boon et al. 1994).

PCBs that are easily metabolized by *meta-* and *para-*epoxidation and possess 2,5 or 2,3,6 chlorine substitution on the same ring, tend to form persistent methylsulfone metabolites through a complex process involving substitution of glutathione at the 3 or 4 positions, excretion of the mercapturic acid, cleavage by gut microflora C-S lyase, resorption, methylation, and hepatic oxidation to 3- and 4-methylsulfone PCBs (Opperhuizen and Sijm 1990). Many of these compounds are resistant to further metabolism and are sufficiently hydrophobic that they are not excreted. The potential therefore exists for biomagnification in food chains, although it has not been demonstrated to date. The concentrations of methylsulfone PCBs relative to remaining unmetabolized PCBs is on the order of 0.01 to 0.10 in the animals studied to date (gray seal, mink, otter, beluga whale, and polar bear). Methylsulfone PCBs have been found to bind to various receptors and induce hepatic enzymes, but very little comprehensive information is available on their toxicology. These compounds should therefore be given careful consideration in PCB risk assessment.

Polychlorinated diphenyl ethers

Polychlorinated diphenyl ethers (PCDEs) compose a group of 209 congeners similar to those of the PCBs. The substitution patterns on the 2 phenyl rings are the same as those of the PCBs. The oxygen bridge of the diphenyl ether leads to a dipole moment of 1.6 Debye (25 °C, benzene). Upon substitution in the 3-position with chlorine, it increases to 1.68 Debye.

The log K_{OW} values for the mono-chloro diphenyl ethers are in the range of 4.4 to 4.7 and increase for the hexachloro diphenyl ether to 6.0 to 6.8 (Becker et al. 1992; Kurz and Ballschmiter 1994). PCDEs have never been produced in substantial amounts per se but are byproducts in the 1 to 10 mg/kg range in chlorophenols (Table 3-3). The degree of chlorination of these byproducts ranges from 5 to 9 chlorines per molecule.

Table 3-3 Major PCDEs and their concentrations in pentachlorophenol

Congener	Substitution	Concentration range (mg/kg)
(PCDE 47)	2,2',4,4'-Cl_4DE	1 to 5
(PCDE 85)	2,2',3,4,4'-Cl_5DE	5 to 10
(PCDE 99)	2,2',4,4',5-Cl_5DE	1 to 5
(PCDE 137)	2,2',3,4,4',5-Cl_6DE	1 to 5
(PCDE 138)	2,2',3,4,4',5-Cl_6DE	1 to 5

PCDEs are also formed in the chemistry of incomplete combustion. They are precursors of the PCDFs. The degree of chlorination has its maximum, in this case (incomplete combustion), of 2 to 3 chlorines per molecule.

PCDE congeners are bioaccumulated by fish (Niimi 1986; Opperhuizen and Voors 1987) and are detected in fish from the Great Lakes (Niimi et al. 1994), with the concentrations in Lake Superior < Lake Huron < Lake Erie < Lake Ontario in the range of 1 to 17 µg/kg for the hexachloro congeners. The hexachloro congeners were present in cod liver oil from the North Atlantic in the 4- to 12-µg/kg range in 1985 and in the 2- to 4-µg/kg range for 1993 (Kurz and Ballschmiter 1995).

The similarity in properties of these chemicals and the PCBs, in terms of hydrophobicity and chemical stability, lead to similar regional and global distribution patterns for PCDEs. The PCDEs also have potential for biomagnification in a food chain. In white tail eagles caught in the Baltic, levels up to 4700 µg/kg lipids have been detected (Koistinen et al. 1993).

Polychlorinated phenoxyphenols

In chlorophenol formulations that were technically produced by chlorination of phenol (rather than by alkaline hydrolysis of chlorobenzene), polychlorinated phenoxyphenols (PCPPs, hydroxylated diphenyl ethers) are found as impurities in the range of 1 to 5% by weight. The most abundant are the 5-*para*- and 4-*ortho*-hydroxylated hexa- to nona-chlorinated congeners (Table 3-4) (Kitunen et al. 1985). They are persistent and have log K_{OW}s between 4.3 and 4.7 (Valo and Salkinoja-Salonen 1986). In historically contaminated sawmill soils, PCPPs have become biomethylated (especially the *para*-hydroxylated congeners), which increases their log K_{OW}s to between 5.0 and 5.5. PCPPs usually occur in contaminated soils together with PCDDs and PCDFs (Kitunen et al. 1985, 1987). Unlike chlorophenols, they bind to soil and show little vertical or horizontal movement in the contaminated sites.

Table 3-4 Polychlorinated phenoxyphenols in technical chlorophenol products and contaminated soil

Polychlorinated phenoxyphenol	Concentration in technical products (mg/kg)	Concentration in contaminated soil (mg/kg dry soil)
4,6-Dichloro-2-(2,4,6-trichlorophenoxy)phenol	500 to 700	1 to 10
2,6-Dichloro-2-(2,4,6-trichlorophenoxy)phenol	600	1 to 10
4,6-Dichloro-2-(2,3,4,6-tetrachlorophenoxy) phenol	1000 to 1200	1 to 10
4,5,6-Trichloro-2-(2,4,6-trichlorophenoxy) phenol	600 to 900	2 to 20
2,3,6-Trichloro-4-(2,4,6-trichlorophenoxy) phenol	600 to 800	1 to 10
2,6-Dichloro-4-(2,3,4,6-tetrachlorophenoxy) phenol	1000 to 1500	1 to 10
4,5,6-Trichloro-2-(2,3,4,6-tetrachlorophenoxy) phenol	1700 to 2300	3 to 30
2,3,6-Trichloro-4-(2,3,4,6-tetrachlorophenoxy) phenol	1900 to 2400	3 to 30
2,3,4-Trichloro-4-(2,3,4,6-tetrachlorophenoxy)phenol	< 10	–

Polychlorinated benzenes, naphthalenes, phenols, and anilines

Polychlorinated benzenes (PCBzs) have been used as insecticides, solvents, or intermediates for synthesizing other chemicals. HCB has been widely used and was found to accumulate in all trophic levels. Lower-chlorinated PCBzs with fewer than 5 chlorine atoms were found to bioaccumulate in lower trophic levels, such as in fish and invertebrates, but not so much in higher trophic levels, such as in birds and mammals, probably due to metabolic breakdown (Mackay et al. 1992a).

Polychlorinated naphthalenes (PCNs) have been produced commercially (e.g., Halowaxes) and are formed and released to the environment through processes in which chlorine is present, such as chlorine production, municipal waste incineration, and copper roasting. They are also present as contaminants in PCB formulations (Brinkman and de Kok 1980). PCNs have been found to bioaccumulate in fish, although the lower-chlorinated congeners (having only 1 or 2 chlorines) were significantly metabolized by fish (Opperhuizen et al. 1985). Some tetra- through hepta-chlorinated naphthalenes have been found in concentrations comparable to PCB concentrations in some wildlife species in Sweden (Järnberg et al. 1993). It has been suggested that octachloronaphthalene is too large to enable uptake through the gills of fish (Opperhuizen et al. 1985).

Of the polychlorinated phenols (PCPs), pentachlorophenol is the best representative, since it was used as a wood preservative that contained many PCDDs as byproducts. In addition, PCPs are byproducts that occur in effluents from bleached kraft mills, where molecular chlorine is used for bleaching. The hydroxy group of phenols is readily attacked by microorganisms that convert them into anisoles through methylation and make them more easily bioaccumulated. In higher organisms, the phenols can undergo conjugation (with glucuronide). The degree of ionization (pK_a) depends on the number and substitution pattern of the chlorines and greatly affects bioavailability.

Polychlorinated anilines (PCAs) are used as intermediates for the production of organic dyes. In addition, they are formed during microbial breakdown of dyes that have been discharged into the environment. They are found to bioaccumulate in fish, but at a slightly lower rate than would be expected based on their K_{OW}s, probably due to metabolic breakdown (De Wolf et al. 1992). Acetyl conjugation of the amino group is found in fish (Sijm and Opperhuizen 1989).

Except for the PCBzs, little information is available about bioaccumulation in animals at lower and higher trophic levels than fish for all these individual classes of chemicals (PCNs, PCPs, and PCAs).

Polychlorinated terphenyls and benzyltoluenes

Polychlorinated terphenyls (PCTs) have been produced as replacements of PCBs. Bioaccumulation of PCTs has been found in saltmarsh cordgrass, oysters, crabs, and mummichog (Jensen and Jørgensen 1983).

Polychlorinated benzyltoluenes (PCBTs) consist of predominantly tetrachlorinated and some trichlorinated congeners (Ehman and Ballschmiter 1989). Bioaccumulation has been found in fish, in both laboratory and field studies (Gallagher et al. 1993). Little quantitative information is available, nor is information about bioaccumulation in animals in lower and higher trophic levels.

Chlorinated aliphatics

Chlorinated solvents, principally formulations of methylene chloride, perchloroethylene, trichloroethylene, and methyl chloroform, have applications in metal cleaning, the manufacture of printed circuit boards, polystyrene foam production, pharmaceuticals, paint removal, and dry cleaning. These solvents have been widely used for 50 years. The monomers vinyl chloride and vinylidene chloride are extensively used in the production of plastics, the worldwide production of polyvinylchloride (PVC) from vinyl chloride being around 20 million metric tons.

Chlorinated C_1–C_3 solvents volatilize easily, and it has been estimated that about 90% end up in the atmosphere (Pearson 1982; Fabian 1986) at 10 to 100 X greater concentrations in air over land than over the ocean. Air concentrations of 1 or several hundreds of ng/m^3 have been reported, usually in the industrial parts of Europe and North America (Frank 1991). They are sorbed from air to plant cuticles, especially those on conifer needles. In the needles, and in air, they may react with atmospheric ozone and form di- and tri-chloroacetic acids, which are found to accumulate in conifer needles to mg/kg levels. Mono-, di-, and tri-chloroacetic acids affect energy metabolism in plant cells, blocking the function of the citric acid cycle, and are suspected of being important contributors to conifer forest decay. Similar concentrations occur in broad-leaved trees, but they appear less lethal because the seasonal dropping of leaves offers a route for elimination of the pollutant (Rönnefahrt 1987; Friege et al. 1989; Wester and van der Valk 1990; Frank et al. 1992; Norokorpi and Frank 1993).

These C_1–C_3 solvents are gases or volatile liquids that rapidly cross biological membranes and concentrate in fatty tissues. They are metabolized by 2 ubiquitous enzyme systems, the cytochrome P450s, principally cytochrome P450IIE1, and the glutathione-S-transferases. Those molecules containing double bonds or protons are oxidized to potential reactive electrophiles that are rapidly eliminated following hydrolysis or conjugation with glutathione. Direct conjugation of these molecules with glutathione similarly leads to rapid excretion. The volatility of these chemicals also facilitates elimination from biological systems in an unchanged form. Of the known metabolites, the only one with a significant half-life in mammalian systems is trichloroacetic acid (36 to 70 h). As a result of the volatility and rapid metabolism of these chemicals, they do not accumulate or biomagnify in biological systems.

Polychlorinated paraffins (CPs) are mainly used as plasticizers and fire retardants, but also as solvents, dielectrics, additives for lubricating and cutting oils, paints, printing inks, rubber, and sealants. The log K_{OW} of a commercial C_{10}–C_{13}–CP mixture ranged for different congener groups between 5.85 and 7.14 (Sijm and Sinnige 1995).

CPs are bioaccumulated by fish. Fish also are capable of partly metabolizing CPs, probably only at the terminal C-atom (Mukherjee 1990). Limited information is available about bioaccumulation in animals in lower and higher trophic levels. A study of Swedish biota found CPs to be among the most prominent chlorinated organic contaminants in 11 representative biological samples, at concentrations ranging from 130 to 4400 ng/g lipid weight (Jansson et al. 1993).

Chlorolignins, chlorohumic matter, and soil-bound residues

Chlorolignins are produced from oxidation and chlorination of lignin and have much less aromatic character than lignin. They are poorly soluble in nonpolar solvents (Banerjee et al. 1984). The main portion of chlorolignin is water soluble at pH > 8 (to ~ 500 mg/L), but below pH 7, it forms colloids and micelles; at lower pH, it precipitates and becomes increasingly soluble into nonpolar solvents (up to 5% in heptane at pH 2), while approximately 70 to 100% is soluble in tetrahydrofuran at pH < 4. Less than 1% is volatile organic halogen.

A small part of the chlorolignin discharge (1 to 5 %) is soluble in nonpolar solvents, < 1% is formed of volatile halogen compounds, and the main part of the AOX is represented by rather hydrophilic compounds, approximately 70 to 100% being soluble in tetrahydrofuran. Of the chlorolignins discharged by bleached pulp mills, 30 to 70% is biodegradable by biological wastewater treatment. The remainder is very persistent, with a half life of > 1 y in water ecosystems (Saski et al. 1996). The chlorolignins adsorb to, but do not greatly accumulate in, sediment and microbial biofilms in the aqueous ecosystem. Despite absorption, no major accumulation occurs, indicating that the surface-attached microbial biomass is a pathway for elimination, mineralizing the chlorolignins (Gribble 1994; Landner et al. 1994).

The organohalogen substances formed during chlorine disinfection of humic-rich raw water is called "chlorohumus." The chlorohumus or chlorolignins are relatively hydrophilic, with no known record of bioaccumulation tendency. Chlorohumus or chlorolignins may gain hydrophobicity by microbial transformations in the aquatic ecosystem, and this may be a reason for the relatively great concentrations (10 to 300 mg Cl/kg fish, fresh weight) of EOX in areas with no other known sources of organohalogen pollution (Håkansson, Sundin et al. 1991). Chlorine disinfection of sewage, practiced in many countries to supplement inadequate sewage treatment, also adds to this load.

Many pesticides form bound residues upon soil application. It is often thought that such residues may be of large molecular weight and poor solubility and therefore not accessible to bioaccumulation, but few studies support this. Bound residues that formed in soil from chlorophenols used for wood, timber, and lumber treatment have been studied (Middeldorp et al. 1992; Laine et al. 1991). The soil-bound organic halogens, extractable into tetrahydrofuran or acetone, occur as halogenated molecules of 200 to 1000 to 200 to 10,000 g/mol, the larger sizes being found in soil high in organic content. When earthworms (*Eisenia foetida andrei*) were exposed to such soils, accumulation occurred to concentrations of 10 to 50 X greater in the worms than in the soil. In the worms, the residues appeared as even larger molecules than in soil, indicating their incorporation into tissue macromolecules (Laine et al. 1991).

Bound-residue bioavailability has been studied with model compounds in which dichloroaniline was linked covalently to glucosides and lignin model compounds, creating organohalogens of 400 to 650 g/mol and chlorine-to-carbon ratios similar to those found in chlorolignins (Sandermann et al. 1990, 1993). It was found that the organohalogen parent compound was released in the rat digestive tract and became bioavailable, possibly as a result of the metabolic attack of gastrointestinal tract bacteria (Sandermann et al. 1990).

Effect of Chlorine Substitution on Partition Coefficient and Bioaccumulation

The effect of chlorine substitution on partition coefficient and bioaccumulation of a number of compounds is shown in Figure 3-2. The data (Table 3-5) are plotted using a log-probability plot (Solomon et al. 1996) that shows the range of values on the x-axis and the probability of observation of a particular value (expressed as rank percent) on the y-axis. This analysis shows that the range of partition coefficients for compounds with no chlorine substitution is very wide and includes many compounds with partition coefficients greater than those of compounds substituted with one or more chlorine atoms. However, there is a general shift of the distributions to greater values with increasing degrees of chlorine substitution. A similar response to increasing degrees of chlorine substitution is seen in the BCF (Figure 3-2), although data are few for some of the higher-chlorinated compounds. These distributions clearly show that degree of chlorination is not the only determinant of physical parameters such as partition coefficient or biological parameters such as BCF, adding further support for a case-by-case assessment of the risks associated with chlorinated compounds.

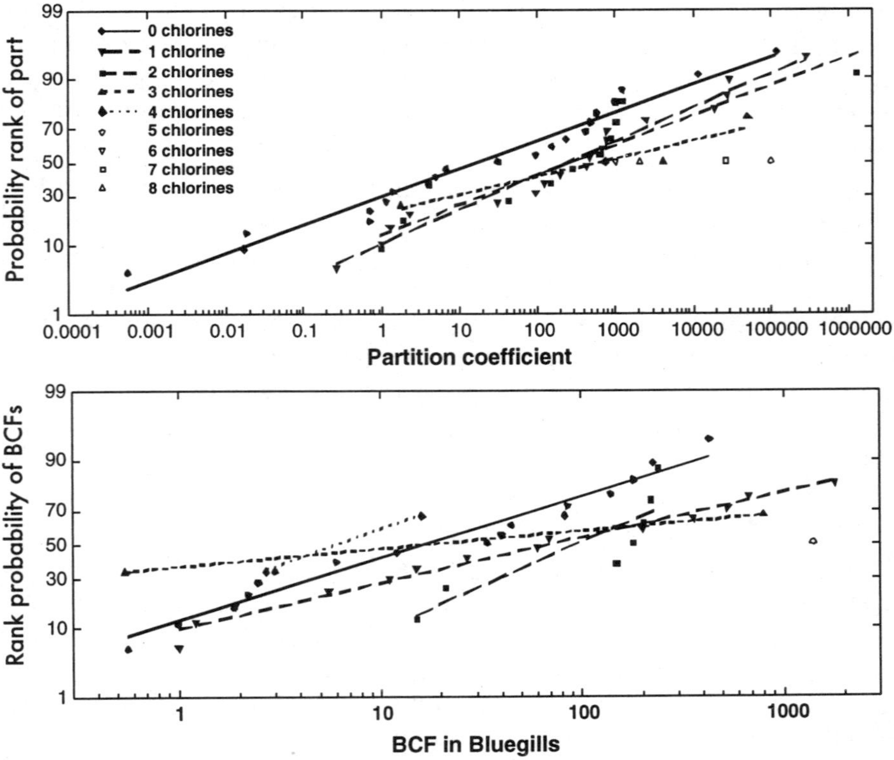

Figure 3-2 Range of partition coefficients and BCF values for a series of compounds with different degrees of chlorination. Plot of rank versus log K_{OW} or log BCF, using a log-normal model.

Table 3-5 Partition coefficient and BAF (BCF) in bluegill and other fish for a series of pesticides with different degrees of chlorine substitution

Substance	Log P[1]	# Chlorines[1]	BCF (Bluegill)[1]	Tissue[1]	Class	Pesticide group
metalaxyl[2]	1.5	0	6	whole	fungicide	misc.
trifluralin	5.07	0	0.567	whole	herbicide	dinitroaniline
glyphosate	−3.25	0	1.87	edible	herbicide	misc.
amitrol	−0.15	0	2.46	whole	herbicide	misc.
primisulfuron[2]	0.059	0	–	–	herbicide	sulphonyl urea
bensulfuron	0.61	0	2.2	–	herbicide	sulphonyl urea
metsulfuron	−1.74	0	0	–	herbicide	sulphonyl urea
nicosulfuron	−1.77	0	–	–	herbicide	sulphonyl urea
sulfometuron	0.84	0	–	–	herbicide	sulphonyl urea
ametryn[2]	2.63	0	83	whole	herbicide	triazine
cyromazine[2]	−0.16	0	1	whole	herbicide	triazine
prometon[2]	2.69	0	3	whole	herbicide	triazine
prometryn[2]	3.1	0	85	whole	herbicide	triazine
fluometuron[2]	2.38	0	40.4	whole	herbicide	urea
aldicarb	0.7	0	–	–	insecticide	carbamate
carbaryl	0.14	0	140	whole catfish	insecticide	carbamate
carbofuran	1.98	0	12	whole	insecticide	carbamate
rotenone	–	0	181	whole	insecticide	misc.
diazinon[2]	3.01	0	34	trout muscle	insecticide	organophosphorus
malathion	2.77	0	–	–	insecticide	organophosphorus
methidathion[2]	2.2	0	45	whole	insecticide	organophosphorus
parathion ethyl	–	0	430	whole	insecticide	organophosphorus
fenoxycarb[2]	4.07	0	277.6	whole	insecticide	phenoxy-phenoxy
baytan	2.9	1	27	whole	fungicide	triazole
myclobutanil	2.84	1	–	–	fungicide	triazole
alachlor	2.64	1	11	whole	herbicide	acetanilide
metolachlor[2]	2.9	1	69	whole	herbicide	acetanilide
butachlor	0	1	–	–	herbicide	acetanilide
propachlor	2.3	1	–	–	herbicide	acetanilide
acifluorfen	–	1	1.9	whole	herbicide	diphenyl ether
oxyfluorfen	4.47	1	1800	whole	herbicide	diphenyl ether
norflurazon	0.36	1	–	–	herbicide	misc.
fenoxyprop ethyl	4.28	1	527	whole	herbicide	misc.
flumetralin[2]	5.45	1	675	whole	herbicide	nitroaniline

Table 3-5 *(continued)*

Substance	Log P[1]	# Chlorines[1]	BCF (Bluegill)[1]	Tissue[1]	Class	Pesticide group
triasulfuron[2]	−0.58	1	1.2	whole	herbicide	sulphonyl urea
chlorsulfuron	–	1	1	edible	herbicide	sulphonyl urea
chlorimuron	0.11	1	–	–	herbicide	sulphonyl urea
thiobencarb	1.5	1	–	–	herbicide	thiocarbamate
atrazine[2]	2.68	1	15	whole	herbicide	triazine
cyanazine	–	1	–	–	herbicide	triazine
simazine[2]	2.1	1	535	whole	herbicide	triazine
diflubenzuron	3.4	1	360	whole	insecticide	benzoylurea
isazofos[2]	1.98	1	200	whole	insecticide	organophosphorus
profenofos[2]	4.44	1	60	whole	insecticide	organophosphorus
vinclosolin	3.02	2	241	whole	fungicide	misc.
iprodione	3.1	2	180	whole	fungicide	misc.
propiconazole[2]	2.95	2	203	whole	fungicide	triazole
dichlobenil	1.64	2	15	guppy whole	herbicide	benzonitrile
paraquat	0	2	–	–	herbicide	misc.
2,4-D	0.27	2	220	muscle catfish	herbicide	phenoxy
dicamba	2.46	2	–	–	herbicide	phenoxy
linuron	2.19	2	–	–	herbicide	urea
diuron	2.81	2	150	minnows	herbicide	urea
permethrin	6.1	2	21	bluefish edible	insecticide	pyrethroid
captan	–	3	–	–	fungicide	cyclohexane dicarboximide
picloram	0.25	3	54	whole	herbicide	misc.
methoxyclor	3.62	3	800	green fish	insecticide	diphenyl chloride
chlorpyrifos	4.7	3	–	–	insecticide	organophosphorus
captafol	–	4	2.7	edible	fungicide	cyclohexane dicarboximide
chlorothalonil	2.88	4	16	whole	fungicide	misc.
lindane	–	6	1400	whole	insecticide	chlorinated hydrocarbon
heptachlor	4.41	7	–	–	insecticide	chlorinated cyclodiene
chlordane	3.32	8	–	–	insecticide	chlorinated cyclodiene

[1] USEPA 1994c.

[2] Unpublished data from Novartis Crop Protection (Gary Dickson, personal communication). Other data from U.S. Dept. of Interior (1980) and Mayer and Ellersieck (1986).

Chapter 4

Modeling the Fate and Transport of Organochlorine Compounds in the Environment

Models enable the efficient assessment of the fate, distribution, and bioaccumulation of a broad range of organochlorines (OCs) and a predictive capability for assessment of potential exposures. Models establish a quantitative relationship between loading of a chemical and its concentrations in various environmental compartments of the system (e.g., air, water, sediments, soil, biota) as a function of chemical and environmental properties. Bioaccumulation models can be linked to the output of chemical fate models to predict OC concentrations in biota. The linkage to biota often aids in the validation of chemical fate model predictions. Modeling of the fate and bioaccumulation of OCs such as polychlorinated biphenyls (PCBs) and chlorinated pesticides has been successfully demonstrated on a site-specific basis (in small watersheds, harbors, bays, and lakes $< \sim 100$ km^2) and is being extended to regional and global scales.

A range of model complexity exists that is governed by the problem statement and available data. The capability to model OCs varies with the chemical and with model type. At the global scale, only chlorofluorocarbons (CFCs) and hexachlorocyclohexane (HCH) can be considered fairly well understood. On a regional scale, polychlorinated biphenyls (PCBs), OC pesticides, and polychlorinated dibenzo-*p*-dioxins (PCDDs) and polychlorinated dibenzofurans (PCDFs) can be modeled with uncertainty of a factor of 10 or less. Further improvements in the modeling of OC chemicals will require better emission or loadings data, more accurate physicochemical and reactivity data, especially for poorly characterized mixtures of OC chemicals, and more "post-auditing" or monitoring to assess the limitations of model predictions.

Chapter Preview

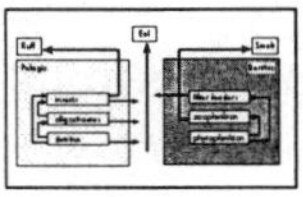

This chapter presents models as tools for exposure assessment. While simple partitioning models are discussed in Chapter 3, the focus here is on mass balance, environmental fate, and bioaccumulation models. The model should be viewed as a tool by which loadings or input rates, measured in kg/d, e.g., are translated into predicted concentrations, such as ng/L in water or μg/g in fish or birds, over time. These concentrations relate directly to exposure and hence to the occurrence and severity of biological effects as discussed in Chapters 5 and 6. Models also satisfy an important scientific purpose by testing quantitatively the ability to describe the complexities of environmental exposure. The importance of expressing such information quantitatively cannot be overemphasized. If assertions are purely qualitative, it is impossible to assess risk or formulate effective regulatory strategies. Without models, however simple, effective environmental risk assessment and management is impossible because of the lack of predictive capability.

In striving to mathematically reproduce observable phenomena in ecosystems, mass balance models have utility in 2 broad areas: 1) for resource management applications and 2) as research tools. Management applications include resource evaluation and protection of ecosystems and, ultimately, human health. A confirmed model can, therefore, be used to compare the response of the system to alternative mitigation or regulatory strategies. There has been an enormous growth in the capability to model environmental processes since the 1920s, when the first models of dissolved oxygen behavior in rivers were formulated. This growth has been catalyzed by the availability of convenient, low-cost computing and by the improved quality of environmental data on the presence and fate of substances.

It is also useful to speculate on the role that models may play a couple of decades from now. This is a "vision" of the future of models as one of the key ingredients of ecosystem management, as well as of ecological risk assessment and management. We believe that there will emerge a modeling capability that is an integral part of regulatory requirements. First, those individuals, municipalities, and industries that discharge chemicals to the environment will disclose information about the nature, relevant properties, and quantities of chemicals emitted. Second, information will be generated (using models) concerning the fate and, especially, the persistence of these chemicals; such information can be used to generate estimates of concentrations in a variety of abiotic and biotic media. Selected results will be tested for validity. Third, these data will form the basis of an exposure estimation that can be examined in concert with known toxicology to better evaluate risk. Ultimately, for many persistent chemicals, risk evaluation must be done globally as well as nationally and locally in site-specific situations.

The concerns expressed by proponents of a "chlorine ban" frequently focus on gaps or inadequacies in this ideal global evaluation capability. For example, chlorinated organics such as the chloroparaffins, toxaphene, and chlorolignins are complex mixtures of uncertain composition, properties, and fate. Global production quantities of many common organochlorine (OC) chemicals such as lindane and dichlorodiphenyltrichloroethane (DDT) are poorly known. Lack of knowledge often results in fear of the unknown, which may lead to irrational responses. This goal of modeling as an integral part of exposure assessment and risk management is regarded as being attainable; indeed, as is discussed later in this chapter, there are several examples of successful components.

This chapter presents a taxonomy of environmental fate and bioaccumulation models, their scales from global to single organisms, an outline of their basic structure, and their data requirements. This is followed by a description of some successful applications of chemical fate models to chlorinated organics. Data gaps, sources of uncertainty, and inadequacies in our understanding of environmental fate are reviewed.

Overview of Existing Model Types

Existing models can be classified by scale, ranging from global to local models. The goal of a model influences the choice of scale. For example, global and regional models aim mostly at the identification of vulnerable areas, while local models and biotic models aim at predicting actual concentrations present in a local environment. Each model type can be run in equilibrium, steady-state (time-independent), or dynamic mode (chemical concentrations vary with time). Four different scales can be identified (Table 4-1).

Table 4-1 Summary of types of models applied to exposure assessment of OCs

Model type	Spatial scale	Example
Global	World	Fugacity model for global distribution
Regional	Country, ocean	Fugacity models, chemical budget models for oceans or countries
Local	Site-specific: river, lake, soil area, point source	Polychlorinated biphenyl (PCB) and hexachlorobenzene (HCB) fate in lakes, air pollution in cities in California
Biotic (aquatic and terrestrial)	Local or regional	PCB accumulation in aquatic food webs

In several cases, especially at the regional and global scales, only chemical budget calculations have been carried out. These calculations are not mass balance models but relate observations of the contaminant output flux to its input for the corresponding time interval. These budget calculations do not provide information on the quantitative transport and fate of the contaminant within a system, but for many systems, it is the only information available.

Input data for chemical-specific models

The use of models to obtain exposure predictions that facilitate policy decisions regarding the management of OC chemicals requires that the models establish a quantitative relationship between chemical loading to the environment and the concentration in various environmental compartments of the system of interest, such as air, water, sediments, soil, aquatic biota, terrestrial biota.

It should be recognized that, within this broad focus, a range of management questions of varying complexity require varying levels of certainty. Examples of OC chemical management information needs that may be provided by exposure models include these:

- applying of current surveillance data to provide useful information on the historical and current loads of pollutants;
- determining the magnitude of changes in loadings required to achieve a given concentration in biota, e.g., at an upper 95% confidence bound;
- determining the time it will take to meet the stated objective with the proposed strategy;
- identification of the optimum source-reduction strategy (i.e., combination of atmospheric, tributary, groundwater, and/or point-source loads) to meet the target load;
- determination of the effects of management programs for eutrophication, fisheries, soil erosion, and exotic species invasions on OC transport, bioaccumulation, residence times, bioavailability, and effects; and
- quantifying the success of regulatory/remediation efforts.

Selection of the most appropriate model can be done only after specification of the problem to be addressed.

Regardless of their complexity, all models require input data. As the complexity of the management question increases, model complexity and sophistication increase. The input data necessary to apply the model also become more complex and must be available in greater resolution. Qualitatively, the relationship between model complexity (defined in terms of spatial, temporal, and kinetic resolution) and model utility (defined in terms of model accuracy and ability to interpret results and information content) follows a relationship like that depicted in Figure 4-1. For a given modeling scenario and available dataset, there exists a level of model complexity that will yield the optimum utility. For a specified problem, if the model is too simple, it may not be capable of answering the question. On the other hand, if the model is too complex, the available data and knowledge will be insufficient to provide accurate inputs for the model, and its utility will be lost. Also, models that are too complex can become difficult to interpret. By way of example, the diagram in Figure 4-1 shows the relative model complexity versus model utility curves for 2 different assessment questions: 1) fish body burden for a site-specific situation in which the chemical of concern has a log

$K_{OW} < 4$ and 2) the same system, but for a chemical with a log $K_{OW} > 4$. In the first case, where accurate solid-phase dynamics and biomagnification are not as critical, a simple transport model and a steady-state bioconcentration factor (BCF) bioaccumulation approach will be adequate. In the second case, a more complex transport model, perhaps containing a dynamic solids mass balance and a dynamic food-chain bioaccumulation model, might be necessary. In this second case, the model is more sophisticated, requires more input data and understanding of the system, but yields more information about the behavior of the system.

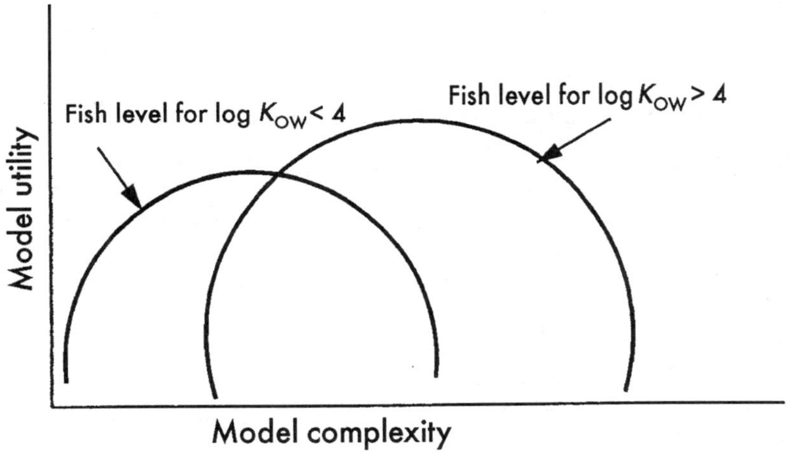

Figure 4-1 Relative model complexity versus model utility for 2 different assessment questions

In addition to model complexity, data intensity, accuracy, and resolution depend on the spatial and temporal scales of the problem, such as local or global, daily, seasonal, or steady-state, and on the desired accuracy of the answer. Nevertheless, in considering the data needs for a given model application, we should consider acquiring the following chemical- and system-specific data (Table 4-2).

At this stage in our theoretical understanding of modeling OC chemicals in environmental systems, system-specific models require calibration and confirmation prior to application. This requires the measurement of all dependent variables, e.g., concentrations in various environmental compartments at some level of spatial and temporal resolution and, if possible, process rates, in order to carry out the calibration and confirmation steps.

Mass balance fate-and-transport modeling

A chemical mass-balance model accounts for the net change in chemical mass over a specified time interval within a specified compartment in terms of all significant

Table 4-2 Data required to model fate and transport of substances in the environment

Process/Property	Input for model
Chemical-specific data	
Chemical entry or removal and current inventory	Location of sources Mass input rate Mode of entry Location, rate of withdrawals, harvest, export Initial conditions in various compartments Boundary conditions
Process-related information	
Chemical-specific properties	Solubility Vapor pressure Octanol/water partition coefficient, K_{OW} Henry' s Law constant (HLC) Soil/sediment adsorption coefficient, K_{OC} Acid/base dissociation
Physicochemical processes	Air–water exchange (volatilization and absorption) Air–land exchange Sorption or partitioning to soils, bed and suspended sediment, aerosols Hydrolysis Photolysis
Biological processes	Biotransformation (rate, extent, products) Uptake rate from appropriate media Chemical assimilation efficiency by relevant organisms Depuration rates Maternal transfer
System-specific data	
Physical properties of the system	Hydrology (water balance) Hydraulics (flows or circulation) Meteorology such as precipitation, wind, solar radiation, and temperature Solids (sorbent) characteristics and dynamics (includes sorbents in all compartments of concern: air (aerosols), water, sediments, soils Temperature profile
General chemical properties of the system	Ph, alkalinity Ionic strength Major ions Dissolved organic carbon (DOC) concentration and characterization Oxygen profile or redox potential
Biological description of the system	Food-web structure, function Organism physiology, metabolism, life history Organism growth dynamics Prey consumption rate and selectivity Population biomass levels such as fish stock and crop cover Biota introductions or harvests such as stocking, planting, fish catch, and crop harvest

inputs, outputs, and transformations (production and decay). This accounting, which is normally formulated as a series of differential equations describing the rate of change with time of each dependent variable within the compartment, attempts to incorporate all of the important transport and transformation processes (see Chapter 2) by expressing them in terms of the appropriate process rate coefficients. The solution of the resulting mass-balance differential equations — ordinary, time-dependent differential equations reduce to algebraic equations under steady-state conditions — provides an estimate of the concentration profile of a contaminant in each environmental compartment of interest, e.g., atmosphere, water, sediment, and algae. Appropriate solutions for the mass balance equations not only relate the concentration of a contaminant in various environmental compartments but also provide quantification of the mass transfer rates among compartments and between spatial segments of the system, thus identifying significant transport-and-fate pathways within the system.

Mass balance models for the aquatic environment have been developed and applied to questions regarding OC chemicals in the Great Lakes by Thomann and DiToro (1983); Mackay (1989, 1991); Bierman et al. (1992); Endicott, Velleux et al. (1992); and DePinto et al. (1994). A general consensus regarding the most important processes has evolved from extensive sensitivity analyses conducted in these modeling studies. In general, the variation in sources, transport, and transformation of particulate organic carbon (POC) sorbents in the system determines the transport and fate of hydrophobic organic chemicals in aquatic systems. Therefore, an accurate solid-phase mass balance is essential to obtaining an accurate chemical mass balance and a workable model. For example, in the Green Bay mass balance modeling study, 5 different polychlorinated biphenyl (PCB) congeners, ranging from 3 to 7 chlorine substitutions, were successfully modeled using the same sorbent dynamics and changing only the chemical-specific parameters, K_{OC} and Henry's Law constant (HLC) (DePinto et al. 1994). In addition to the solids dynamics, the dissolved phase of the chemical in the water column and the flux of chemical to the bottom sediments is quite sensitive to the partitioning of particulate matter in the water column. For less hydrophobic OC chemicals, transformations in the water column, such as biodegradation, hydrolysis, and photolysis, may be significant (see Chapter 3).

Modeling bioaccumulation in aquatic biota

Bioaccumulation models can be linked to the output of chemical fate models in order to predict concentrations of OC chemicals in biota from predicted concentrations in air, surface waters, suspended solids, sediment, and soil. Because aquatic and, to a lesser extent, terrestrial biota have been monitored for persistent OC chemicals, the linkage to biota often aids in the validation of chemical fate model predictions. Simple partitioning models such as BCF and biota–sediment accumulation factor (BSAF) have been used to predict concentrations in aquatic biota.

Toxicokinetic models

Toxicokinetic models describe the bioaccumulation of chemical substances and the distribution and metabolism within organisms. They include a series of rate constants that represent uptake, elimination, and metabolic transformation through various mechanisms such as food consumption, inhalation, and gill ventilation. An example of this kind of model is a toxicokinetic bioaccumulation model for fish:

$$dC_f / dt = k_1 \bullet C_{WD} + \alpha \bullet F \bullet C_{FOOD} - C_f(k_2 + k_E + k_M + k_G) \qquad \text{(Equation 4-1)},$$

where C_f = concentration in fish, k_1 = uptake rate from water (L/kg•d), C_{WD} = dissolved concentration in water, α = chemical assimilation efficiency, F = feeding rate (kg food/kg fish/d), C_{FOOD}= concentration in food, k_2 = elimination through the gills (1/d), k_E = elimination through fecal egestion, k_M = metabolic transformation rate, and k_G = growth rate (1/d). At steady state, $dC_f / dt = 0$, and the mass balance of chemical in the fish can be written as

$$C_f = (k_1 \bullet C_{WD} + \alpha \bullet F \bullet C_{FOOD}) / (k_2 + k_E + k_M + k_G) \qquad \text{(Equation 4-2)}.$$

This type of model differs fundamentally from the simple partitioning models because it views bioaccumulation as the net result of the rates of uptake and elimination. This approach is essential to describing bioaccumulation in some organisms such as birds and mammals. Because of this model's focus on rates, the relationship between internal concentrations and bioaccumulation factors (BAFs) and time can be described. It can also include the processes of biomagnification through food and inhalation. In addition, the methodology allows the inclusion of any number of uptake and elimination processes such as fecal and urinary excretion, metabolic transformation, and reproductive elimination through breast milk and eggs. The models can be applied to a larger range of chemicals, i.e., chemicals with log K_{OW} between 2 and 8.5 and possibly greater. For chemicals with log K_{OW} between 2 and 5, the toxicokinetic models often result in estimates of bioaccumulation that are comparable to those of the partitioning models. The only exception is for chemicals that are readily metabolized. If metabolic transformation rates can be estimated or measured and then introduced into the model, toxicokinetic models will produce better estimates than partitioning models. A major difficulty for the application of toxicokinetic models is the fact that metabolic transformation rates are often unknown and largely unpredictable from the chemical structure.

Bioenergetics-based bioaccumulation models such as Food Gill Exchange of Toxic Substances (FGETS) (Barber et al. 1991) differ from toxicokinetic models by including physiological characteristics of the organism (e.g., gill morphometry) and fractional aqueous, lipid, and nonlipid composition of the fish. Unlike the food-chain bioaccumulation models of Thomann (1989) and Gobas (1993), FGETS assumes that assimilation of OC chemicals from food by fish decreases with increasing concentration in the fish tissue and with exposure time. Bioenergetics-based toxicokinetic models are also available for birds and marine mammals. Clark et al. (1987) developed a 2-compartment model consisting of plasma and seasonally variable lipid to interpret temporal trends in the bioaccumulation of OC compounds in herring gulls and their eggs (Figure 4-2). The dynamics of OCs in herring gulls were shown to be highly dependent on seasonal fluctuations of the lipid compartment size. Whole-body half-lives ranged from 48 d for *p,p′*-dichlorodiphenylethane (DDD) to 740 d for *p,p′*-DDE. Hickie et al. (1995) developed a bioenergetics-based model to predict PCB concentrations in marine mammals and the maternal transfer to the fetus and, through lactation, to the neonate. Model predictions show good agreement with observed age-related accumulation of recalcitrant PCB congeners in fat of male and female beluga whales (*Delphinapterus leucas*).

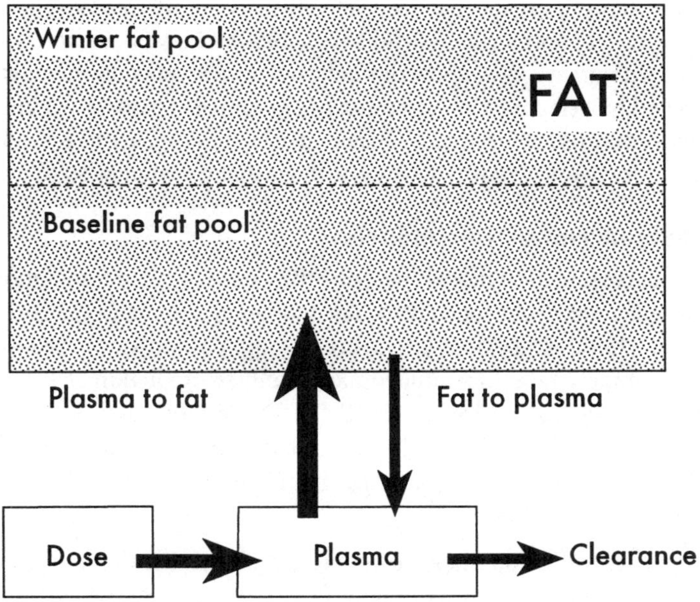

Figure 4-2 Schematic model of herring gull pharmacokinetics. Box sizes are scaled to show relative size of plasma and fat compartments (Clark et al. 1987).

Physiologically based toxicokinetic models

Physiologically based toxicokinetic (PBTK) models have been used to assess the internal distribution and bioaccumulation of chemicals in specific organs following exposure (Andersen et al. 1987; Nichols et al. 1990). These models consist of a series of differential equations, i.e., one for each organ, including an equation for the blood compartment, that distributes the chemical between the organs. These equations are usually solved numerically with the help of computers to give estimates of chemical concentrations over time. The PBTK models often require a considerable effort in terms of defining values of each input parameter ("parameterizing") the model. Model parameterization is specific to the organism in question, and in many cases, laboratory experiments are necessary to find appropriate values for many of the model parameters such as blood flow, organ size, and organ-to-blood partition coefficients for the chemical. The models are particularly useful for chemicals that preferentially absorb in certain organs, e.g., due to metabolic transformation. These models are most relevant for chemicals that are rapidly eliminated by processes such as metabolic transformation and urinary excretion and in situations where exposure conditions are short and/or highly variable. There is considerable evidence from field studies that chemicals that are eliminated relatively slowly and not metabolized, such as PCBs, tend to reach concentrations in organs that are relatively constant when expressed on a lipid-weight basis within the organism (Nichols et al. 1990). Many of the nonmetabolizable OC chemicals with greater K_{OW} show this behavior, and as a result, the internal distribution and bioaccumulation can often be simplified considerably and the detail of PBTK model may not be necessary. Hence, application of the PBTK models may be more relevant for chemicals with low K_{OW} or those that metabolize significantly.

Food-chain bioaccumulation models

Food-chain bioaccumulation models describe the extent of bioaccumulation as a result of trophic interactions in the food chain. The models of Thomann (1981, 1989), Thomann et al. (1992), and Gobas (1993) are examples of kinetics-based food chain models. These models have been applied successfully to prediction of PCBs and 2,3,7,8-tetrachlorodibenzo-*p*-dioxin (2,3,7,8-TCDD) concentrations in Lake Ontario lake trout (Endicott and Cook 1994), kepone in fish in the James River (Connolly and Tonelli 1985), and PCBs in New Bedford Harbor (Connolly 1991) and Green Bay (Connolly and Parkerton 1994).

The food-chain accumulation model of Gobas (1993) includes the same food chain relationships as the Thomann and Connolly (1984) model. It derives internally the rate constants for chemical uptake from water, uptake from food consumption, elimination through the gills to water, elimination in the feces, and growth dilution, based on fish weight, lipid content, and K_{OW}. Trophic transfer and biomagnification are then calculated at steady state from feeding preferences that are also entered into the model. The model consists of 8 compartments: water, sediment, phytoplankton, benthic invertebrates, and 4 fish weight classes. Each fish compartment can feed on various fractions of each compartment, i.e., phytoplankton, benthos, or other fish. At steady state, the concentrations in fish are given by

$$C_F = (k_1 \bullet C_{WD} + k_D \bullet \Sigma P_i \bullet C_{D,i}) / (k_2 + k_E + k_G) \qquad \text{(Equation 4-3)},$$

where $k_D = \alpha \cdot F$ = rate constant for chemical uptake from food (kg food/kg fish/d), P_i = fraction of the diet of the fish consisting of prey I, and $C_{D,i}$ = the chemical concentration in prey I (µg/kg). The Gobas model as well as the recent Thomann et al. (1992) model includes a link to the benthic food chain. They assume that benthos are in chemical equilibrium with sediments and pore water. Concentrations in phytoplankton or detritus in the overlying water are calculated from the water concentration by assuming that the water/algae partition coefficient can be approximated by the K_{OW} of the chemical and the lipid content of the organism.

These food chain models are data intensive. They depend on empirical relationships between log K_{OW} and uptake and elimination rates derived mainly from PCBs and other persistent OC chemicals in laboratory studies. Other key parameters are growth rates and feeding rates for organisms at each trophic level. To apply these models on a seasonal basis requires knowledge of bioenergetics, i.e., respiration rates of fish at various temperatures as well as variations in feeding rates and food assimilation efficiency that could lead to different lipid levels in the animal. Differing diets and feeding rates for different age classes of fish are needed to model trends in levels in the population as a whole. Sijm et al. (1992) found that growth, biotransformation, and mother-to-young transfer were the key processes in elimination of PCBs by guppies over an entire 2-y life cycle.

Sensitivity analyses have shown that critical parameters are fish growth rates and assimilation efficiency (Thomann 1989). In the Thomann model, partitioning to phytoplankton is a critical step in the transfer of contaminants along the pelagic food web. This partitioning varies seasonally. Log K_{OC} values for PCBs were found to be constant over a range of log K_{OW} in the "growth" period (summer data). Under low water-temperature conditions, K_{OC} values of PCBs varied with K_{OW} (Skoglund and Swackhamer 1992; Swackhamer and Skoglund 1992). Thus, models may be good predictors in some seasons but not in others.

Modeling the terrestrial environment

Models for the terrestrial environment have been developed to a lesser extent than those for the aquatic environment. In principle, terrestrial models must deal simultaneously with the air, soil, and plant compartments as well the transfer of OC chemicals to animals and humans. Thus, a multimedia approach is needed. However, most fate modeling of OC chemicals has dealt only with soil-to-water transfer, or with air–plant or plant–animal accumulation, rather than with all processes together. There have been several recent initiatives to adopt a multimedia modeling approach. For example, an exposure model for polychlorinated dibenzo-*p*-dioxins and -furans (PCDD/PCDFs) has been developed to calculate exposure concentrations for all media (ambient air, soil, crops, meat, milk) resulting from the additional loadings of PCDD/PCDFs when a municipal waste incinerator goes into operation (Fries and Paustenbach 1990; USEPA 1992a). Based on these results, as well as on measured environmental levels, the additional daily intake can be calculated for humans. Scheil et al. (1994) have described the Chemical Exposure Modeling System (CemoS), a multimedia model that allows users to define various terrestrial fate and bioaccumulation scenarios for OC chemicals in the terrestrial environment. The following is intended to provide background on the various components of a multimedia model for partitioning and uptake of airborne or soilborne OC chemicals by soil, plants, and biota.

Modeling fate and bioavailability in air

Input of OCs to the atmosphere can occur through combustion processes and volatilization of chemicals from surfaces such as soil, vegetation, and water, or from other surfaces such as buildings, roofs, and streets. Input may be either continuous or discontinuous, as in the case of accidents from point sources or diffuse multiple sources, as well as by wind erosion of dust or soil particles. Atmospheric deposition of OC chemicals is discussed in more detail in Chapter 2. Semivolatile organic chemicals, including many OC chemicals, can be present in air either in the gaseous phase or adsorbed to a particle. Their concentrations in the atmosphere are dynamic, due to air mass movements and temperature changes (Eitzer and Hites 1989; Hoff et al. 1992). For example, for PCDD/PCDFs, greater air concentrations are found during the winter months than in summer (Eitzer and Hites 1989), while the reverse is true for PCBs and OC pesticides (Hoff et al. 1992). The OC chemicals are removed from the air either through "extraction" from the gas phase, mainly to vegetation surfaces, or through dry and wet deposition on soil, vegetation, surfaces, and waters. Based on actual TCDD measurements, McLachlan (1994) has calculated the contribution of the 3 pathways, gaseous phase:dry deposition:wet deposition, into grass as being 10:1:0.5. Eisenreich and Strachan (1992) found wet deposition of PCBs and PCDD/PCDFs predominated over dry deposition in the Great Lakes region but did not estimate gaseous-phase deposition.

Atmospheric transport-and-deposition models for OCs, and for other semivolatile organics, are gaining increased sophistication. An atmospheric deposition model recently developed by the Netherlands Organization for Applied Scientific Research calculates deposition flux of PCBs, chlorinated pesticides, and chlorinated solvents to the Rhine River basin from estimates of emissions of these chemicals in Europe (Baart and Diederen 1991). The model was based on Photochemical Oxidant Acid Deposition Model (PHOXA 1989). PCB emission estimates were made on a grid (resolution 55 × 55 km) by assuming that they were related to population density, while pesticide emissions were assigned to areas of arable land. A long-range transport model was used to calculate the dispersion of the OCs from the source to the receptor area and to estimate dry and wet deposition. Uncertainties in the model included the lack of emission data, especially for nonpoint-source, semivolatile compounds such as PCBs and chlorinated pesticides, and limited information on dry and wet deposition parameters.

Modeling fate and bioavailability in soil

Surface soils, and more precisely their organic carbon content, are the sinks for airborne, lipophilic OC chemicals. Inputs to soils also occur (and often to a larger extent) directly by application of chemicals such as pesticides or indirectly by application of sewage sludge or compost. In the soil compartment, OC chemicals can undergo transformation processes, especially biodegradation and hydrolysis (Chapter 2). Direct and indirect photolysis in the upper layers (< 1 mm) may be important especially for newly applied pesticides (Hebert and Miller 1990). The fate of OC chemicals in soil has been modeled at the local or site-specific scale, but to date, no models have been developed that quantitatively describe the transport and transformation processes of chemicals in soil on a regional or global scale.

The extent of volatilization of an organic chemical from soil will depend upon its physical properties, such as vapor pressure and soil adsorption coefficient (K_{OC}), as well as on environmental factors such as properties of the soil, soil moisture, and temperature. The Jury et al. (1983) model assumes

1) uniform soil properties of constant water content,
2) linear soil adsorption isotherms,
3) linear HLCs,
4) uniform initial incorporation of pesticide, and
5) loss of pesticide and water to the atmosphere, limited by gaseous diffusion through a stagnant air-boundary layer above which the pesticide has 0 concentration.

Jury et al. (1983) present the mass balance and flux equations of the model and apply it to the volatilization of lindane and 2,4-D from soil. Spencer et al. (1988) found that volatilization of chemicals with low HLC is controlled by the air-boundary layer above the soil surface, while loss of higher HLC compounds is controlled within the soil. Boundary-layer thickness and water evaporation rate control volatilization of chlorine-containing pesticides with low HLC, such as atrazine and 2,4-D acid. These chemicals generally have greater water solubility and can accumulate at the soil surface as soil water moves up the capillaries to replenish water evaporated at the surface. Although such compounds are not expected to volatilize significantly, they may do so when they are deposited at the soil surface through transport in soil water.

Organochlorine chemicals in soil can be transported with the water flux to ground water or lost through overland runoff of soil particles. Widely used models to predict losses of pesticide to streams flowing through agricultural watersheds and to ground water include Chemicals, Runoff and Erosion from Agricultural Management Systems (CREAMS) and Agricultural Runoff Model (ARM), developed for the U.S. Department of Agriculture (Knisel 1980), Pesticide Root Zone Model (PRZM) (Carsel et al. 1984; Mullins et al. 1993), and Simulator for Water Resources in Rural Basins (SWRRB) (USEPA 1985). All of these models depend upon well-developed models of the hydrologic cycle. Transport of pesticides in water or sediment is simulated using physical properties, partition coefficients, and degradation rates for each compound. All models calculate a water balance for a defined area and for several soil depths within the area, using equations for runoff, infiltration, evapotranspiration, and snowmelt. A watershed is frequently divided into areas or segments of different soil type and cropping practices. The sediment loss portions of the models are generally based on the Universal Soil Loss Equation (Knisel 1980). Model inputs are similar to those for any mass-balance chemical fate model (Chapter 4, "Input data for chemical specific models"). They include

1) a meteorological dataset describing rainfall, insolation, temperature and rainfall intensity;
2) a pesticide dataset consisting of physical properties (e.g., soil adsorption coefficient) and dates of application; and
3) a site hydrology-and-crop dataset describing the soil, topography, and vegetation characteristics that influence the water balance.

These site-specific data are harder to obtain and are available for only a limited number of sites at which experimental runoff studies have been performed (all in the U.S., at present). Many pesticide fate simulations can be run with the same meteorological and hydrological or crop datasets. In principle, these models can be used to predict the fate of atmospherically deposited OC chemicals, although their application to date has been limited to pesticide spray deposits.

Some OC chemicals in soils can be taken up by plant roots, with subsequent delocalization of the substance into the plant, such as in fruit or new growth. This process is dependent on the water solubility of the compound and its ability to pass through membranes. Root uptake is negligible for very lipophilic compounds (i.e., octachlorodibenzo-dioxin (OCDD). OCs in soils may be available to organisms such as earthworms and insects to form the first step of a soil organism-based, terrestrial food chain with subsequent bioaccumulation to birds. OC chemicals in soil can also enter the terrestrial food chain directly through ingestion of soil by humans or animal species. Simple bioaccumulation models for PCDD/PCDFs from soils to chickens have been developed by Petreas et al. (1991). In controlled exposure studies, these authors found that chickens rapidly accumulated PCDD/PCDFs from ingested soil and reached steady-state levels in less than 30 d.

Air–plant transfer

Plants represent a large reservoir of biomass, with greater lipophilicity on their surfaces (waxes). Thus, the waxy layers of plants can "extract" gaseous OC chemicals as well as those on airborne particles (Reischl et al. 1987). From the waxy layers, the OC chemical either can be transported to the inner part of the leaf or needle or can be revolatilized, depending on the ambient air temperature, humidity, and its chemical-physical constants, especially vapor pressure. The physiology of the plant and the time of exposure play important roles when plant uptake is modeled. For example, long exposure times offer better possibilities for an effective uptake of gaseous pollutants than do short exposure times. In most cases, plants represent a temporary sink because deposited OCs will enter a new compartment after harvest or because of seasonal changes; leaf fall in the autumn will result in transfer to the soil; harvest will transfer an OC chemical as foodstuffs or enter the food chain through animal fodder, resulting in a potential contamination of meat, milk, and eggs.

Bacci et al. (1990), Paterson et al. (1991), and Müller et al. (1994) derived air-to-plant bioconcentration factors (BCF_V) for hydrophobic OC chemicals using azalea leaves that enabled plant concentrations (C_{VPA}) to be estimated as a function of vapor-phase air concentrations (C_A):

$$C_{VPA} = BCF_V \bullet C_A \bullet VG / d \qquad \text{(Equation 4-4)},$$

where VG = a correction factor to correct concentrations in leaves to whole plant, and d = density of air (1.19 kg/m^3). The BCF_V for a series of chlorinated organics was found to be a function of the air/water partition coefficient (K_{AW}) and the octanol/air partition coefficient (K_{OA}) (Paterson et al. 1991):

$$BCF_V = 0.19 + 0.7K_{AW} - 0.05K_{OA} \qquad \text{(Equation 4-5)}.$$

This indicates that leaves exhibit partitioning behavior equivalent to 19% air, 70% water, and 5% octanol.

Deposition of particle-borne TCDD in vegetative matter has been modeled by several authors using the following equation:

$$C_{PPA} = F \bullet I / k_W \bullet Y \qquad \text{(Equation 4-6)},$$

where C_{PPA} = concentration in plant due to settling of contaminated particles, F = deposition flux (ng/m^2/y), I = fraction intercepted, k_W = weathering dissipation half-life of the plant (1/y), and Y = dry matter yield of the crop (kg/m^2). The USEPA has used Equations 4-4 and 4-6 in the exposure assessment of vapor-phase and particle-borne 2,3,7,8-TCDD resulting from municipal-waste incinerator emissions (USEPA 1992a).

Transfer factors for PCDD/PCDFs from soil to edible subsurface parts of plants, such as peeled potato tubers, have been determined to be < 0.001 on the basis of toxic equivalents (TEQs) of 2,3,7,8-TCDD (Hülster and Marschner 1992). Similar low-transfer efficiency has been seen with other hydrophobic organics (Wild and Jones 1991). The main pathway of most OC chemicals into leaves is through air. A maximum transfer factor of 0.01 has been determined for PCDD/PCDF from soils to leaves (McLachlan 1994). The transfer has been shown to be dependent on the concentration in the soil, indicating that increasing concentrations result in lower transfer rates.

However, no translocation of the dioxin contamination was observed for soil levels below 1000 ng TEQ/kg (dry weight) soil.

Higher trophic levels in the terrestrial food chain

Contamination of cow milk with lipophilic OC chemicals can occur through

1) atmospheric input to (leafy) fodder;
2) local contamination during storage and handling of the fodder, such as from wood preservatives and other pesticides; and
3) ingestion of soil particles (green fodder may contain up to 10% soil by weight).

The transfer of chlorinated pesticides dieldrin and heptachlor from the soil–pasture system to grazing animals has been modeled by Cullen and Connell (1994). The model assumes an equilibrium between lipid-based concentrations in the food and the cow. Biomagnification of dieldrin and heptachlor was overpredicted by the model, possibly because of nonequilibrium conditions for pesticide absorption in the stomach, losses to metabolism, and milk excretion.

Grass-to-milk transfer of PCDD/PCDF is highly dependent on the degree of chlorination in the molecule (Fürst et al. 1992). Thus, 2,3,7,8-TCDD is more bioavailable than is OCDD, with a transfer efficiency of 3.5 compared to 0.1 for the latter compound. The accumulation of OCs through ingestion of plants, as well as the metabolism and excretion by the organism, can be modeled with toxicokinetic or PBTK models. McLachlan (1994, 1996) has developed such a model for a lactating cow for PCDD/PCDF and PCB. The first comprehensive food-chain model for human intake of 2,3,7,8-TCDD through the terrestrial route of exposure has been described by Fries and Paustenbach (1990), who estimated the additional contamination caused by the operation of a municipal waste incinerator.

Examples of Successful Model Applications

Several examples of local-scale models that have been successfully applied to local water systems exist. These models combine a chemical fate model that links sources with concentrations in water, suspended sediment, and bed sediment with a food-chain bioaccumulation model.

Exposure modeling framework for PCB congeners in Green Bay, Wisconsin

The overall goal of the Green Bay Mass Balance Study (GBMBS) was to evaluate the feasibility of mass balance modeling of toxic substances as a planning and management tool for the Great Lakes. Specific objectives included

1) assessing the technical and economic feasibility of the mass balance approach for use in the management of pollutant loadings and impacts on Great Lakes ecosystems;
2) calibrating the mass balance modeling framework for understanding sources, transport routes, and fates of pollutants in the Green Bay ecosystem;
3) identifying the major sources of PCBs entering the Green Bay ecosystem and ranking their relative significance; and
4) demonstrating methods and priorities for further studies of toxic contaminants in the Great Lakes (Great Lakes National Program Office [GLNPO] 1992).

A series of models were developed that simulated the transport and fate of PCBs within the Fox River system (Fox River between Lake Winnebago and the DePere dam by Steuer et al. 1994; Fox River between the DePere dam and Green Bay by Endicott, Velleux et al. 1992) and within Green Bay itself (water column and sediment mass balance by Bierman et al. 1992; food chain bioaccumulation by Connolly and Parkerton 1994; hydraulic transport by Martin et al. 1995).

The structure of the integrated exposure-modeling framework for Green Bay itself, named GBTOX (Figure 4-3), was driven primarily by the realization that particle dynamics, through sorption and subsequent particle transport and transformation, was instrumental in controlling the transport and fate of hydrophobic organic chemicals. To capture this important aspect of chemical modeling, a series of linked models were developed to permit accurate simulation of sources, transport, and fate of sorbents such as POC and DOC in the system. In essence, there is a water balance, a sorbent mass balance supported by a separate eutrophication model that generates autochthonous loading of sorbents (phytoplankton production), and finally, a toxic-chemical mass balance that integrates all three.

GBTOX consists of a sorbent dynamics model that, unlike its predecessors, explicitly includes 3 sorbent-state variables: biotic carbon (BIC; live phytoplankton), particulate detrital carbon (PDC; all other particulate matter), and nonsettleable organic carbon (DOC) on the basis of their organic carbon concentrations. In addition to the conventional transport dynamics such as advection, dispersion, settling, resuspension, diffusion, this model includes both allochthonous and autochthonous inputs of organic carbon sorbents as well as various transformation pathways among these sorbent compartments (DePinto et al. 1994).

The final, coupled, sorbent-chemical mass balance model is depicted in Figure 4-3. This model was calibrated to GBMBS field data collected during the field monitoring program in 1989 to 1990. Having calibrated the model to Green Bay on a spatial and temporal basis, the computational framework was then used as both a research and management tool.

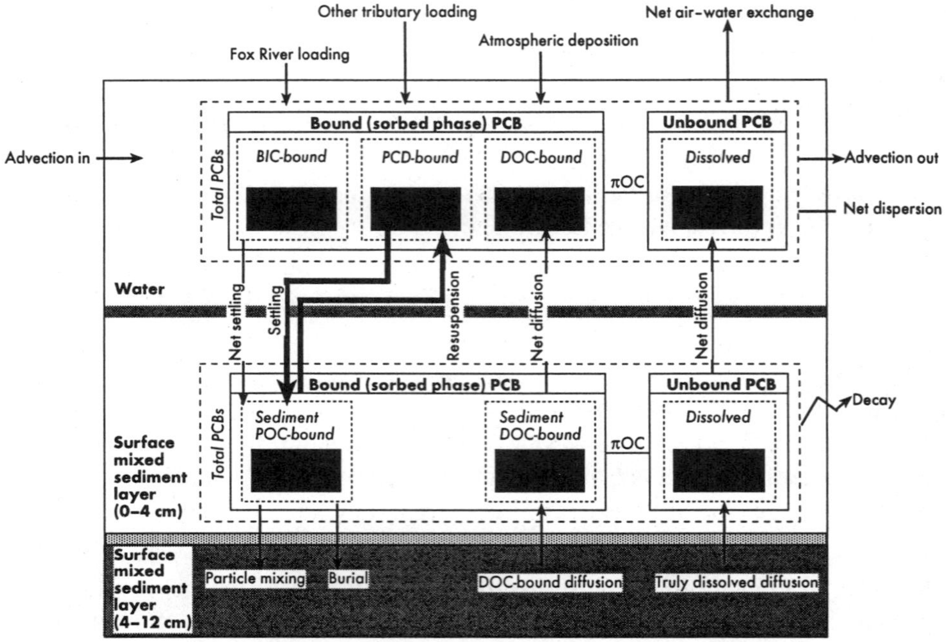

Figure 4-3 Conceptual diagram of GBTOX, the mass balance model for transport and fate of PCBs in Green Bay, Lake Michigan

GBTOX as a management tool

As a demonstration of how GBTOX can be used as a management tool, the model was applied to estimate the long-term response of the system to potential PCB-loading scenarios. For a number of reasons, the uncertainty in model predictions increases as long-term responses are extrapolated on the basis of short-term calibration data. The more a model can be validated with additional data, the more confidence there is in its long-term management predictions. Nevertheless, significant insights can be gained about the relative value of alternative remediation or regulatory actions and about the time it will take the system to respond to those actions by using the model in a management mode.

The long-term response of PCBs in the water column, the upper mixed-sediment layers (top 4 cm), and the entire modeled sediments (12 cm) for a range of loading scenarios are shown in Figure 4-4. The first diagram shows the trajectory of PCBs for repeating 1989 loadings and forcing functions on an annual basis, i.e., the entire time-series of loads were repeated every year, thus preserving the seasonal dynamics of PCBs. This scenario represents, in effect, the approach of the system to steady state. The whole-bay, annual average, water-column steady-state value is about 0.9 ng/L.

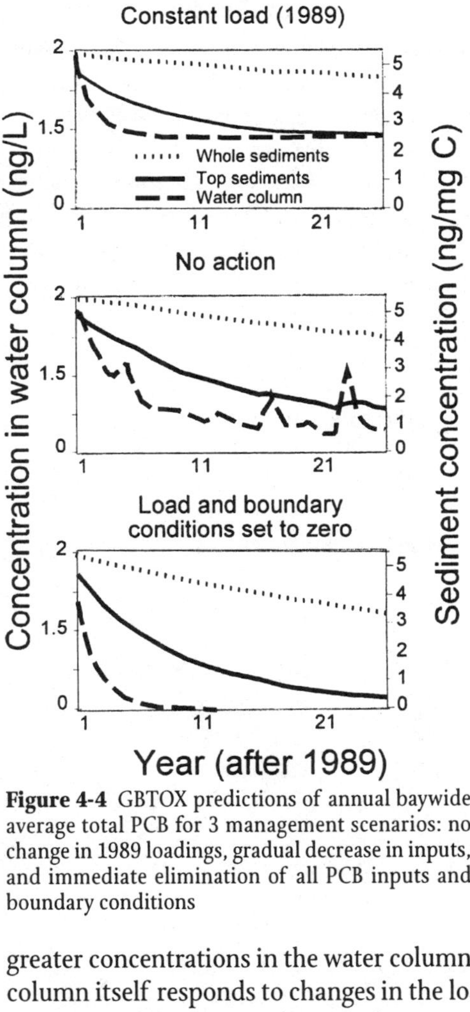

Figure 4-4 GBTOX predictions of annual baywide average total PCB for 3 management scenarios: no change in 1989 loadings, gradual decrease in inputs, and immediate elimination of all PCB inputs and boundary conditions

The second diagram in Figure 4-4 shows the response of the bay to a no-action scenario for the Fox River loads; "no action" in this context refers to no further remediation or regulatory action beyond what has already taken place in the system. These declining loads were provided by the lower Fox River model developed by Endicott, Velleux et al. (1992). These loads represent the natural washout and burial of PCBs in the Fox River as a function of 25 y of actual flow records in the river.

The third diagram in Figure 4-4 shows the "washout" of the bay if all loads, forcing functions, and boundary conditions are set to 0 at the start of the simulation. This simulation represents a bounding of the possible system response between no change in the loading, as shown in the first run, and complete elimination of all inputs, as shown in this run. It shows that, if all external loads are eliminated, the water column concentrations will still take about a decade to decline to near-0 values. This is due to the sediment–water interaction, which can sustain greater concentrations in the water column after loadings terminate. The water column itself responds to changes in the loads within the first year. The long-term trajectory of the water column is thus controlled by the concentrations in the active sediment layers. The deep sediments lose PCBs, primarily by net deposition of solids from the water column (burial). Thus, the vast reservoir of PCBs in the bay sediments continue to be buried deeper into the sediments. These PCBs are unavailable to the active Green Bay system, barring a catastrophic event remobilizing them.

The long-term behavior of PCBs in the Green Bay system shows that, if present loading levels continue, significant contamination of sediments and water throughout the system will be present for several decades. Total elimination of all external loads to the system will still require nearly a decade before the PCB concentrations in the water column are below detection limits. The Green Bay system shows a remarkable tendency to retain PCBs discharged into the system.

Application of DELWAQ-IMPAQT to HCB in lakes in the Netherlands

The model DELWAQ-IMPAQT describes the transport, distribution, and degradation of organic and heavy metal pollutants in the aquatic environment. This model was developed for the National Institute for Inland Water Management and Waste Water Treatment, the Netherlands. IMPAQT (De Vries 1987) simulates the total, dissolved, and particulate concentrations of micropollutants in the water column and sediment of a lake (Figure 4-5). DELWAQ calculates the transport of water, particulates, and modeled compounds. A food-chain model (UPTAQE) simulates the bioaccumulation of organic micropollutants and heavy metals in a food chain of the eel, using the results of IMPAQT calculations (De Vries and Goossens 1987). Bioaccumulation equations for this model were identical to Equations 4-1 and 4-2.

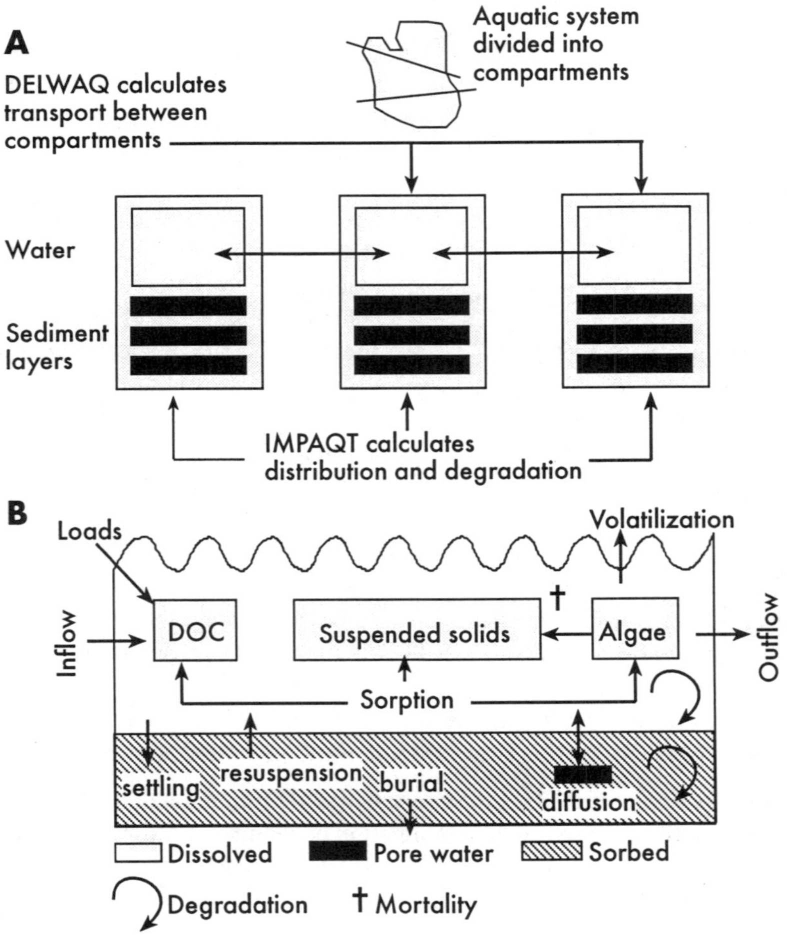

Figure 4-5 (A) Relationship between DELWAQ and IMPAQT chemical fate-and-transport models. (B) Major processes in the IMPAQT model.

The models were applied to calculate the distribution, fate, and bioaccumulation of HCB in 2 interconnected lakes in the Netherlands: Lake Ketelmeer (water residence time = 3 d, mean depth = 2.7 m, area = 35 km^2) and Lake Ijsselmeer (water residence time = 3 mo, mean depth = 4.2 m, and area = 1150 km^2). Water entering Lake Ketelmeer originates from the Rhine River; Lake Ijsselmeer receives its water from Lake Ketelmeer. Lake Ketelmeer was modeled as 1 segment while Lake Ijsselmeer was divided into 3 parts: south, north, and deep areas (> 5 m). Results of the model calculations were verified using the available measured concentrations in water, suspended sediment, sediment layer, and eels in the lakes. The measured concentrations in water were considered to be less reliable than those measured in the suspended sediment in water. The results of the IMPAQT model for Lake Ijsselmeer are shown in Figure 4-6.

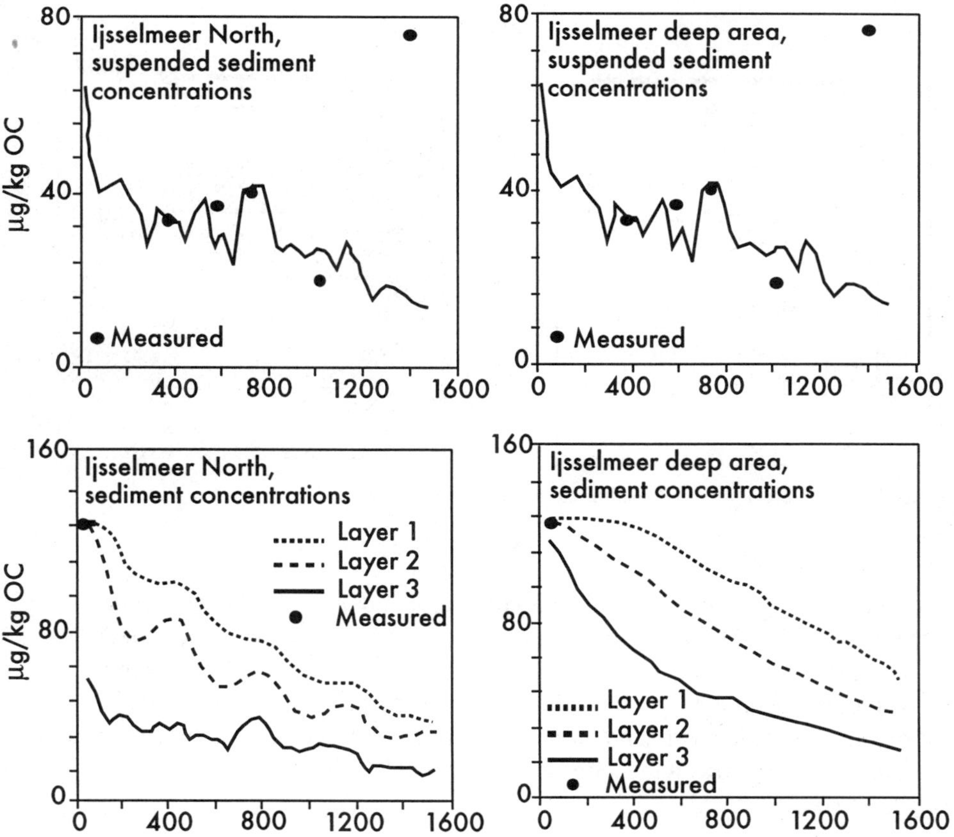

Figure 4-6 Results of IMPAQT model for Lake Ijsselmeer for HCB concentrations in suspended and bottom sediments caused by inputs from Lake Ketelmeer upstream

The application of UPTAQE to estimate HCB concentrations in eels used 2 different classes of eel (23 cm and 36 cm) because of the different food preferences (Figure 4-7). Results are presented for the 2 age classes in a contaminated system (Lake Ketelmeer; Figure 4-8 A) and for an individual eel entering the system without pollution history (Lake Ijsselmeer; Figure 4-8 B). Eels entering Lake Ketelmeer, which receives input from the Rhine River, are predicted to have increased concentrations of HCB, with the larger age class achieving greatest levels. Eels entering Lake Ijsselmeer from Lake Ketelmeer are predicted to show declining concentrations of HCB in the short term. Field measurements and model results show a good agreement for concentrations of HCB in both suspended sediments and accumulated levels in eel. Field observations of concentrations in eels could only be reproduced by assuming a food chain from benthic and pelagic organisms to smelt/ruff to eel.

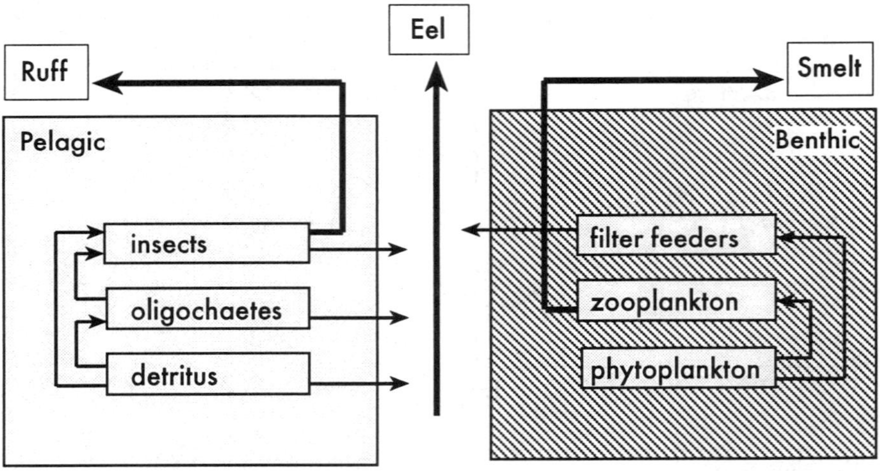

Figure 4-7 Structure of food web leading to eel used in UPTAQE model

The conclusions from modeling of HCB contamination of the 2 lakes are similar to those from the GBTOX study: concentrations in the water column are dominated by the suspended sediment dynamics and input load. If loads are stopped, the time needed for the sediment layer (5- to 10-cm depth) to reach half its initial value is approximately 4 y. Removing polluted sediment without reduction of the loads will not help to decrease pollutant concentrations in the lakes significantly.

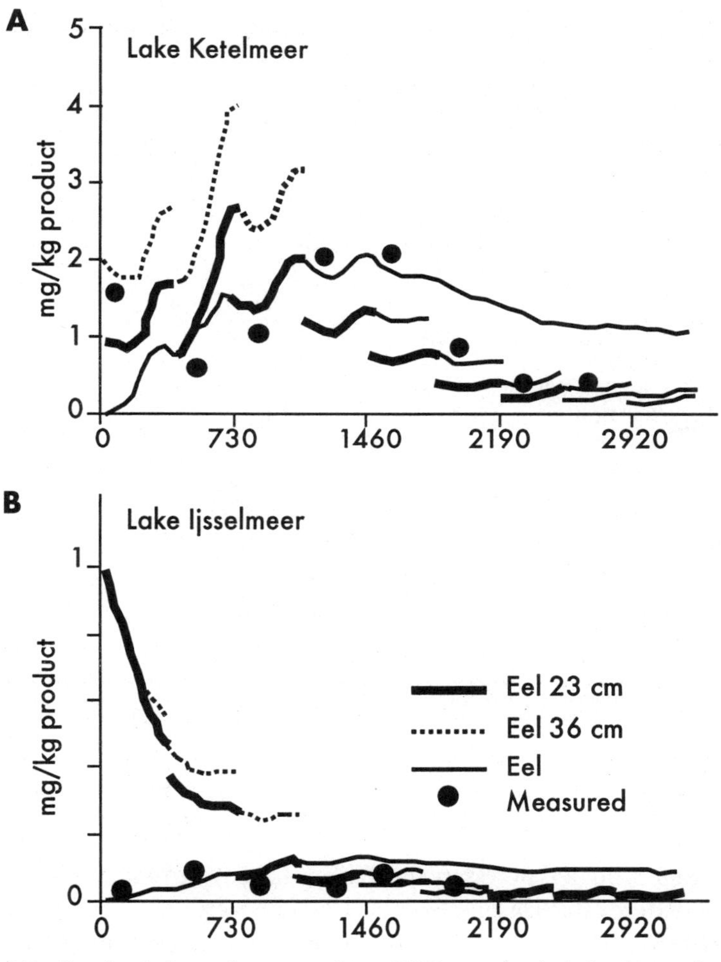

Figure 4-8 Predicted and observed concentrations of HCB over time in 2 size classes of eels in Lake Ketelmeer and Lake Ijsselmeer

Uncertainty and the Use of Models to Predict Environmental Concentrations

Table 4-3 describes the current state of knowledge in the application of chemical fate and food-chain modeling for various OC chemicals. For some chemicals that are well understood, the distribution in various environmental compartments can be predicted within a factor of 3. For other substances that are fairly well understood, the corresponding uncertainty is about 10, while the environmental concentrations of some poorly understood compounds have been predicted only within a factor of about 100.

Table 4-3 Summary of modeling capability for OC chemicals

Type of model	Compounds		
	Well-simulated	Moderately well-simulated	Poorly simulated
Global		CFC HCH[1]	PCB DDT
Regional	CBz[2]	PCB HCH PCDD/PCDF DDT	CP[3] Toxaphene
Local and site-specific	PCB CBz PCDD/PCDF HCH DDT C_1–C_3 (in air)[4] Chloro phenoxy acids	CP	
Biotic	PCB CBz PCDD/PCDF DDT HCH	CP	

[1] hexachlorocyclohexanes, including lindane (γ-HCH)
[2] chlorobenzene
[3] chlorinated paraffins
[4] chlorinated alkanes

chlorinated organics, but it is necessary to establish and adhere to some ground rules. First, it is important to establish an acceptable level of uncertainty. This level will depend upon the way in which the results will be used. For example, uncertainty is more tolerable in screening exercises aimed at prioritizing lists of hundreds or even thousands of chemicals than in a situation in which hazards are known and significant, but where the economic stakes are higher. Second, the scale of the model that is needed must be known. Scale means not only local versus regional versus global but also involves a decision as to whether a single-medium or multimedia fate model is more appropriate. Finally, a sensitivity analysis can help to identify critical and noncritical parameters. This is important because the impact of missing data is chemical- and site-specific and varies with the situation being modeled.

Release data

Environmental exposures occur only if a chemical is released as a result of its manufacture, processing, or use. This means that the use of models to predict environmental concentrations is possible only if data on emissions are available or can be estimated with acceptable accuracy. The quantities emitted, the chemical composition of the emissions, and their geographical location are fairly well defined for some OCs, but for most OCs, emissions are poorly characterized. Emissions quantities are, therefore, often the most critical data gap and largest source of uncertainty in exposure modeling. Special efforts should be made to address this data gap.

Physicochemical properties

Missing data on chemical properties are also a principal source of uncertainty. Properties can often be estimated with acceptable accuracy, but reliable measured values are always preferred. Large databases of measured values exist for log K_{OW} and to a lesser extent for water solubility, but for other parameters (such as vapor pressure), data are relatively sparse. Unfortunately, these key property data are often missing for the so-called "gray area" chemicals. The properties of these chemicals are of particular interest because they are less well characterized than well-studied OCs like PCBs.

Environmental transformations

As a general rule, reliable data on transformation rates are more limited than data on properties, and the ability to estimate these rates on a site-specific basis is not well developed. Compilations of transformation rate constants and other fate data do exist and could provide a convenient source of model input data. Biodegradation rates are especially important in determining long-term environmental accumulation, but environmentally relevant data are in short supply. Parameter uncertainty may be compounded by site-to-site variation, which may or may not be significant, depending upon the scale of the modeling exercise. Another source of potential uncertainty is the extent to which the basic transformation process is understood. For example, photo-oxidation catalyzed by hydroxyl radicals is relatively well understood, but the factors determining rates of indirect photolysis are not. This uncertainty may impact a model's validity at the most basic level.

Process descriptions

In addition to problems that arise from missing data, aspects of the partitioning process itself may not be well understood. Examples of processes that may be important for some chemicals, but that may not be correctly described by models, include partitioning between snow or ice and surrounding media, between atmosphere and plants, and between surface microlayers on water bodies and the atmosphere and water that they contact.

Environmental data

Models vary in their requirements for data on the physical and chemical properties of the system being modeled, due to variation in the complexity of those systems. If important data are not available — e.g., data on the hydrology or meteorology of the system — it may not be possible to run the model. However, the more significant problems from the standpoint of uncertainty analysis arise when most of the needed data exist but some model inputs must be estimated and/or when distributions (such as of stream flow data) must be captured as a single number or range of values for input to the model. Here again, a sensitivity analysis can reveal the impact of these parameters on model results.

Complex mixtures and effluents

Mixtures and effluents often contain many components and may be poorly characterized. This problem is not merely a source of uncertainty but in fact often leads to a judgment that modeling is not feasible. Therefore, it is important to consider 2 kinds of multicomponent systems: one that consists of chemicals with somewhat similar properties and behavior and another that is more heterogeneous. In the former case, it may be possible to characterize the behavior of a mixture by reference to one or more representative chemicals.

Confirmation of model results

Confidence in models is strengthened to the extent that predictions compare favorably to actual environmental concentrations. It is a given that any model being considered for application to OC chemical compounds has been fitted to some dataset and is therefore valid for those conditions. However, the ability of a model to accurately predict concentrations using data not in the calibration set helps to define the scope of the model's applicability as well as increase confidence in its accuracy. Beyond validation, it is useful to conduct post-auditing of model predictions, when this is feasible. Post-auditing involves making measurements of environmental concentrations after some action has been taken to reduce levels of contamination (e.g., site remediation). The model predictions are then compared to the actual outcome of the corrective action.

Chapter 5

Effects of Organochlorine Chemicals on Organisms

Assessing the potential effects of organochlorine (OC) compounds on organisms in the environment has typically been performed using laboratory studies, usually to determine acute and/or (sub)chronic toxicity to organisms specifically selected for the purpose. The biologically meaningful effects or endpoints of concern for such tests have included mortality, effects on reproduction, development and growth, and others such as neurotoxicity, immunotoxicity, and endocrine effects. This chapter focuses on the effects of OC compounds in relation to their degree of chlorination and mechanisms of action, from nonspecific narcosis (baseline toxicity) to more specific mechanisms such as respiratory uncouplers, inhibitors of acetylcholine esterase (AChE), aryl hydrocarbon receptor (AhR)-mediated mechanisms, endocrine toxicants, metabolically activated reactive intermediates, immune suppressants, and neurological and behavioral toxicants.

Narcosis can be caused by many organic chemicals and is observed in all organisms when sufficient concentrations in target tissues are reached. Chlorine does not necessarily contribute to or determine narcotic potency. Although well described in laboratory toxicity testing, and important in some site-specific situations and spill events, OC-induced narcosis does not appear to be a serious environmental problem.

Chapter Preview

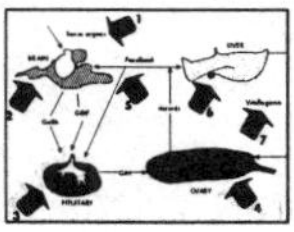

The actions of toxicants (such as pesticides or drugs) with specific sites of action are dependent on stereospecific interactions with a specific target system or receptor. Chlorinated chemicals in this group cannot be treated as a class, as their actions are generally specific. Risk assessments on a case-by-case basis are more appropriate.

There is growing evidence that a broad range of environmental chemicals may have the potential to disrupt reproduction in vertebrates by altering reproductive and endocrine function. These endocrine toxicological effects are not restricted to OCs. To date, studies have indicated that hormone-modulating actions of certain OCs may be of significant concern. Early life-history stages, e.g., during sexual differentiation, appear to be particularly sensitive. However, the endocrine effects of the majority of OCs and other substances have not been investigated, and more information is needed.

Toxicity associated with metabolic activation has been observed in a variety of chemical classes, both with and without chlorine. Chlorine does not appear to play an essential role in this process. Metabolic activation is most extensively studied in mammals, but examples have been seen in fish, birds, and bacteria as well. Compounds that are subject to metabolic biotransformation reactions are less likely to be environmentally persistent, in general, than are compounds that do not have this property.

AhR-mediated responses include lethality, reproductive and developmental toxicity, immunotoxicity, and cancer. To date, the group of chemicals producing the full range of toxic responses mediated by the AhR is dominated by certain halogenated (mainly chlorinated) organic compounds. While some polynuclear aromatic compounds also act through the AhR, they do not show all the AhR responses, as a result of their greater rates of metabolic clearance. Continuous exposure studies have not been conducted to confirm this hypothesis. Exposure to mixtures of AhR agonists appears to cause responses that are additive, based on AhR binding and potency.

Laboratory effects of AhR agonists have been observed at environmentally relevant levels. For certain locations in the industrialized world, there is probably not an adequate margin of safety for current maternal body burdens of tetrachlorodibenzo-*p*-dioxin (TCDD) and related chemicals in some mammals and for egg burdens of TCDD and related chemicals in some fish and birds.

Immunotoxicity has been observed in a variety of classes of chemicals, both with and without chlorine. Chlorine does not appear to play a specific role in the mechanism. Some of the older OC pesticides are immunotoxic at high doses, but data on low-dose exposure or dose-response studies are lacking. As with most substances, the immunotoxicity of OCs has not been well studied and good data are not available.

In general, the addition of chlorine to organic molecules increases the lipophilicity and decreases the rate of degradation of these compounds. The pronounced toxicological effects of some OCs, such as some PCBs and some older chlorinated pesticides, are influenced by their degree of chlorination as well as by the inherent toxicity of the basic molecular structure. However, there are OCs that do not share the adverse toxicological properties of the PCBs and chlorinated pesticides. Conversely, there are nonchlorinated compounds that act through mechanisms that are commonly observed in OC compounds.

Assessing the effects of chemicals on living organisms has typically been performed using laboratory studies. Laboratory studies of ecotoxicity have usually been conducted on laboratory organisms as surrogates to determine the acute toxicity and/or chronic toxicity of a substance to organisms in the environment. The biologically meaningful endpoints of concern for such tests can consist of mortality and/or effects on reproduction, development, and growth. Other endpoints, such as neurotoxicity, immunotoxicity, and endocrine effects, are also important but are more difficult to interpret in terms of ecological significance. Basic, acute toxicity data for many organochlorine (OC) chemicals are available for a few standard aquatic species, and examples of such data have been extracted from the large pool of information available for organic chemicals (see Chapter 6).

It should be noted that the testing procedures and protocols employed in generating these data are standardized, and no differentiation is made between tests of chlorinated chemicals as compared to nonchlorinated chemicals. Where some substances may have their impacts through persistence in the environment and bioaccumulation in organisms, longer-term ecotoxicity testing must be considered, e.g., whole life-cycle or multiple-generation testing. Testing of this type is currently required only for pesticide registration. In contrast, good mammalian toxicity data are generally available for many widely used OC compounds, such as solvents, pesticides, and drugs.

Laboratory tests alone seldom adequately describe what is likely to occur in the environment. Nevertheless, laboratory testing is often the first step taken in assessing ecotoxicity, for several reasons. Laboratory studies usually are easier to develop and control than are field studies, and laboratory results are easier to interpret. However, environmental realism is often compromised or sacrificed. Extrapolation of laboratory data to the field may overestimate or underestimate field responses. In addition, modifying factors such as physical and chemical fate and biological interactions can influence the results of both laboratory and field research, e.g., alterations in bioavailability of the chemical, variation in physicochemical parameters of the system under consideration, variable exposures over time, behavioral effects or avoidance of the contamination, and nutritional and reproductive status of the exposed organism. An important but often under-appreciated factor is that toxicity testing results will vary, depending upon the time of exposure to the toxic agent in relation to the life cycle and the diurnal cycle. Toxic agents may have more or less influence, depending on the time of day or even the time of year. This is related to the natural circadian, life-cycle, and seasonal rhythms inherent in organisms (Zeeman 1986).

While substances produced and used in large quantities can be and, in some cases, have been well characterized, data gaps for OC compounds, (e.g., physicochemical properties, environmental fate, loadings, and potential ecological effects) frequently exist. With some substances potentially subject to regulation, lack of data has prompted the grouping of chemicals by various criteria into classes that are then regulated based on properties of selected members of the class. Chemical-structure-based groupings are often used, and the USEPA, e.g., has successfully employed toxic-effects-based groupings in the regulation process for new organic chemicals (Zeeman, Nabholz, Clements 1993; Zeeman, Auer et al. 1995; USEPA 1994b; Zeeman 1995).

Mechanisms causing ecotoxicity include narcosis (nonspecific toxicity) and more specific mechanisms that result in enhanced toxicity, such as respiratory uncouplers, acetylcholine esterase (AChE) inhibitors and central nervous system convulsants (Lipnick 1993; McCarty and Mackay 1993). Sections of this chapter focus on various mechanisms such as AhR, steroid receptors (estrogenic chemicals), metabolic activation, immune suppression, and neurological and behavioral effects. Discussions in this chapter examine mechanisms of action for chemicals because the nature of the toxic action of a chemical is valuable for the assessment of its potential ecological impacts. Although this type of information is scanty, some generalizations are possible for classes of OC compounds. It is important that the mechanisms of action be identified for compounds that may be persistent, bioaccumulative, and toxic, so appropriate generalizations can be made for homologous series and quantitative structure-activity relationships (QSARs) can be developed.

The OC chemical groups being considered in this chapter are diverse (Table 5-1); they can therefore be expected to act by a variety of the previously mentioned mechanisms and may, in fact, have different mechanisms of action in different circumstances. This chapter tries to relate the mechanisms of action to these chemical classes. The authors realize that there are exceptions to any of the mechanisms stated or known for any chemical class. Furthermore, classification of mechanism of action can be by biological response characteristics (see McCarty and Mackay 1993 for a summary and explanation of the categories proposed by workers at the USEPA-Duluth laboratory) or by similarities in chemical reactivity (Lipnick 1993 and Table 5-1). Potential target organisms are grouped as bacteria, plants, invertebrates, fish, birds, and mammals. Although the term "mechanism of action" is used here, it is more accurate to term these groupings "modes of action," since the actual detailed mechanism of action often is not known. As noted by Drummond and Russom (1990), more than one specific biochemical mechanism can be associated with each whole-organism response mode.

In the case of OCs, the addition of chlorine to organic molecules generally results in greater lipophilicity of the chlorinated molecules as well as in decreasing rate of breakdown of these compounds by various biological and chemical processes (Chapter 2). It is clear that the pronounced environmental effects of some OCs, such as some polychlorinated biphenyls (PCBs) and some older chlorinated pesticides, are influenced by their degree of chlorination as well as the inherent toxicity of the basic molecular structure. Recognized toxicological modes or mechanisms of action are not unique to OC chemicals. However, the environmental effects of some of the most prominent members, such as PCBs and the older chlorinated pesticides, require specific examination. As further data are developed on this group of chemicals, a better delineation of the environmental impacts of members of these chemical classes will result.

The problem of considering the toxicity of mixtures, which is most often the case in real-world exposures, is a longstanding one for which there is no general solution. Fortunately, synergistic interactions of chemicals in mixtures are less likely at typical environmental concentrations (Heindel et al. 1994). Such interactions are seen more often in pharmacological studies and in human and animal testing, in which doses are greater, often near or at those associated with overt toxicity at the maximum tolerated dose. In the absence of data to the contrary, simple additivity of exposures is generally an appropriate assumption for assessing mixtures with similar modes of action.

Table 5-1 Summary of mechanisms of action by OC chemical class

Chemical class	Nonspecific toxicity (narcosis)	Toxicants with specific site of action	Endocrine toxicology	Metabolism activation, covalent binding, lipid peroxidation	AhR-mediated responses	Immunotoxicology
Dioxins	BA,P,I,F,B,M	NA	F,M	—	F,B,M	F,B,M
PCBs	BA,P,I,F,B,M	NA	F,B,M	—	F,B,M	F,B,M
Chlorinated pesticides	BA,P,I,F,B,M	P,I,F,B,M	F,B,M	—	—	I?,F,B,M
Old	BA,P,I,F,B,M					
New	BA,P,I,F,B,M	P,I,F,B,M	—	—	—	
Chlorinated aromatics	BA,P,I,F,B,M	P	—	M,F	—	M
Neutral (benzenes, naphthalenes)						
Polar (phenols, anilines, diphenyl ethers	BA,P,I,F,B,M	P,I,F,B,M	—	—	—	
Chlorinated aliphatics	BA,P,I,F,B,M		—	M	—	
C_1–C_3						
C_3–C_{10}						
> C_{10}						
Cyclic chlorinated aliphatics	BA,P,I,F,B,M	I,F,B,M	—	F,M	—	
Chloroacids and chloroaldehydes	BA,P,I,F,B,M	—	—	—	—	
Chlorinated polymers chlorolignins, humics, and soil-bound residues		—	—	—	—	
Nonchlorinated chemicals	BA,P,I,F,B,M	P,I,F,B,M	P,I,F,B,M	I,F,B,M	F,B,M	I?,F,B,M

BA bacteria; P plants; I invertebrates; F fish; B birds; M mammals; ? probable but not reported yet

There are some modest successes in assessing toxicity of mixtures, such as in dealing with mixtures of narcotic chemicals where the indirect toxic unit approach or its direct equivalent, the critical body residue (CBR) approach, provides a reasonable understanding. The toxic equivalency factor (TEF) approach relating relative potency for tetrachlorodibenzo-*p*-dioxin (TCDD)-like toxicity, which is exhibited by many polychlorinated dibenzo-*p*-dioxins (PCDDs)/polychlorinated dibenzofurans (PCDFs) and PCBs, is another potentially useful strategy. However, the TEF approach has yet to be fully experimentally validated. Although there are likely other possibilities within other specific modes of action, the methodologies are not as well developed. It is noteworthy that the existing successes reflect the advantages of classifying chemicals based on their modes of action rather than the more traditional, chemical-structure-based groupings.

Nonspecific Toxicity (Narcosis)

Narcosis can be caused by most organic chemicals, and this mode of action can be found in all organisms when sufficient exposure occurs. Assessment of narcotic chemicals is based primarily on laboratory testing. Chlorine neither contributes to nor determines narcotic potency. Because narcotic chemicals are essentially equipotent when expressed as molar concentration in organisms, toxicity of mixtures can be adequately estimated by addition of residue-based internal doses. Although well described in laboratory toxicity testing, and important in some site-specific situations and spill events, OC-induced narcosis does not appear to be a serious environmental problem.

Mechanism of action

Simple, reversible narcosis resulting in general, nonspecific toxicity is a characteristic of all organic chemicals that can traverse a biological membrane and enter the body of a living organism. Narcosis is often referred to as "baseline toxicity" or as "anesthesia" (Lipnick 1993) and is generally the toxicity observed most frequently with high-level exposures. Chemicals expressing greater toxicity than this generally will be eliciting such toxic effects by specific modes or mechanisms of toxicity.

All organisms — bacteria, plants, invertebrates, and vertebrates — can exhibit narcotic effects when exposed to organic chemicals, and such responses have been found in laboratory experimentation. The actual mechanism of action is not known, the 2 leading theories being the "membrane swelling" and the "protein-pocket" theories. Both theories are related to general disruption of membrane-associated metabolic activities (Abernethy et al. 1988). Acute narcotic effects are similar in most organisms, with an initial excitatory phase, followed by an activity depression phase and, with sufficient dose and duration, death. Many humans are familiar with the narcotic effect of organic chemicals by virtue of their intoxication experiences with beverage alcohol. The universality of narcotic action resulting from exposure to organic chemicals is reinforced by the fact that specifically acting chemicals, such as pesticides, often exhibit narcotic activity in nontarget species, but at much higher concentrations than those at which they cause specific, receptor-mediated effects (Figure 5-1).

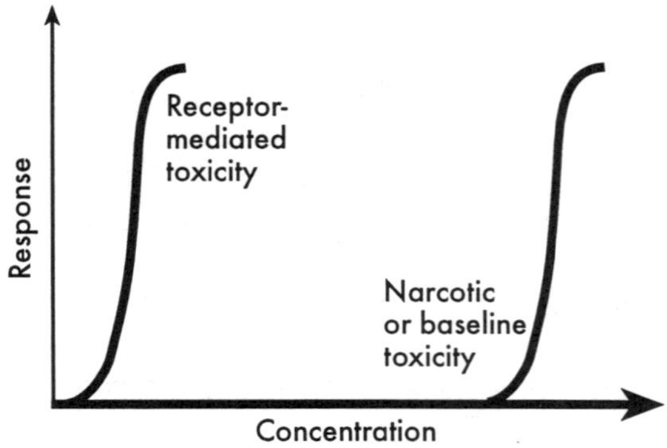

Figure 5-1 Difference between toxicity of substances in organisms with the receptor and narcotic toxicity in those without

Although chlorine does not affect the potency of the effective dose in an organism, it has substantial influence in the partitioning characteristics between the exposure medium and the organism by increasing hydrophobicity relative to the nonsubstituted form. Thus, variation in expressed toxicity is controlled primarily by partitioning behavior, the toxicokinetics processes that control the rate of accumulation and concentration in the membrane. In general, chlorination results in greater molecular volume and greater hydrophobicity. For example, in toxicity tests with fish, this hydrophobic influence can produce estimates of lethal concentrations to 50% of the population (LC50s) that vary by 6 orders of magnitude, while the concentrations in affected organisms are approximately the same. Some well-known narcotic OCs include many members of the chlorobenzenes, chloroethanes, and chloroethenes.

In addition to narcosis caused by neutral organic compounds, narcosis may be associated with organic chemicals that are polar in their molecular structure and may ionize in solution. This is known as "polar narcosis." There is some debate about whether this is a completely distinct mode of action or a minor complication in the general narcosis group that is associated with variability introduced by increased molecular polarity. Although chlorine may produce polarity, it is not unique in this regard, as other substituents can create polarity in organic molecules. Some of the well-known, polar-narcotic OCs can be found in the chlorinated phenols and chloroanilines.

Chemical classes

Narcotic potency is related to molecular volume and partitioning properties. It is not a function of the structure of the chemical, nor has it anything to do with the substitution of chlorine or any other substituent group unless this changes the partitioning properties. Narcosis occurs when the dose of compound in the organism reaches a threshold concentration, the CBR (Figure 5-2). Thus, narcotic effects can be observed in all of the chemical groups listed in Table 5-1. However, as noted earlier, chlorine substitution may have substantial influence on the partitioning characteristics between the exposure medium and the organism by increasing hydrophobicity relative to nonsubstituted forms. Thus, variation in expressed toxicity of narcotic OCs is controlled primarily by partitioning behavior (toxicokinetics). Although the amount of chemical in the exposure medium may vary dramatically for a given response, the amount in the organism (CBR) is relatively constant. As seen in Figure 5-2, this level is on the order of 1 to 10 mM in the whole organism, in this case for small fish (< ~ 5 g) and invertebrates, although this is modified somewhat by body lipid levels (Ferguson 1939; McCarty et al. 1992). When considered on a body-residue basis rather than on an exposure-concentration basis, mixtures of narcotics appear to be simply additive on the basis of molar concentration.

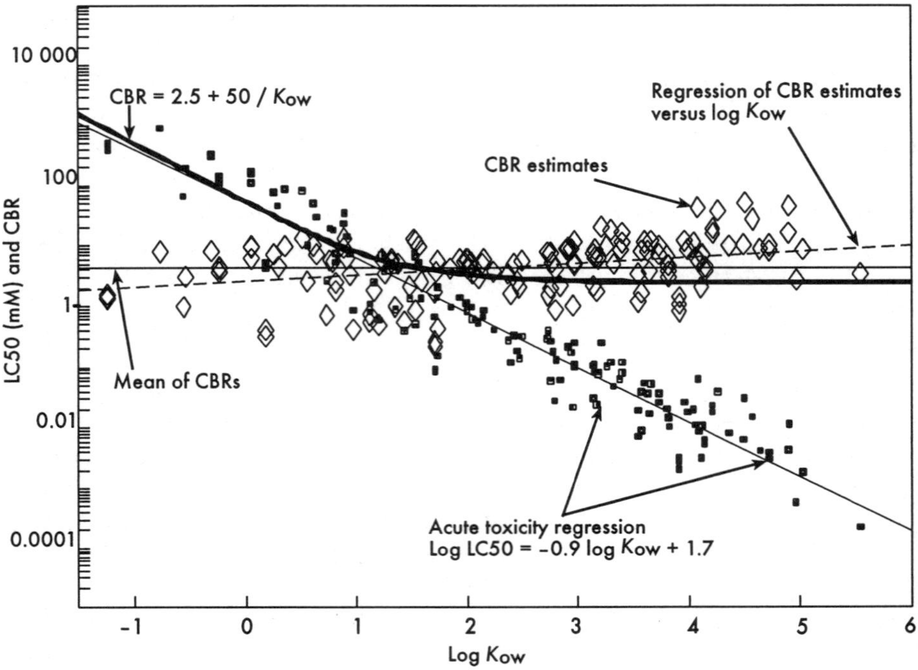

Figure 5-2 Relationship between octanol/water partition coefficients, acute toxicity, and critical body residues (CBRs) for fathead minnows exposed to narcotic organic chemicals (redrawn from McCarty et al. 1992)

Effects

In addition to direct acute mortality, chemicals causing narcosis can decrease long-term survival, development and growth, and reproduction. Chronic lethal effects of narcotic chemicals typically occur at levels about an order of magnitude below those of acute effects (McCarty and Mackay 1993; Zeeman, Clements et al. 1993; Zeeman 1995; Zeeman et al. 1995). Such effects are reversible unless death has occurred.

Toxicants with a Specific Site of Action

The actions of toxicants with specific sites of action, as is often the case with pesticides and drugs, are dependent on stereospecific interactions with a specific target system or receptor. The size and conformation of the molecule, regardless of the presence or absence of chlorine substituents, determines the toxicant's action.

The presence of chlorine in the molecule may affect the reactivity of the compound or its resistance to degradation by biota and in the environment. These properties may change concentration at the target site but do not appear to influence the inherent toxicity of the molecule. Chlorinated chemicals in this group cannot be treated as a class because their actions are generally specific. Risk assessments on a case-by-case basis are more appropriate.

Mechanism of action

Specific toxicity is associated with compounds that, inter alia, include pesticides such as the AChE inhibitors, neurotoxic agents, photosynthetic blockers, respiratory blockers, drugs, and uncouplers of oxidative phosphorylation. A number of groups of chemicals have mechanisms of action that involve binding to specific sites on cell membranes, enzymes, or proteins and are thus receptor-mediated. These include a range of pesticides, drugs, and other compounds.

Neurotoxic pesticides and drugs are specifically toxic to animals. Neurotoxic pesticides are most commonly used as insecticides. Differences between the nervous systems of vertebrates and invertebrates may allow some specificity of action. This specificity results from differences in the target nervous systems (receptors) as well as the distribution, transport, and metabolism (pharmacokinetics) of these compounds. In general, these compounds act through 2 main mechanisms: either as an agonist or antagonist of the natural compound that binds to the receptor or as substances that interfere with the synthesis, release, or degradation of endogenous neurotransmitters and that inhibit enzymes critical to cell functions.

Receptor interactions

Dichlorodiphenlytrichloroethane (DDT) (Figure 5-3) is understood to act on the nerve axon by binding to a receptor associated with the sodium or the potassium gate in the membrane. The role of chlorine substitution in the binding of the molecule to the receptor appears to be unimportant, as evidenced by the equally potent analogs in which the chlorine groups have been replaced by organic moieties of similar "volume," such as in methoxychlor and the substituted ethane analogs of DDT (Figure 5-3) (Kapoor et al. 1972, 1973; Coats et al. 1977). The natural and synthetic pyrethroids both interact with nervous system receptors in a way similar to DDT. The addition of chlorine or other halogens to the basic (natural) pyrethroid molecular structure of pyrethrin makes these compounds more useful as agricultural pesticides because it increases

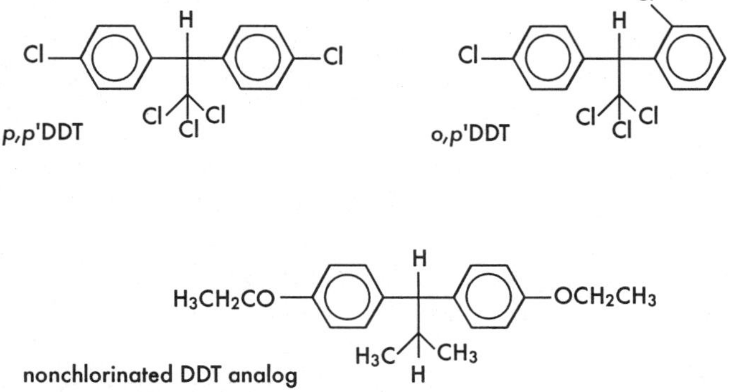

Figure 5-3 DDT and a nonchlorinated, biologically active analog

their persistence both in the organism and in the environment (Elliott et al. 1974; Briggs et al. 1976; Elliott 1976, 1989). However, halogen substitution makes little difference to their inherent toxicity, in the absence of metabolism, in target and nontarget organisms.

Several neurotoxic insecticides, such as cyclodienes, toxaphene, and lindane, also contain chlorine, some in considerable quantity. Although these compounds are not as widely studied as DDT and its analogs, there is no reason to believe that these compounds require the presence of chlorine to interact with their respective receptors. Many of these are understood to interact with the γ-amino butyric acid (GABA) receptor in the nervous system. This GABA-receptor binding property is also displayed by other naturally occurring neuroactive compounds, such as the avermectins (Pong et al. 1982; Mellin et al. 1983), which do not contain chlorine.

Specific inhibitors of enzymes

The AchE-inhibiting insecticides are perhaps the best known and most widely used group of pesticidal chemicals. The mechanism of action of these compounds is also understood (Figure 5-4). AchE has been isolated, and the enzyme kinetics are defined. The esteratic site contains a basic group (an N+) and an OH group (from serine). In the case of organophosphorus pesticides, the chemical mimics the normal substrate (acetyl choline), binds to (k+1), and phosphorylates the enzyme (k2), as, e.g., in the case of paraoxon (Figure 5-4). The rate of enzyme reactivation (hydrolysis, k3) of phosphorylated enzyme is approximately 10^9-fold slower than natural acetylated enzyme. Thus, phosphorylated AChE is effectively blocked from acting in the normal way for as long as 12 h. As a result, acetylcholine accumulates in the synapse, and the postsynaptic membrane is stimulated to initiate a continuous series of action potentials in the nervous system. This disruption results in the appearance of cholinergic toxicity.

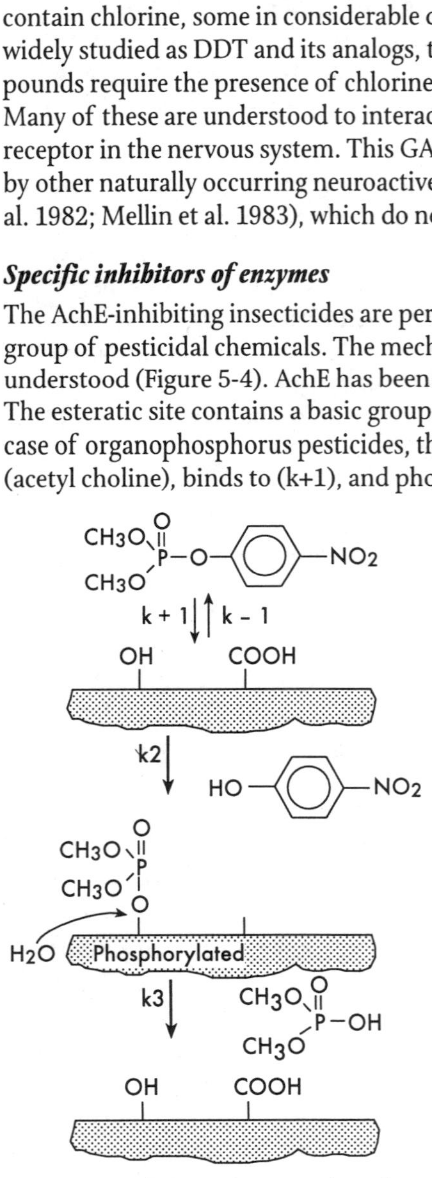

Figure 5-4 Hydrolysis of paraoxon by AChE

While some compounds that inhibit AChE contain chlorine, its presence is not necessary to the action of the compounds. However, chlorine substitution may change the reactivity of the compound (k2) and thus increase the rate of the phosphorylation. In general, the organophosphorus compounds are proportionally more active, with greater electron withdrawal from the phosphorous atom. Substitution with chlorine atoms may be used to manipulate this electron withdrawal, but the same effect can be obtained with other substituents such as nitro groups.

Another group of compounds that inhibit biochemical processes by specifically binding to a receptor are the herbicides that act as photosynthetic inhibitors. Photosynthesis is the process whereby light energy is converted to chemical energy (adenosine triphosphate NADPH) for plant growth. Photosynthetic inhibitor herbicides such as atrazine competitively inhibit photosynthesis through binding to the plastoquinoine receptor site in the D-1 protein of the photosynthetic complex (Fuerst and Norman 1991; Devine et al. 1993). Photosynthetic inhibitors have a range of chemical structures, some of which contain chlorine. These include the chlorinated phenyl ureas, 1,2,4-triazinones, chlorinated s-triazines, azido-s-triazines, thiocarbamates, uracils, carboxamides, pyridazinones, alkyl anilides, diphenylethers, and p-alkyl thioanilides. All of these compounds appear to have the same mechanism of action: deactivation of the enzyme system responsible for cleaving the water molecule in the photosynthetic pathway. As with the insecticides, the presence of chlorine in these molecules is not necessary for biological activity. However, substitution with chlorine atoms may play an important role in modifying their pharmacokinetic properties in the target organism and their persistence and fate in the environment.

Chlorinated phenols can act as uncouplers of bacterial and mitochondrial oxidative phosphorylation. This property is thought to be related to the degree to which these compounds can accept electrons and interfere with the electron transport system in the mitochondrion. Again, this property is not exclusive to chlorine substitution; rather it is a function of the properties that the substituent groups or atoms confer on the molecule. For example, 2,4-dinitrophenol, which contains no chlorine atoms, is a more potent uncoupler of oxidative phosphorylation than is pentachlorophenol (PCP).

Chemical classes

It is not possible to classify all OC compounds in classes by mechanism of action. Specific mechanisms of action in target sites are not general to the chlorinated compounds (Table 5-1). Receptor-mediated neurotoxicity is found across most of these groups, except for the PCDD/PCDFs, and the chlorinated polymers. Inhibition of photosynthesis is caused by chlorinated aromatics and some chlorinated aliphatics but is not caused by PCBs, PCDD/PCDFs, and the C_{1-10} aliphatics.

Effects

Toxic potencies of chlorinated compounds with specific sites of action vary greatly and can cause lethality at small environmental concentrations. Effects of neurotoxicants are confined to organisms with nervous systems. In general, specificity seems to be a function of pharmacokinetics and distributions in the environment. The systems initially affected are generally the nervous system or energy production. However, secondary effects also may result from this interference. Disruption of the neuroendocrine system and homeostasis has been reported as the actual cause of death in insects treated with insecticides (Casida and Maddrell 1971; Maddrell and Casida 1971; Maddrell and Reynolds 1972). It has been reported that neurotoxicants may interfere with normal behavior in organisms such that they may be exposed to greater risks of predation or be unable to hunt for food (Morgan and Kiceniuk 1991). This disruption is secondary to the initial toxic action and is unrelated to the chemistry of the molecule.

Some of these compounds and their metabolites may have effects on reproduction (see "Endocrine toxicology," this chapter). Developmental effects have been reported in some studies with OC compounds (methoxychlor; Holdway and Dixon 1986; Holdway et al. 1987). However, the exact mechanisms of action are not well understood.

Endocrine Toxicology

There is growing evidence that a broad range of environmental chemicals may have the potential to disrupt reproduction in vertebrates by altering reproductive and endocrine function. These endocrine toxicological effects are not restricted to OCs. To date, studies have indicated that hormone-modulating actions of certain OCs, such as PCBs, DDT analogs, PCDD/PCDFs, and kepone, may be of significant concern. Potential neuroendocrine effects of certain OCs with central nervous system (CNS) actions are also of concern. Early life-history stages, e.g., sexual differentiation, may be particularly sensitive. However, the endocrine effects of the majority of OCs, and of other substances, have not been investigated and more information is needed.

The complexity of endocrine systems in vertebrates makes the investigation of the sites and mechanisms of toxic action difficult. Nevertheless, evidence from laboratory and field studies indicates adverse effects on reproduction in vertebrate populations from exposure to xenobiotic OCs. In some cases, population-level effects have been observed. Certain OC compounds are among those chemicals with the capability to disrupt the endocrine or reproductive systems in a range of species, including humans[1]. These and other reproductive toxicants can also impair reproduction by mechanisms other than endocrine-system (see "Toxicants with a specific site of action," this chapter).

[1] The authors and editors recognize that endocrine toxicology and the possible effects of endocrine-disrupting substances in the environment have been the subjects of considerable recent activity in the research community. A number of scientific publications as well as popular press and media attention to the topic have made them the subjects of intense debate on many levels. The importance of this subject and its relevance to some of the OCs is acknowledged, but the editors judged that adding an extensive discussion of the topic to this book at this time, without the participation of the workshop attendees, would be inappropriate. Several workshops have been held on the topic recently, and the interested reader is referred to these (Kavlock et al. 1996; Kavlock, Ankley 1996; Kendall et al. 1998) for further information.

Hypothalamus-pituitary-gonadal axis

In vertebrates, reproduction is controlled by hormones secreted by the hypothalamus-pituitary-gonadal (HPG) axis (Figure 5-5). Environmental stimuli, such as photoperiod and temperature, are detected by sense organs (site 1) that relay this information to the hypothalamus (site 2). The hypothalamus is the principal region of the brain that controls pituitary-hormone secretion and hormone processes such as reproduction. Signals from the sense organs and internal sensory system are integrated in the hypothalamus to control the release of neurotransmitters that, in turn, regulate the production and secretion of specific neurohormones termed "releasing hormones." The releasing hormones then act on specific cell types in the pituitary gland (site 3) to regulate the production and secretion of pituitary hormones into the blood for circulatory distribution. In general, the secretion of each class of pituitary hormones is regulated by a different releasing hormone. For example, the secretion of gonadotropins (GTHs), the primary pituitary hormones controlling the reproductive cycle in vertebrates, is regulated by gonadotropin-releasing hormone (GnRH). Inhibitory factors or pathways may also be present, such as the dopaminergic-inhibitory mechanism for the control of reproduction in some teleost species (gonadotropin release-inhibiting factor) (Figure 5-5). Gonadotropins are released into the blood in response to GnRH. Gonadotropin acts at the ovary or testes (site 4) to stimulate gonadal development and the production of sex steroid hormones, mainly estrogens, by the ovaries and androgens by the testes. During pregnancy, the placenta is also a major source of GTH and steroid hormones (progesterone) in mammals. Steroid hormones have multiple sites of action. They act within the gonad to influence gonadal development and gamete production (site 4). A major site of action of estradiol in egg-laying vertebrates such as birds, reptiles, amphibians, and fish, is the liver (site 5), where the precursor of egg yolk, vitellogenin, is produced in response to estrogen stimulation. In mammals, steroids of placental origin act locally to maintain pregnancy. Sex steroid actions on other components of the HPG axis, sexual organs, and secondary target tissues result in the broad range of sex differences in axis morphology, function, and behavior observed in vertebrates. The mechanism of action of these compounds is through intracellular steroid receptors that modulate deoxyribonucleic acid (DNA) transcription. The brain is especially sensitive to steroids or foreign compounds with steroidal activity during the critical period in the early life-history cycle of vertebrates when sexual differentiation occurs. Thus, transient alterations in brain steroid concentrations can cause permanent reproductive dysfunction, although not at all stages of the life cycle. The brain and pituitary are also important sites for steroid action (site 5) in adults, as part of the steroid feedback mechanism modulating the production and release of GnRH and GTH.

Mechanisms of toxicant action

While it is recognized that some chemicals may be toxic directly to the reproductive tissues, many others may act through the endocrine system. Toxicants can act at any of the multiple sites in the HPG axis to impair reproductive or endocrine function (Figure 5-5). Toxicants may act on the peripheral sense organs (site 1) or directly interfere with neural activity by altering neurotransmitter function (site 2) to influence GnRH or gonadotropin release-inhibiting factor (GRIF) secretion. Alterations in the secretion of these neurohormones would affect GTH secretion from the pituitary, or alternatively, certain chemicals could act directly on the pituitary (site 3; Thomas 1993) to alter GTH secretion. Changes in GTH secretion would secondarily influence gonadal function. Chemicals could exert direct effects at the gonad to alter gonadal development and steroid secretion (site 4; Singhal et al. 1985). Changes in steroid secretion would alter steroid feedback of GTH secretion (site 5) or other steroid-dependent processes, such as vitellogenesis (site 6). Increased metabolic clearance of steroids due to induction of hepatic P450 enzymes (site 5) would also alter circulating steroid levels and steroid function. Steroid function can also be altered directly by chemicals that can compete with the natural steroid ligand for binding to its receptor (sites 4 and 6; Kupfer and Bulger 1980; Huang and Nelson 1986; Thomas and Smith 1993).

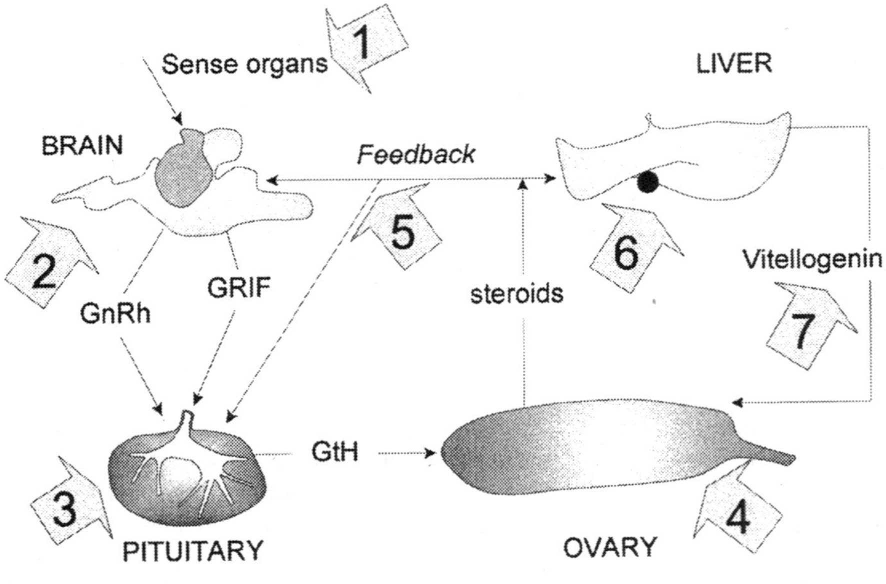

Figure 5-5 Hypothalamus–pituitary–gonadal–liver axis in a female teleost, showing possible sites of chemical interference (redrawn from "Teleost model for studying the effects of chemicals on female reproductive endocrine function," P.T. Thomas, *J Experimental Zool Suppl* 4:126–128, © 1990 Wiley-Liss, Inc. Adapted by permission of Wiley-Liss, Inc., a subsidiary of John Wiley & Sons, Inc.)

Chemical-induced alterations in the function of one level of the HPG axis will ultimately result in alterations throughout the HPG axis. Therefore, it is frequently difficult to determine the precise sites and mechanisms of action of chemicals on reproductive endocrine function. In some organisms, certain stages of reproductive development and the differentiation of the gonads are especially sensitive to disruption of the endocrine system by toxicants. This may result in feminization or masculinization. However, it is frequently difficult to identify these effects because they may become manifest later when the offspring reaches sexual maturity (e.g., the effects of diethylstilbestrol (DES) in humans, such as infertility, cancer, morphological abnormalities, and behavior).

Reproductive or endocrine toxicants can be classified as hormonal mimics and blockers with direct actions and as indirect disruptors such as the CNS-active compounds. Both mechanisms of action have been observed in OC chemicals. It should be recognized that alterations in the concentrations of hormones may not necessarily result in adverse effects: the endocrine system is homeostatic and many compensatory responses are possible.

Hormone mimics and blockers

Hormone action is mediated by hormone binding to specific hormone receptors in target tissues. Estradiol, e.g., binds to its nuclear receptor in an estrogen target tissue such as the liver, causing a conformational change termed "activation" that enables the receptor to bind to specific acceptor sites on the DNA in the nucleus. This results in alteration of gene transcription, such as in the vitellogenin gene. A variety of nonsteroidal estrogenic compounds, such as the synthetic estrogen DES, the phytoestrogen coumestrol, and the OCs *o,p'*-DDT and kepone, have been shown to exert their estrogenic effects by the same mechanism and are therefore termed "hormonal mimics" (Figure 5-6). Potency of the compounds may vary considerably. For example, DES is 10 times more potent than estradiol, while *o,p'*-DDT is 10,000 times less potent. PCB mixtures, particularly those with lower chlorine contents (up to 48%, such as Aroclor 1221) have weak estrogenic activity (Bulger and Kupfer 1985). PCBs show weak binding affinity to estrogen receptors with a potency 10,000- to 100,000-fold less than estradiol (Nelson 1974). Compounds that displace the natural ligand from its receptor but fail to induce a hormonal response (gene transcription), and thereby antagonize the actions of the hormone, are termed "antihormones" or "hormonal blockers" (Figure 5-6).

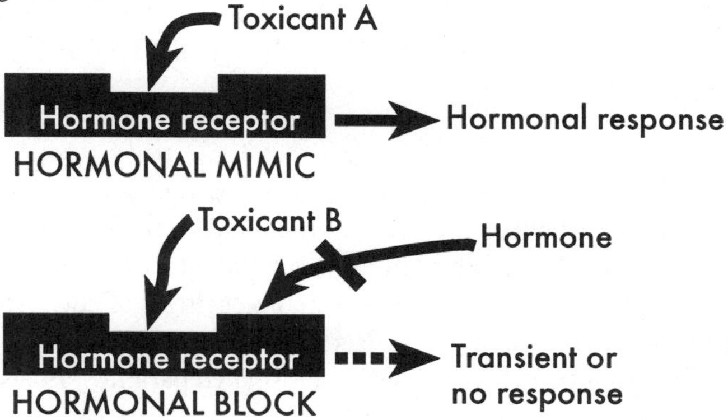

Figure 5-6 Exogenous chemical activity at hormone action sites (redrawn from McLachlan 1993)

Some OCs, such as *o,p′*-DDT, kepone, methoxychlor, dibromochloropropane, and PCBs, have estrogenic activity in wildlife. These compounds can have both agonistic or antagonistic activities at a number of sites on the HPG axis (Figure 5-5), depending on the circulating concentrations of the natural estrogen and the stage of the reproductive cycle. Many of these have been shown to act by binding to the estrogen receptor, such as *o,p′*-DDT and kepone. However, other OCs can exert their estrogenic or antiestrogenic effects by other mechanisms, such as TCDD. In addition, there is preliminary evidence that certain OCs also bind to other steroid hormone receptors, such as *o,p′*-DDT to the progesterone receptor, atrazine to the androgen receptor, and *o,p′*-DDT and kepone to the progestin membrane receptor (Korach and McLachlan 1995). However, nonestrogenic steroid actions of these compounds have received little attention.

Ortho-para*′-DDT*

The estrogenic actions of DDT have been investigated extensively in mammals and birds (Kupfer and Bulger 1980). *Ortho-para′*-DDT was identified as the major component of technical-grade DDT with estrogenic activity. Several estrogenic and antiandrogenic actions of *o,p′*-DDT (dose 50 to 200 mg/kg) have been observed in female rats, including uterine growth, increases in ornithine decarboxylase activity, changes in glycogen and progesterone receptor levels and RNA synthesis, and a decline in circulating GTH levels. Feminizing effects of *o,p′*-DDT and methoxychlor have been observed in male rats and male gull embryos (Fry and Toone 1981; Bulger and Kupfer 1985). In neonatal rats, *o,p′*-DDT (300 µg/rat) caused premature vaginal opening and persistent vaginal estrus, leading to sterility. Effects also have been seen with estradiol and DES treatments (Bulger and Kupfer 1985). A direct relationship between the estrogenic activity of the DDT isomers and their binding to the estrogen receptor has been demonstrated. Moreover, treatment with *o,p′*-DDT causes activation of the estrogen receptor resulting in binding to DNA in the cell nucleus, mimicking the actions of estrogen. However, binding of the receptor is maintained longer after treatment with *o,p′*-DDT than with estradiol, similar to the receptor dynamics effects observed with synthetic antiestrogens such as tamoxifen (Kupfer and Bulger 1980). The results suggest that *o,p′*-DDT may have some antiestrogenic properties through its sustained binding to the receptor.

Indirect disruptors

Organochlorines such as PCBs and some OC insecticides as well as nonchlorinated compounds may alter reproductive endocrine function directly by altering the function of neurotransmitters involved in the regulation of GTH secretion (Thomas 1989). However, it is often difficult to distinguish direct effects of chemicals on neurotransmitter function from indirect effects due to altered steroid feedback. Specific information for other classes of OCs (chlorinated paraffins) is lacking.

Enhanced metabolic clearance of steroids and a decline in steroid levels, due to induction of cytochrome P450, has been shown for AhR-active PCBs. However, compensatory mechanisms of the HPG axis may eventually overcome this clearance, so that reproduction is delayed but not prevented (Thomas 1989).

Other possible modes of actions have not been investigated, including direct effects on hormone secretion by disruption of hormone-receptor or second-messenger systems such as cyclic adenosine monophosphate (AMP), calcium, and polyphosphoinositol turnover. Kepone increases progesterone concentrations in plasma, although the mechanism is unknown. OC-induced alterations of other endocrine systems, such as the thyroid and adrenal function in fishes and marine mammals, could secondarily disrupt reproductive endocrine functions.

Chemical classes

Interaction of chemicals with the endocrine system is not specifically related to chlorination. A number of nonchlorinated compounds, such as the natural phytoestrogens and synthetic compounds like DES, also have potent endocrine activity. Even among chlorinated compounds, potency varies. DDT isomers have weak affinity for the estrogen receptor, compared to estradiol (see "Hormone mimics and blockers," this chapter). *Ortho-para* derivatives of DDT have greater estrogen receptor-binding affinity and estrogenic potency than do *p,p′*-DDT derivatives. For example, the rank order of potencies of DDT analogs (10 mg/100 g body weight [bw]) on uterine ornithine carboxylase activities in immature rats were *o,p′*-DDT > *o,p′*-dichlorodiphenylethane (DDD) > *p,p′*-DDT > *p,p′*-dichlorodiphenyldichloroethylene (DDE) > *p,p′*-DDD (Kupfer and Bulger 1980), which agrees with their relative binding affinities for the estrogen receptor (Nelson 1974). Kepone has a binding activity and an estrogenic potency similar to those of *o,p′*-DDT. Methoxychlor causes estrogenic effects in vivo over the same concentration range, but it requires metabolic activation to the didemethoxylated metabolite of methoxychlor to display estrogenic activity and binding to the estrogen receptor (Kupfer and Bulger 1980). Some currently used pesticides, such as endosulfan, have been shown to have estrogenic activity in in-vitro assays (Soto et al. 1994); however, their mechanisms of action are unclear.

Effects

Reproduction may be affected by hormone modulators. Hormonal modulators can interfere with hormone action at a variety of target tissues, such as estrogen receptors in ovary, brain, pituitary, and liver in oviparous vertebrates or uterus in mammals (sites 4, 5, 6, Figure 5-5), resulting in altered ovarian function, neuroendocrine function, and yolk production. Interference with membrane progesterone receptors can affect oocyte maturation (Ghosh and Thomas 1995). Interference with androgen, progesterone, glucocorticoid, and vitamin D actions and their receptors has not yet been investigated. Hormonal cycles may be disrupted by indirect interference with neuroendocrine function or circulating hormone concentrations. Also, general toxic effects may result in secondary responses in the endocrine system.

Development may also be affected by hormone modulators. Alterations in hormone binding to receptors caused by hormone modulators during critical stages of a life cycle (sexual differentiation) and their potential effects on sex ratios and sexual behavior are not fully understood. However, *o,p'*-DDT, which has estradiol receptor-binding activity, causes feminization in birds and reptiles in the laboratory and also as a result of food-chain biomagnification. Significant deposition of persistent, lipid-soluble OCs in developing oocytes may occur. As a result, the subsequent course of development of the embryo can be affected but may not be manifested until late in life (Fry 1995).

Likely target species

Direct evidence for hormonal modulators has been obtained in mammals (Nelson 1974), birds (Turner and Eliel 1978), and fish (Thomas and Smith 1993); indirect evidence has been obtained in reptiles. Although invertebrates have some steroid hormones (e.g., ecdysone, the molting hormone) similar in structure to those identified in vertebrates, information is lacking on xenobiotic steroid–receptor interactions. The freshwater crustacean, *Daphnia magna*, has been shown to have enzymes capable of metabolizing steroids (Baldwin and LeBlanc 1994) and also to be responsive to DES (Baldwin et al. 1995), although exposures were conducted at the relatively high concentration of 0.5 mg/L, at which narcosis might be expected. Direct evidence of effects on the endocrine system secondary to neurotoxicity has been found in all groups of animals, but indirect reproductive responses through neuroendocrine effects have been demonstrated only in mammals, fish, and birds.

Methyl sulfone polychlorinated metabolites and adrenotoxicity

Nonpolar, methylsulfonated metabolites of DDE and various nonplanar PCBs have been found in various wildlife species (up to 10% of the total amount of DDT and PCBs respectively). The greatest concentrations were found in the liver, fat, and adrenals in mammalian species studied. In vitro and in vivo studies on 3-methylsulfonyl-DDE ($MeSO_2$-DDE) in mice have shown a strong adrenocorticolytic effect causing necrosis of *zona fasciculata* (Brandt et al. 1992). The metabolite is covalently bound to the *zona fasciculata* in both birds and mammals. The covalent binding and toxicity were mediated by cytochrome P450. Effects were found at a single dose of 3 mg/kg (lowest dose examined). This dose level is comparable to the greatest concentrations found in wildlife (6 mg/kg in seal blubber); however, $MeSO_2$-DDE stored in fats may not be available to the adrenal tissue in the same way as oral doses in mice.

Metabolism, Activation, Covalent Binding, and Lipid Peroxidation

Toxicity associated with metabolic activation has been observed in a variety of chemical classes, both with and without chlorine. Chlorine does not appear to play an essential role in this process. Metabolic activation is most extensively studied in mammals, but examples have been seen in fish, birds, and bacteria as well. Because of their ability to be biotransformed, compounds exhibiting this mechanism are less likely to be environmentally persistent, in general, than are compounds that do not have this property.

Mechanism of action

Some OC compounds, such as tertiary amines and polycyclic aromatic hydrocarbons, are electrophilic and react to form covalent bonds with biologically important molecules, including both low- and high-molecular-weight compounds such as DNA, proteins, and polypeptides. On the other hand, some OC compounds are not directly reactive with macromolecules unless they have first undergone metabolic transformation by enzyme systems to form an electrophilic species. In either case, the electrophilic species are capable of inducing toxic effects in biological systems through reaction with critical cofactors or biological macromolecules.

Role of the chlorine atom

The presence of chlorine in a molecule, however, is neither necessary nor sufficient to imply that a compound will undergo activation. Many compounds that contain chlorine are both unreactive themselves and resistant to metabolic activation. On the other hand, many chemicals that do not contain chlorine are reactive enough, either by themselves or after metabolic activation, to induce toxicity in a variety of organisms.

Selected examples

Reactive metabolites of chloroform covalently bind to proteins in the liver and kidney and induce necrosis and even death associated with renal failure in mammals (Davidson et al. 1982; Corely et al. 1990). Chloroform is metabolized to phosgene by reductive dechlorination mediated by cytochrome P-450 in the kidney tissues. The phosgene formed in this process may react with glutathione to 2-oxothiazolidine-4-carboxylic acid or with nucleophilic macromolecules to further cause cellular damage (Smith 1986; Smith et al. 1983). This metabolic pathway is specific to some animals and in laboratory mice there are strain and sex differences (Lock 1993). Chronic exposure to chloroform has also been shown to increase the frequency of certain spontaneously occurring liver kidney tumors in mammals (IARC 1979a), possibly through binding of reactive metabolites to DNA. Similar mechanisms have been demonstrated in the case of nonchlorinated compounds such as benz(a)pyrene which is metabolized via the mixed-function oxidase (MFO) enzyme system to the epoxide that then binds covalently to DNA, thus causing mutations that could lead to carcinogenesis.

Less frequently, metabolic activation, which normally takes place in the cytoplasm, produces a reactive intermediate that is stable enough to be transported to the nucleus of the cell and react with the nucleophilic sites in DNA. In this case, the chemicals may be capable of inducing somatic mutations and/or cancer in susceptible species. An example of this type of material is vinyl chloride, which is carcinogenic in mammals and mutagenic in bacteria and mammalian cell lines (IARC 1979b).

Similar mechanisms have been demonstrated in the case of nonchlorinated compounds such as benz(a)pyrene, which is metabolized through the MFO enzyme system to the epoxide that then binds covalently to DNA, thus causing mutations that could lead to carcinogenesis.

Another mechanism involving metabolic activation is demonstrated by carbon tetrachloride. When carbon tetrachloride is metabolically activated, free radicals are produced that react with a variety of cellular macromolecules (IARC 1979c). This has particularly serious consequences for organisms because these radicals can participate in chain reactions (e.g., lipid peroxidation) where one activated molecule can cause adverse effects in multiple target molecules.

Chemical classes

Organochlorines, and many molecules that do not contain chlorine, undergo metabolic activation. However, while the reactive metabolites may result in acute toxicity to mammalian, avian, or aquatic organisms, susceptibility to this mechanism also decreases the likelihood that metabolizable compounds will persist in the environment long enough to biomagnify in the food chain.

Metabolic activation has been shown to occur in some, but not all, members of the following chemical groups:

1) chlorinated aliphatics C_1–C_3, C_3–C_{10}, > C_{10};
2) organophosphorus pesticides;
3) chlorinated aldehydes;
4) chlorinated aromatics (neutral and polar); and
5) nonchlorinated polycyclic aromatic hydrocarbons (PAHs).

On the other hand, this mechanism does not occur to a significant extent in most members of the following groups of OCs :

1) PCDDs,
2) PCDFs,
3) PCBs, and
4) chloroacids.

It is not clear whether metabolic activation plays a role in mediating the effects (if any) of complex chlorinated polymers such as chlorolignins, soil-bound residues, and synthetic chlorinated macromolecules.

Effects

Chemicals acting through this mechanism of metabolic activation have caused mortality or decreased growth rates, mostly as a result of liver or kidney failure, at greater, nonenvironmentally relevant concentrations in mammals. Most chemicals acting through this mechanism do not appear to be potent reproductive toxicants or immunotoxicants, at least at levels below those causing mortality or growth retardation. Many of the chlorinated aliphatics cause transient, narcotic effects, especially among the chlorinated solvents (see "Nonspecific Toxicity [Narcosis]," this chapter).

Organisms exhibiting activation mechanisms

Because enzymatic metabolism to reactive intermediates is required for expression of toxicity for compounds acting through this mechanism, differential enzymatic activity likely accounts in part for species differences in toxicity. No generalizations about types of organisms affected by different chemical classes are possible without knowledge of the relative activities of the enzyme systems responsible for metabolic activation in those species. Compound-specific information must be gathered, as in the examples given above. Enzymes to activate these chemicals exist in fish, birds, and mammals, but it is likely that the toxicity to aquatic organisms cited elsewhere in this book (see Chapter 6, for example) is mediated through other (unknown) mechanisms.

AhR-mediated Toxic Responses

Several OC compounds, such as PCDD/PCDFs and planar PCBs, act through the AhR common to fish, birds, and mammals. AhR-mediated responses appear to include lethality, reproductive and developmental toxicity, immunotoxicity, and cancer.

To date, the group of chemicals producing the full range of toxic responses mediated by the AhR is dominated by certain halogenated (mainly chlorinated) organic compounds that bind with the AhR and also are resistant to metabolism. While some polynuclear aromatic compounds also act through the AhR, they do not show all the AhR responses, usually as a result of their greater rates of metabolic clearance. Continuous exposure studies have not been conducted to confirm this hypothesis. Exposure to mixtures of AhR agonists appears to cause responses that are additive, based on AhR binding and potency.

Laboratory effects of AhR agonists have been observed at environmentally relevant levels. For certain locations in the industrialized world, there probably is not an adequate margin of safety for current maternal body burdens of TCDD and related chemicals in some mammals and for egg burdens of TCDD and related chemicals in some fish and birds.

Mechanism of action

The dioxin, 2,3,7,8-TCDD is the prototype compound for structurally similar PAHs that act by an AhR mechanism of action (Poland and Knutson 1982; Whitlock 1990, 1993). The initial step in the action of TCDD is stereospecific binding with greater affinity to the cytosolic AhR. Endogenous ligands for the AhR have not been identified. After binding, the TCDD–AhR complex is translocated to the nucleus, where it dimerizes with the AhR (Figure 5-7) nuclear translocator protein (Arnt). The TCDD-AhR-Arnt complex binds to dioxin response elements on DNA, causing the transcription of such genes in the AhR gene battery as cytochrome P450A1. Binding with the AhR is a necessary step toward, but is not sufficient to produce, toxicity. The TCDD-responsive genes that cause the various signs of toxicity in a highly tissue-specific and species-specific manner currently are unknown.

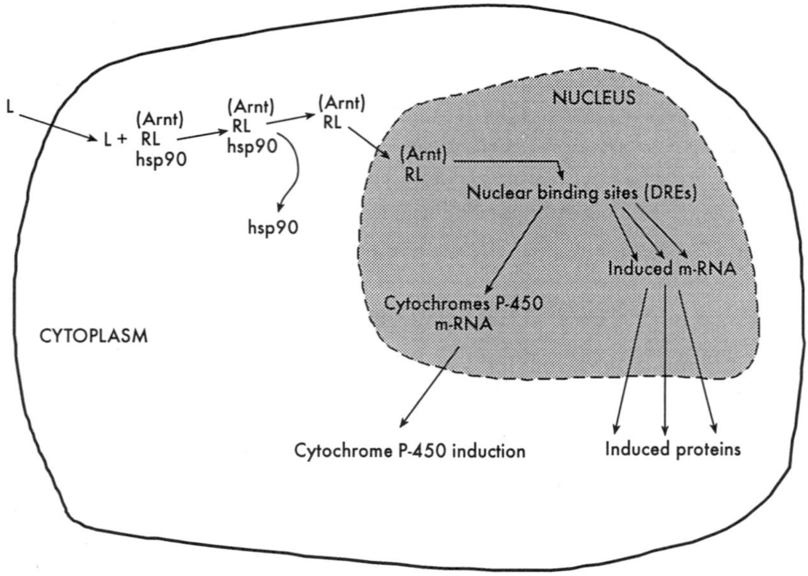

Figure 5-7 Proposed mechanism of TCDD toxicity. L = TCDD or other ligand.

Certain nonchlorinated environmental pollutants, such as PAHs, bind to the AhR and elicit the prototypical, nontoxic, dioxin-like response: induction of cytochrome P4501A1 (Eisen et al. 1983). Naturally occurring compounds, such as the indole-carbinols, have similar transient AhR-mediated effects typified by induction of monooxygenase activity (Bjeldanes et al. 1991). In general, both the PAHs and the indolecarbinols have lower binding affinity for the AhR than TCDD and fail to produce a sustained effect on gene expression because they are rapidly biotransformed to nonbinding metabolites. Consequently, they fail to elicit the entire spectrum of dioxin-like toxic effects. TCDD and related persistent coplanar chemicals cause, in addition to P451A1 induction, delayed-onset toxic effects that are characteristic of sustained-action, polychlorinated AhR agonists. These toxic effects are highlighted below.

Functional AhR has been detected in teleost and elasmobranch fishes, birds, and mammals (including humans) but not in plants, 9 species of invertebrates representing 8 classes of 4 phyla, or in primitive fishes such as hagfish, lamprey, or fishes in the class Agnatha (USEPA 1993a). Depending on dose and species sensitivity, the symptoms of toxic response caused by TCDD and related compounds can include the following: lethality, body weight loss, immunosuppression, edema, dermal toxicity, developmental toxicity, reproductive toxicity, neurobehavioral toxicity, liver toxicity, endocrine toxicity, alterations in lipid metabolism and gluconeogenesis, modulation of responsiveness to hormones and growth factors, alterations in circulating levels of hormones, thymic atrophy, cancer, tumor promotion and induction of various enzyme activities, including aryl hydrocarbon hydroxylase (AHH) and ethoxyresorufin-*O*-deethylase (EROD) that are associated with induction of cytochrome P4501A1 (Poland and Knutson 1982).

Chemical classes

A number of chemical classes of compounds show binding to the AhR. This is related to the size and shape of the molecules. Polychlorinated dibenzo-*p*-dioxins and dibenzofurans (PCDD/PCDFs) substituted in at least the 2,3,7,8-positions occur in the environment and act by an AhR mechanism in fish, birds, and mammals. Similarly, laterally substituted PCB congeners (International Union of Physicists and Chemists [IUPAC] nr. 77, 81, 126, and 169) occur in the environment and act by an AhR mechanism in fish, birds, and mammals. Mono-ortho-substituted PCB congeners occur in the environment and act by an AhR mechanism in birds and mammals but not in fish.

Other classes of polychlorinated aromatics also act through binding to the AhR. Certain congeners in the following classes also act by an AhR mechanism: polychlorinated naphthalenes, azoxybenzenes, chlorinated diphenylethers (Howie et al. 1990; Harper et al. 1993), and chlorinated benzyltoluenes (Addison et al. 1991; Murk et al. 1991). Congeners in other related classes may act by an AhR mechanism of which we are currently unaware. For example, structure-activity relationships for AhR-mediated toxicity caused by polybrominated dibenzo-*p*-dioxins, furans, and biphenyls appears to be similar to that for PCDDs, PCDFs, and PCBs.

Several classes of chlorinated chemicals and one class of a nonchlorinated chemical (PAHs) may have the potential to cause toxic effects by an AhR mechanism. However, the PAHs, which bind to the AhR and induce cytochrome P4501A1, do not cause the full spectrum of TCDD-like signs of toxicity. This failure of PAHs to produce all endpoints of toxicity is generally used to support the conclusion that binding to the AhR is necessary but not sufficient to cause toxicity. However, the PAHs are more rapidly metabolized; they would thus have shorter durations of exposure, possibly explaining the observed differences in mode of action.

Effects

Because TCDD is the most potent substance that acts by an AhR mechanism, it is used as the prototype to describe 4 types of ecologically relevant effects: mortality, reproductive toxicity, developmental toxicity, and decreased growth. All of these effects are considered ecologically significant because they have the potential of ultimately reducing populations of fish, birds, and mammals exposed to TCDD and related compounds. In general, AhR agonists, which produce sustained AhR-mediated responses, will cause the same profile of dose-related, specific effects as TCDD on mortality, reproduction, embryonic and fetal development, and growth. Thus, no-observed-adverse-effect levels (NOAELs) and lowest-observed-adverse-effect levels (LOAELs) for these endpoints in fish, birds, and animals can be determined to assess ecotoxicological risk.

Mortality

Following exposure to high concentrations of TCDD, delayed-onset mortality occurs in most species. In freshwater fish species, yellow perch and carp are among the most sensitive. For avian species, the chicken is among the most sensitive, and for mammals, the guinea pig and mink are highly sensitive (Peterson and Walker 1992; USEPA 1993b). Wide species differences in susceptibility to mortality are seen. For example, the lethal dose to 50% of the population (LD50) of TCDD in the hamster is more than 3 orders of magnitude greater than that in the guinea pig and mink. This may be related to the amount of body fat in the organisms, as male Syrian hamsters carry 17 to 20% body lipid and male guinea pigs have only 4.5% body lipid (Geyer et al. 1990). The ultimate cause of death due to TCDD exposure has not been fully elucidated, but among mammalian species, it is invariably preceded by a wasting syndrome in which animals can lose up to 50% of their body weight (Seefeld et al. 1984).

Reproduction

Tetrachlorodibenzo-*p*-dioxin exposure causes female reproductive toxicity in various mammalian species (Morrissey and Schwetz 1989; Peterson et al. 1993), including reduced fertility, reduced litter size, effects on the gonads of females, alterations in menstrual or estrus cycles, and alterations in responsiveness to estrogen. Species in which one or more of the above TCDD effects have been observed include the rat, mouse, guinea pig, rhesus monkey, hamster, mink, and rabbit. In female birds and fish, TCDD exposure reduces egg production. However, the most sensitive sign of toxicity is embryo mortality (Peterson and Walker 1992; USEPA 1993b).

In male animals, TCDD and related compounds decrease testis and accessory sex organ weights, cause abnormal testicular morphology, decrease spermatogenesis, and reduce fertility when given to adult animals in doses that are in the overtly toxic range. Such effects have been reported for chickens, rhesus monkeys, rats, guinea pigs, and mice. Effects of TCDD exposure in adulthood on the male rat are believed to be due, in part, to an androgenic deficiency (Moore et al. 1985). This deficiency is characterized by decreased plasma testosterone and dihydrotestosterone concentrations. Although plasma-luteinizing hormone (LH) concentrations and plasma clearance of androgens and LH are unaffected, the decrease in circulating androgen levels in TCDD-treated rats appears to be due to a combination of decreased LH-stimulated testicular steroidogenesis and altered regulation of pituitary LH secretion (Bookstaff, Kamel et al. 1990; Bookstaff, Moore, and Peterson 1990; Kleeman et al. 1990; Moore et al. 1991).

Development

Fetal and embryo toxicity. Early life stages of fish are more sensitive to TCDD-induced mortality than are adults. Therefore, greater concentrations of TCDD and related compounds in fish eggs may pose the greatest risk to wild fish populations. Species suffering early life-stage mortality include the zebrafish, Japanese medaka, northern pike, rainbow trout, brook trout, and lake trout (Peterson and Walker 1992; USEPA 1993b). In salmonid sac fry, the classical TCDD toxicity syndrome is characterized by pericardial and yolk-sac edema, subcutaneous hemorrhages, craniofacial malformations, and arrested growth and development, culminating in death (Spitsbergen et al. 1991; Walker et al. 1991). This syndrome in rainbow, brook, and lake trout sac fry exposed as fertilized eggs to TCDD is indistinguishable from blue sac disease. Similar signs of developmental toxicity have been observed in Japanese medaka, zebrafish, and northern pike (Peterson and Walker 1992; USEPA 1993b).

In birds, the LD50 of TCDD in the chicken embryo is less than 1/100 of that needed to cause mortality in the adult chicken (Peterson et al. 1993; USEPA 1993b. Species in which TCDD and related compounds cause embryo mortality include the chicken, ring-necked pheasant, turkey, eastern bluebird, mallard duck, domestic duck, goldeneye, herring gull, and black-headed gull (Brunstrom 1988; USEPA 1993b). Birds exhibit developmental toxicity in different ways (Peterson and Walker 1992; Peterson et al. 1993). In the chicken embryo, pericardial and subcutaneous edema, liver lesions, lymphoid toxicity, structural malformations, and mortality occur. However, in other birds, many of these embryotoxic effects are not seen; the only toxic effect that appears common to all bird species is embryo mortality.

Tetrachlorodibenzo-*p*-dioxin can also adversely effect pregnancy maintenance, embryo or fetotoxicity, and postnatal development in mammals. When TCDD exposure occurs during pregnancy, prenatal mortality has been observed in the mouse, rat, hamster, guinea pig, rabbit, mink, and rhesus monkey. In addition, in most mammals, gestational exposure to TCDD produces a characteristic pattern of fetotoxic responses that consist of thymic hypoplasia, subcutaneous edema, and decreased fetal growth (Couture et al. 1990; Peterson et al. 1993).

Certain fetal and neonatal effects of TCDD are species-specific (Peterson et al. 1993). Intestinal hemorrhage is observed in the rat fetus and neonate following in-utero and lactational exposure to TCDD. In monkeys and humans, tissues derived from ectoderm, such as the skin, conjunctiva, nails, teeth, and central nervous system, are sites of action (Rogan et al. 1988).

Structural malformations. In birds and fish, TCDD-induced malformations include cardiovascular malformations in the chicken (Cheung et al. 1981) and craniofacial malformations in sac fry of various salmonid fish species (Walker et al. 1991). Various deformities have been observed in colonial fish-eating water birds exposed to TCDD and related chemicals (Peterson and Walker 1992; Ludwig, Auman et al. 1993; Peterson et al. 1993; Giesy, Ludwig, Tillitt 1994; Giesy, Verbrugge et al. 1994).

In-utero and/or lactational exposure to TCDD causes structural malformations in various mammalian species (Morrissey and Schwetz 1989; Couture et al. 1990; Peterson et al. 1993). Cleft palate formation occurs in the mouse, hydronephrosis in the mouse and hamster, extra ribs in the rabbit, and at the time of puberty, vaginal malformations are seen in the rat and hamster. Susceptibility to cleft palate in the mouse depends on AhR genotype of the fetus (low-affinity AhR strains are not affected, whereas greater-affinity AhR strains respond) and stage of development at the time of exposure.

Postnatal functional alterations. In-utero and lactational exposure to TCDD causes postnatal alterations in neurobehavioral and male reproductive function in the rhesus monkey and rat, respectively (Peterson et al. 1993). The alterations in postnatal male reproductive function in the rat consist of a spectrum of effects, including reduced anogenital distance, delayed time to testis descent, reductions in sex organ weights, decreased daily sperm production, decreased epididymal and ejaculated sperm numbers, demasculinization and feminization of sexual behavior, and feminization of the regulation of pituitary LH secretion (Mably, Bjerke et al. 1992; Mably, Moore, and Goy, Peterson 1992; Mably, Moore, and Peterson 1992).

In the male rat, the vast majority of the postnatal signs of male reproductive toxicity can be caused by the in-utero route of TCDD exposure only (Bjerke and Peterson 1994). Alterations in sexual behavior and reduced epididymal sperm numbers appear to be permanent consequences of in-utero and lactational exposure to TCDD, which to date have been observed in 2 rat strains and in the hamster (Gray et al. 1996).

Growth

Decreased growth is generally observed in all classes of vertebrates that possess the AhR after exposure to overtly toxic doses of TCDD. This response tends to be more apparent in mammals, possibly because the presence of edema in birds and fish masks the reduction in body tissue mass. It appears that the wasting syndrome is caused, in large part, by dose-related decreases in feed intake (Seefeld et al. 1984). However, alterations in intermediary metabolism are also involved. In young animals, TCDD toxicity is manifested by decreases in growth rate, whereas in sexually mature animal, whose body weight has stabilized, the wasting syndrome is characterized by loss of body weight.

TCDD toxic equivalents

Because AhR agonists act by a common mechanism, the effects of exposure to complex mixtures of these compounds are generally considered to be additive, on the basis of toxic equivalents (TEQs). Thus, although individual compounds in a mixture may not be present at a toxic dose, the combined subtoxic doses of individual AhR agonists could result in toxicity.

The potential toxicity of mixtures of AhR agonists in environmental samples has been evaluated using the concept of "TCDD equivalents," in which each compound is assigned a TEF, indicating its toxic potency relative to TCDD. The value of the TEF usually is determined based on short-term toxicity tests and sometimes even on in-vitro tests. The concentration of each individual compound in a sample is then multiplied by its appropriate TEF to give its concentration as a TCDD TEQ. The TEQs of all substances are then added to give the total TEQ concentration of the mixture.

There are certain limitations to the TCDD-equivalency concept. The advantages and limitations of the use of TCDD equivalents as risk management tools have been extensively discussed (Safe 1990; Ahlborg et al. 1992). As research continues, it is becoming clear that TEFs can be most useful for predictive purposes when they are developed for specific endpoints. Internationally agreed-upon TEFs have been developed for PCDD/PCDFs (CEPA 1990) and for PCBs (Ahlborg et al. 1994) for use in human risk assessment.

Toxic equivalency factors used for human risk assessment can also be applied to mammals but not to fish or birds. This is because TEFs, particularly for PCBs in fish, are significantly different from those proposed for human health risk assessment (Walker and Peterson 1991; Parrott et al. 1995; Zabel et al. 1995). Fish appear to be less sensitive to nonortho-substituted PCBs, whereas the same congeners cause toxicity in mammals.

Lowest-observed-adverse-effects levels for developmental and reproductive toxicity

Cross-species comparisons of NOAELs and LOAELs for reproductive and developmental toxicity of TCDD and related chemicals in mammals, birds, and fish were recently reviewed (Peterson and Walker 1992; Ludwig, Auman et al. 1993; Peterson et al. 1993; Giesy, Verbrugge et al. 1994). In fish, birds, and mammals, the embryo or fetus is more sensitive to the lethal effects of TCDD than is the adult. LOAELs for endpoints that are relevant to the ecotoxicological risk posed by bioaccumulative, persistent, AhR agonists in these 3 classes of vertebrates are given in the following paragraphs.

The LOAEL for craniofacial malformations in lake-trout sac fry is in the same range as that for blue-sac syndrome-associated mortality, 40 to 55 ng TCDD/kg egg (Spitsbergen et al. 1991; Walker et al. 1991).

In birds, the LOAEL based on egg dose is 240 ng/kg in the chicken for embryo mortality and 40 ng/kg in lake trout for sac-fry mortality (Spitsbergen et al. 1991). LOAELs for cardiovascular malformations in chickens (Cheung et al. 1981) or embryo mortality in double-crested cormorants is less than 100 ng/kg egg (Ludwig, Auman et al. 1993; Giesy, Verbrugge et al. 1994).

In mammals, the LOAEL based on maternal body burden for prenatal mortality in the rhesus monkey is 97 ng/kg (Schantz and Bowman 1989). At low TCDD exposures, structural malformations are not seen in mammalian species. However, postnatal functional alterations occur following in-utero and lactational exposure to TCDD in mammals and are considered to be ecologically relevant. In adult male rats, the LOAEL for decrease in epididymal sperm numbers is 64 ng/kg of TCDD administered to the dam on gestation day 15 (Mably, Moore, Goy, and Peterson 1992).

In rats exposed to TCDD in adulthood, the most sensitive toxic responses have been seen following long-term, low-level exposure. In a 3-generation reproduction study (Murray et al. 1979), the steady-state body burden of TCDD in female rats chronically exposed to 10 ng/kg/d was estimated to be 290 ng/kg (Peterson et al. 1993), which is the LOAEL for reduced reproductive capacity in the female rat. Based on concentration of TEQs in the diet, the LOAEL for reproductive effects in ranch mink that were fed Saginaw Bay-contaminated fish was 19 pg TEQs/g diet (Heaton et al. 1991; Ludwig, Giesy et al. 1993).

Other effects

Cancer and tumor promotion

TCDD has been demonstrated to be a potent animal carcinogen (Kociba et al. 1978), and a mixture of hexachlorinated dibenzo-*p*-dioxins has also been shown to be carcinogenic (National Toxicology Program [NTP] 1980). In addition, mixtures of higher-chlorinated PCBs (Aroclor 1260) are carcinogenic in rodent bioassays. Subsequently, TCDD, several other PCDD/PCDFs, and non- and *mono-ortho*-substituted PCBs have been demonstrated to act as tumor promoters (Pitot et al. 1980; Wœrn et al. 1991; Hemming et al. 1993). Although carcinogenicity is of human health relevance, its significance for assessing ecotoxicological risk of TCDD and related chemicals is less clear.

Vitamin A and thyroid hormones

In several mammals, PCDDs, PCDFs, and PCBs have interfered with the turnover of vitamin A (Brunström et al. 1991; Håkansson, Johansson et al. 1991; Håkansson et al. 1992), and this effect is related to the toxic potency of these compounds (Håkansson et al. 1990, 1992; Brunström et al. 1991; Chu et al. 1994, 1995). Mammalian studies (Brouwer and van den Berg 1986; Lans et al. 1993) indicate that metabolites of the lower-chlorinated PCBs and dioxins can interact with the thyroid-hormone transport protein (transthyretin). This protein is involved with vitamin A transport. As a result of this interaction, changes in thyroid hormone as well as vitamin A metabolism occur. Although reductions in thyroid hormone and vitamin A levels have been observed in marine mammals and fish-eating birds, the toxicological relevance of this mechanism of action should be validated. In addition, interference with thyroid-hormone metabolism can also be caused by induction of the thyroid-hormone glucuronidation. Laboratory studies with rodents indicate that increased glucuronidation might also be AhR-mediated.

Vitamin K

Several studies have shown that, after exposure to dioxins and PCBs, mammals, birds, and fish can hemorrhage. Laboratory studies with rats show that TCDD as well as 2,2′4,4′,5,5′-hexachlorobiphenyl (HxCB) interferes with several enzymes in the hepatic vitamin K cycle. As a result, vitamin K-dependent, blood-coagulation factors are decreased in plasma. These effects have been found to be compound-, sex-, and age-dependent. Dose-response studies with TCDD and 2,2′4,4′,5,5′-HxCB indicate that a nonAhR mechanism might be involved, as the latter compound, an extremely weak AhR agonist, was more potent than TCDD (Bouwman et al. 1992, 1993; Ivens et al. 1993).

Immunotoxicity

Immunotoxicity has been observed in a variety of classes of chemicals, both with and without chlorine. Chlorine does not appear to play a specific role in the mechanism. Among OC compounds tested to date, PCBs and dioxins are the most potent immunotoxicants (active at low doses, causing a range of effects and multispecies toxicity). Polychlorinated biphenyls and dioxins appear to be immunotoxic in several species at doses that may be environmentally relevant. Many of the older OC pesticides are immunotoxic at high doses, but data on low-dose exposure or dose-response studies are lacking. As with most chemicals, the immunotoxicity of OCs has not been well studied, and good data are not available.

General concepts

"Immunotoxicity" has been defined as the adverse effects on the immune system of foreign substances (Luster et al. 1988; Trizio et al. 1988) and is classified in 2 main categories:

1) the alterations of normal immune responses through an immunodepression or an immunopotentiation that will consequently compromise the host's defense mechanisms and
2) the induction of abnormal immune responses such as various types of hypersensitivity (allergy and autoimmunity, e.g.) (Bennett and Mercier 1987).

These chemical-induced immunotoxic effects can be organ-specific, cell-specific, immune function-specific; can be related to the intoxication of other organs; or can be nonspecific. They have been described by many authors (Zeeman and Brindly 1981; Descotes 1988; Luster et al. 1988; Trizio et al. 1988; Gleichmann et al. 1989; Tryphonas et al. 1991; Wong et al. 1992; Wood et al. 1992; Kzrystyniak et al. 1995). These immunotoxic effects can occur through various mechanisms, such as modification of the -SS/SH ratio, activation of the AhR, and lipid peroxidation. In this section, we will focus primarily on OC chemicals.

Chemical classes

Polychlorinated biphenyls

The results of in-vivo and in-vitro experimental studies indicate that commercially available PCB mixtures alter several morphologic and functional aspects of the immune system. These effects have been reproduced in several species: fishes, birds, rodents, and nonhuman primates (reviewed in Descotes 1988; Dunier 1994; Tryphonas 1995). Furthermore, limited data from studies of rodents that were fed fish from selected areas of the Great Lakes (Tryphonas, Fournier, Lacroix et al. 1998; Tryphonas, Fournier, McGuire et al. 1998) and seals that were fed fish from selected areas of the Baltic Sea suggest that their immune systems may be affected by contaminants in the fish (Ross et al. 1995).

High PCB exposure resulted in a) structural alterations of immune system organs (loss of thymic cortical lymphocytes, reduction of germinal center size, and reduction of leucocyte and T-lymphocyte counts in peripheral blood) and b) altered functional reactivity of the immune system, characterized by reduced antibody production against foreign antigens and reduced skin reactivity to specific antigens. Functional defects were also observed in the mononuclear phagocytic cells and natural killer-cell activity and in increased susceptibility to normally tolerated doses of bacterial, viral, and parasitic infestations. For example, in monkeys chronically exposed to PCBs, levels of antibodies to the T-dependent antigen sheep red blood cells were reduced by doses as low as 5 μg/kg body weight/d (Tryphonas et al. 1989, 1991). In general, higher-chlorinated PCB mixtures (Aroclors 1260, 1254, and 1248) were more immunotoxic than lower-chlorinated Aroclors 1232, 1016, and 1242.

The mechanism of PCB-induced immunomodulation has not been completely elucidated. Many of the immunotoxic effects of PCB congener mixtures depend on the presence of the AhR and on the ability of the PCBs to bind to this receptor. The molecular events following binding of PCBs to the receptor are believed to be similar to those of dioxin. It should be noted, however, that other nonAhR-interacting PCB congeners also cause immunotoxicity, suggesting a different mechanism of action.

*Chlorinated dibenzo-*p*-dioxins*

TCDD (2,3,7,8-tetrachlorodibenzo-*p*-dioxin) is probably one of the most widely known immunotoxic xenobiotics. Immunotoxicity is expressed as lymphoid tissue depletion, especially in thymus and bone marrow, functional alteration in immune responsiveness, and increased susceptibility to infectious disease. Some studies suggest that prenatal exposure to TCDD is more immunosuppressive than comparable exposure in adults. This may be related to effects on the thymus, critical for the development of T cells (review in Descotes 1988; Wong et al. 1992; Kerkvliet 1995; Dunier 1994).

The immunotoxicity of TCDD depends on the AhR. In mice, the generation of a primary antibody response appears to be particularly sensitive to suppression when mature animals are exposed to TCDD. The effect on antibody production may result from effects on both T and B lymphocytes in rats (Birnbaum 1994). However, these findings cannot be generalized to other species, since the rat is less sensitive (Smialowicz et al. 1994). Although most macrophage functions appear to be resistant to TCDD-induced alterations, recent studies indicate that macrophage production of inflammatory cytokinins is enhanced following TCDD exposure. Natural killer-cell cytotoxicity and mitogen-induced blastogenesis of lymphocytes in adult mice are not altered by doses of TCDD that significantly suppress antibody production. Delayed type-hypersensitivity (DTH) responses and T-cell-mediated cytotoxic responses are suppressed by TCDD but require greater doses when compared to antibody production.

Organochlorine pesticides

Several pesticides are known to have immunomodulatory effects. These include hexachlorobenzene (HCB), mirex, lindane, chlordane, dieldrin, and DDT and its metabolites. The immunotoxicity of these compounds has been demonstrated in several species and includes the loss of resistance to infections. Dieldrin is a potent immunotoxicant, suppressing natural resistance to certain viral infections, although doses in test organisms were orders of magnitude greater than relevant environmental levels. In most cases, the mechanisms of action are still not well defined, even if some studies suggest that some mechanisms may counteract the cell activation process (reviewed in Descotes 1988; Wong et al. 1992; Dunier 1994; Thomas 1995).

Other chlorinated chemicals

Among other chlorinated chemicals, few data are available on immunotoxic potential, except for PCP (reviewed in Wong et al. 1992). In rodents as well as in several fish species, PCP interferes with phagocytic activity of macrophages and reduces antibody production. The mechanisms of action and the possible involvement of dioxin contaminants in PCP are still speculative.

Nonchlorinated compounds

All the immunotoxic effects observed with exposure to chlorinated compounds have been observed in nonchlorinated chemicals and in many different species. These chemicals include organo-tin compounds, antibiotics, hormones and their mimics, mycotoxins, food additives and phenols (Zeeman and Brindley 1981; Anderson and Zeeman 1995), and metals (Bernier et al. 1995). There is no evidence that chlorinated compounds, in general, possess any special characteristics that confer immunotoxic properties.

Effects

Certain OC compounds may have deleterious effects on the immune system. These effects may vary from the simple change of cellularity in a lymphoid tissue to a loss of resistance to infections. These effects and endpoints do not all carry the same biological significance. The detrimental health effect as a result of a loss of resistance to infection is clear and valuable for a risk assessment. But change in cellularity of the thymus is difficult to associate with any specific detrimental effect, especially in adults. In order to differentiate the significance of immunotoxic potential of compounds, the following classification of effects, based on their increasing biological significance (from A to D), is proposed:

- Type A: Deleterious effects on cellular and molecular parameters after in-vitro exposure. These data are helpful for assessing mechanisms but have no predictive value.
- Type B: Modification on general parameters, such as tissue cellularity and histology after in-vivo exposure. Those are clearly toxic effects, but biological significance, when considered alone, is equivocal.
- Type C: Demonstrated effects on immune reactivity tested after in-vivo exposure. These represent clear immunotoxic potential for the chemical tested.
- Type D: Demonstrated detrimental effects mediated by immune response, such as decrease in the resistance to infections and cancers, autoimmunity, allergy, as well as other forms of hypersensitivity.

Immunotoxic effects of OC chemicals

Based on the ranking of effects (A to D), the toxicities of OC and other chemicals to the immune system have been summarized (Table 5-2). Data presented are based on published observations and on the highest effect recorded. Therefore, e.g., although 2,4-D presents a Type-B effect in fishes, this may not represent its maximal immunotoxic potential because a resistance-to-infection assay may not yet have been conducted.

Table 5-2 Summary of immunotoxic effects of OC chemicals in several species

Chemical	Fishes	Birds	Rodents	Primates	Other mammals
PCBs	B	D	D	C	C
PCDD/PCDFs	B	—	D	C	—
Atrazine	B	—	B	—	—
2,4-D	B	—	C	—	—
Diuron	B	—	B	—	—
DDT	C	C	C	A	C
Endosulfan	B	—	C	—	—
Endrin, dieldrin	C	D	D	A	—
Mirex	B	C	—	—	—
Lindane	D	—	B	A	C
Chlordane	_	—	D	—	—
HCB	—	—	D	—	—
PCP	C	—	C	—	C
Nonchlorinated chemicals	D	—	D	D	D

— lack of data
A deleterious effects on cellular and molecular parameters after in-vitro exposure
B modification on general parameters after in-vivo exposure
C demonstrated effects on immune reactivity tested after in-vivo exposure
D demonstrated detrimental effects mediated by immune response such as allergy, autoimmunity, or other forms of hypersensitivity as well as decrease in resistance to infections and cancers

Because of the magnitude of immunotoxic effects and the range of species affected (Table 5-2), compounds such as PCBs should be of greatest concern from the point of view of environmental and human health. Moreover, effects noted in primates have been obtained at exposure doses that are environmentally relevant. PCDD/PCDFs are also of concern. Highly immunotoxic in some rodents and in primates, their effects in other organisms remain to be confirmed. However, like PCBs, the dose factor is crucial because TCDD is toxic at extremely low doses. For most of the OC pesticides, immunotoxic properties vary from weak to strong, depending on species and type of effect. However, most of the overtly toxic effects were recorded only after exposure doses were increased. Additivity of effects of immunotoxic chemicals that are persistent in the environment is poorly understood; however, food-chain biomagnification may result in exposures that reach biologically active levels. Therefore, for all those chemicals, it is important to take the following actions:

- Complete their toxicological evaluation in order to determine characteristics such as LOAEL and NOAEL, especially for immune parameters, because they appear to be more sensitive (as demonstrated for PCBs).
- Complete the effects database by further determination of immunotoxic effects in several species, in order to establish conditions for interspecies extrapolation.
- Describe the uncertainties linked to the use of immune parameters in the risk assessment process, and proceed to the quantitative determination of the associated risk to the environment.

Chapter 6

Assessment of Organochlorine Chemicals for Terrestrial and Aquatic Animals

Chlorinated compounds cause a spectrum of effects on wildlife (terrestrial and aquatic organisms), but the nature of their effects and their potencies overlap those of nonchlorinated compounds. Neither the nature of effects nor the methods of assessment should be dictated strictly by the presence or absence of chlorine. In both cases, the responses of organisms depend upon the magnitude of exposure, the duration of exposure, the potency of the chemical, the tolerance of the organism, and the interactive effects of other chemicals or stressors.

The history of chemical effects on wildlife demonstrates that adequate hazard evaluations, specific to the use and expected production volumes of the chemical, are required before chemicals are released to the environment; field observations and biological monitoring are invaluable indicators of unexpected effects; and managing the risk of chemicals requires institutional frameworks for re-evaluating use and release of chemicals if new evidence for adverse effects is found.

Modern chlorinated pesticides do not have the persistence, bioaccumulation potential, or broad-spectrum toxicity typical of older pesticides that caused widespread effects on wildlife.

The long-term adverse effects on wildlife of chemicals other than pesticides are due primarily to polyaromatic organic compounds that are persistent, bioaccumulative, and highly toxic. These properties are typical of industrial compounds and byproducts containing chlorine, and effects on wildlife are associated with bioaccumulation to high concentrations. However, labile nonchlorinated polyaromatic hydrocarbons that accumulate in sediments are also highly toxic to wildlife continuously exposed to these sediments.

Populations of several species of birds, fish, and mammals have apparently recovered from the effects of such persistent chlorinated compounds as polychlorinated biphenyls (PCBs) and dichlorodiphenyltrichloroethane (DDT), as concentrations of these compounds have decreased in the environment. However, recovery may be incomplete because some populations are maintained through replacement or immigration of individuals from unaffected areas (birds) or through stocking (some fish species).

Chapter Preview

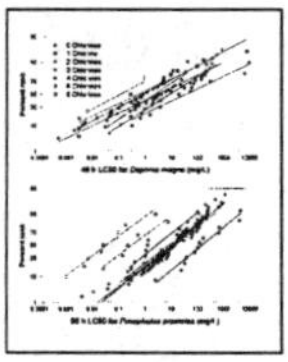

Laboratory approaches are available for characterizing most effects of chemicals on aquatic biota. However, their successful application requires that exposure methods and test endpoints suit the characteristics of the chemicals and ecosystems of concern. The ecological risks of recently discovered endocrine modulators and disruptors are poorly characterized. To reduce uncertainty in risk assessments, there is an urgent need to evaluate the hazards of these compounds using existing tools and to develop new tools appropriate for this new class of effects. Immunotoxicity is poorly characterized, and suitable methods for assessing this hazard in the laboratory or in the field must be developed. To characterize hazard, uncertainty could be reduced by expressing exposure in terms of tissue dose and the specific congeners causing toxicity.

Factors that contribute to the ecological risk of chlorinated compounds derive primarily from the effect chlorine may have on controlling the dose accumulated by biota, not the hazard or effects of that dose. The characteristics of chemicals that control dose are polarity, water solubility, and resistance to degradation, which in turn affect persistence and bioaccumulation. For persistent bioaccumulative compounds, the characteristics of wildlife that modify the accumulated dose include lipid levels, longevity, preferred habitat, food-web structure, and food-web position.

In light of the emerging awareness of previously unexpected chemical hazards, such as reproductive, immunotoxic, and multigeneration effects, the data required for pesticide registration and for new chemicals notification should be re-evaluated.

Biomonitoring of chemical effects on wildlife provides a critical check on the success of the regulation of chemical use and discharge to the environment. While laboratory methods for hazard assessment are relatively well developed, approaches for characterizing effects in natural ecosystems are much less so. The establishment of routine biomonitoring requires an improved capability to detect effects and to establish cause–effect relationships that will speed mitigation. A better integration of the measurements at the chemical, biochemical, population, and ecosystem levels is required, as are better sampling designs to apply the measurements.

As noted in Chapters 2, 3, 4, and 5, the presence of chlorine in organic chemicals can affect their reactivity, persistence, and bioaccumulation. Chlorine does not confer unique chemical or biological characteristics; other compounds lacking chlorine can exhibit similar properties over similar concentration ranges. Like other chemicals, organic chemicals containing chlorine exhibit a spectrum of toxicological attributes and can exert a range of effects in both aquatic and terrestrial wildlife. Thus, hazard evaluations for chlorine-containing compounds should be treated no differently than for other chemicals. The critical issue for all chemicals is whether their effects on natural populations and communities can be assessed with enough certainty to support informed decisions.

Effects of chemicals in aquatic ecosystems can be characterized by extrapolating from laboratory hazard evaluations or by evaluating effects on natural populations in situ. In either case, it is critical to establish a relationship between endpoints of interest in ecosystems (assessment endpoints) and test endpoints in the laboratory (measurement endpoints). The results of field-based assessments must be evaluated in controlled laboratory experiments to determine mechanisms and to understand causality. The results of laboratory measurements must be validated in the field to establish their relevance to natural ecosystems. The appropriate methods depend on the nature of the chemical and on the characteristics of the ecosystem. Current techniques for characterizing the ecological hazards of mixtures must be improved in both laboratory and field settings.

An important consideration in risk assessment is whether exposure is to distinct, identifiable compounds or to complex mixtures typical of effluents and of environmental transformations of compounds. Although complex mixtures are not unique to chlorinated compounds, bleaching and disinfection with chlorine increases the complexity of effluents by increasing the numbers of chemicals in the mixture.

Compounds that Affect Wildlife

Chlorinated compounds cause a spectrum of effects on wildlife (terrestrial and aquatic organisms), but the nature of their effects and their potencies overlap those of nonchlorinated compounds. Neither the nature of effects nor the methods of assessment are dictated strictly by the presence or absence of chlorine. In both cases, the responses of organisms depend upon the magnitude of exposure, the duration of exposure, the potency of the chemical, the tolerance of the organism, and the interactive effects of other chemicals or stressors.

The association between the hazard presented by certain persistent, bioaccumulative, and toxic chlorinated chemicals and the effects on wildlife has been deduced from

1) chemical analyses of tissues from numerous species,
2) well-defined dose–response relationships for adverse effects,
3) a decline in adverse effects in parallel with a decline in chemical concentrations, and
4) the measurement of similar adverse effects under controlled laboratory conditions.

Chlorinated compounds are acutely lethal at a wide range of doses in aquatic biota. The acute lethalities of various classes of low-molecular-weight aliphatic and aromatic compounds span several orders of magnitude, but the degree of chlorination explains only some of this variation (Figure 6-1). Toxicity increases with increasing chlorination, particularly for compounds acting by nonspecific narcosis; increasing the number of chlorine atoms increases their hydrophobicity and accumulation by lipids and membranes, the site of narcotic action (see Chapter 5). Nonetheless, the number of chlorine atoms per molecule is not a direct and consistent index of potency. Acutely toxic compounds (LC50 < 1 mg/L) may contain no, few, or many chlorine atoms; some chemicals with the greatest degree of chlorination may still be less toxic than the median for all OC compounds. The range of toxicities for nonchlorinated compounds is also similar to that of chlorinated compounds; e.g., the botulinum toxin (not shown) is more acutely toxic than 2,3,7,8-tetrachlorodibenzo-*p*-dioxin (TCDD). Acute mortality of wildlife due to exposure to chemicals is uncommon. The major impacts of chlorinated compounds are associated with the sublethal effects of chronic exposure, either to nonpersistent compounds (generally of low chlorine content) released continuously in effluents or to persistent compounds that bioaccumulate (generally of high chlorine content). Therefore, the data in Figure 6-1 cannot be generalized to chronic toxicity, as will be shown in later sections of this chapter.

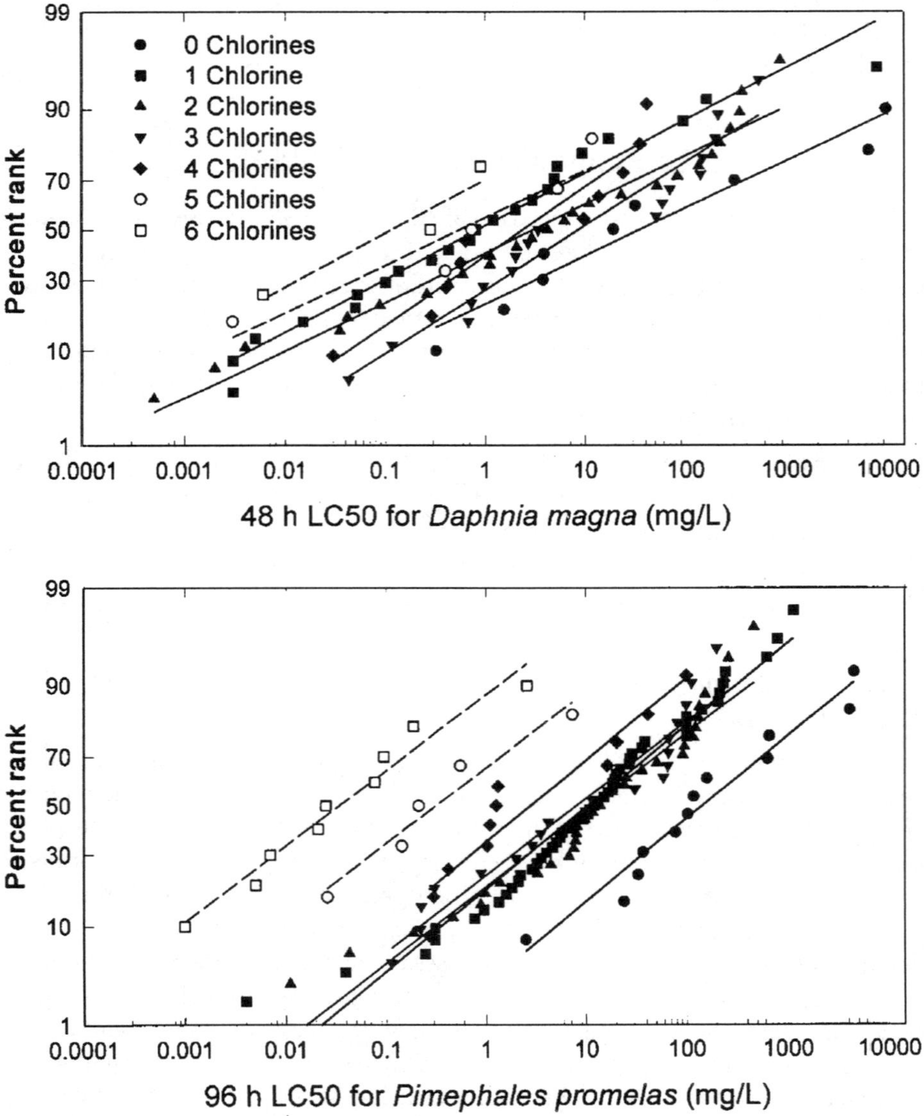

Figure 6-1 Toxicity of pesticides versus chlorine content (data provided by R. Kimerle, Monsanto Co., St. Louis MO). Plot of toxicity rank versus log LC50 using a log-normal model. (See the appendix to this chapter for the raw data.)

Persistent chlorinated pesticides: history of dichlorodiphenyltrichloroethane (DDT)

The history of chemical effects on wildlife demonstrates that adequate hazard evaluations, specific to the use and expected production volumes of the chemical, are required before chemicals are released to the environment; field observations and biological monitoring are invaluable indicators of unexpected effects; and managing the risk of chemicals requires institutional frameworks for re-evaluating use and release of chemicals if new evidence for adverse effects is found.

The history of dichlorodiphenyltrichloroethane (DDT) use, effects, and prohibition in the United States illustrates many of the basic challenges inherent in discovering and correcting ecological effects of chemicals, both chlorinated and nonchlorinated. Beginning in the 1940s, DDT was produced and released into the environment in the absence of a basic understanding of environmental persistence, long-range transport, and bioaccumulation; the discovery of DDT in Antarctic biota by Peterle (1969) provides a good example of the subsequent problems. Toxicological testing of DDT and other pesticides was perfunctory for wildlife prior to the 1970s. Subsequently, many countries enacted legislation that required more complete testing for pesticide registration, such as the Federal Insecticide, Fungicide and Rodenticide Act (FIFRA) of the U.S.

Initial findings of adverse effects in the field raised concerns about unanticipated effects on nontarget organisms (Cottam and Higgins 1946). These concerns were summarized by Couch (1946):

> It is disturbing to find DDT released for public use long before adequate research has revealed its limitations. Its promise as the ultimate Utopia in insecticides seems to overlook dangers to other forms of animal life. The effectiveness of DDT for household use in curbing flies, bedbugs, and mosquitoes, military use in controlling lice that spread typhus, and against insects of veterinary importance, has been demonstrated. In agriculture and forestry, potential dangers have not been fully appreciated. In fact, applications of DDT in the outdoors, whether on marshes, lakes, streams, on crop and pasture lands and in forested areas, present a serious threat to fish and wildlife species. Only now are we becoming aware of the possible misuse of this powerful chemical in the hands of the uninformed public.

Biologists reported a variety of suspicious changes in terrestrial biota but were unsuccessful in identifying ecologically significant effects in aquatic species until 1951 (Robbins et al. 1951). Aquatic invertebrates were affected, but these results were attributed to similarities among target and nontarget species. Concurrent laboratory tests were conducted using procedures adapted from pharmaceutical test methods. Even though effects were observed, they occurred at unrealistically high exposures (Coburn and Treichler 1946).

It remained for field studies in the early 1950s by Broley (Carson 1962) and Robbins et al. (1951), among others, to identify the first ecologically significant effects on wildlife: reproductive failure and population decline in birds. During the early 1960s, a primary mechanism causing these effects was eggshell thinning, as shown by Ratcliffe (1967, 1970). Despite these early reports, the lack of an adequate legal framework slowed the response of regulatory agencies. DDT was restricted in Canada in 1970, followed by the U.S. and Germany in 1972, Sweden in 1975, the Netherlands in 1981, and subsequently, many other countries. Recovery of many wildlife populations in North America, such as American robins (Beaver 1980), is now substantial, but lingering effects persist in some locations and species. Even now, there is only limited global agreement on the regulation of such compounds (see Chapters 1 and 7).

The history of DDT provides several valuable principles:

1) Prior to the environmental release of chemicals, adequate chemical and biological characterization are necessary, specific to the use and expected production volumes of the chemical;
2) field observations and biological monitoring are invaluable indicators of unsuspected effects; and
3) managing the risk of chemicals requires institutional frameworks that can apply scientific knowledge prior to a chemical's use and following its release, if new evidence for adverse effects is found.

Even with well-studied compounds such as DDT, new understanding of environmental hazards is still emerging: e.g., methylsulfonated metabolites of DDE have been discovered, and they are more toxic than the parent compound (Lund et al. 1988; Brandt et al. 1992; Jonsson et al. 1993).

It has been more difficult to demonstrate or document widespread effects of chlorinated compounds on aquatic organisms. OC insecticides have caused fish kills and impaired reproduction of adult fish (Niimi 1983; Eisler and Jacknow 1985; Hansen et al. 1985; Giesy et al. 1986; Vonwesternhagen et al. 1987; Panday 1988; Knuth 1990; Gorge and Nagel 1990; Anon. 1991); DDT, toxaphene, lindane, and dieldrin are most frequently associated with adverse effects on fish. Of the OC insecticides still used in North America, endosulfan is associated with most acute poisonings in fish (Graney et al. 1994).

Modern chlorinated pesticides

While not all of the toxic chemical effects observed in wildlife have been caused by chlorinated compounds, a significant number of manufactured chlorinated pesticides had a high toxicity to wildlife, including DDT, aldrin, and dieldrin (Rosene 1965; Stickel 1975; Blus et al. 1979; Flickinger 1979; Clark et al. 1983). Most of the persistent chlorinated insecticides that historically caused adverse effects are no longer used in North America or Western Europe. However, some, such as DDT, are still manufactured in North America and used in other regions of the world (Raloff 1996). While a number of "modern" pesticides contain chlorine, chlorine alone does not determine their toxicological properties or persistence, as it did in older pesticides (Delzell et al. 1994); they are designed to be less environmentally persistent, to bioaccumulate less, and to have greater target organism specificity. Within each of several classes of modern pesticides, including both insecticides and herbicides, there are both chlorinated and nonchlorinated compounds (Potts 1977; White et al. 1979; Flickinger et al. 1980; White, King et al. 1982; White, Mitchell et al. 1982; Grue et al. 1983; Rands 1985; Stone and Gradoni 1985).

Nonpesticidal chemicals

The long-term adverse effects on wildlife of chemicals other than pesticides are due primarily to polyaromatic organic compounds that are persistent, bioaccumulative, and highly toxic. These properties are typical of industrial compounds and byproducts containing chlorine, and effects on wildlife are associated with bioaccumulation to high concentrations. However, labile nonchlorinated polyaromatic hydrocarbons that accumulate in sediments are also highly toxic to wildlife continuously exposed to these sediments.

Aromatic compounds

A number of chlorinated hydrocarbons not manufactured as pesticides have caused widespread toxicity to wildlife, the best known of which are the polychlorinated diaromatic hydrocarbons (PCDHs) such as polychlorinated biphenyls (PCBs). PCBs have caused widespread toxic effects in wildlife because of their often indiscriminate use and release to the environment (Jensen 1972; Hoffman et al. 1987; Kubiak et al. 1989; Tillitt et al. 1992; Bosveld et al. 1995). Localized impairment of reproductive performance, deformities, and developmental effects still persist at some locations. These effects are linked primarily to PCB and/or dioxin exposure in North America and Western Europe (Gilbertson et al. 1976; Aulerich et al. 1986; Bellward et al. 1990; Fox et al. 1991; Tillitt et al. 1992; Yamashita et al. 1993; Van den Berg, Craane et al. 1994; Van den Berg, DeJongh et al. 1994; Giesy, Ludwig et al. 1994; Giesy, Verbrugge et al. 1994; Bosveld et al. 1995). For example, a causal link between tissue concentrations of PCBs and reproductive impairment of seals in the Baltic Sea and the Dutch Wadden Sea has been proposed (Helle et al. 1976a, 1976b; Reijnders 1976, 1980, 1986; Bergman and Olsson 1986, 1989; Olsson et al. 1992, 1994a). PCBs are suspected of causing a disease syndrome (hyperadrenocorticism) associated with bone erosion, renal glomerulopathy, chronic intestinal ulcers, uterine tumors (leiomyomas), arteriosclerosis, hyperkeratosis, thin epidermis, cystic dilation of hair follicles, and bilateral adrenocortical hyperplasia in Baltic gray and ringed seals (Bergman and Olsson 1986). Relative to the late 1970s, an increase in the number of gray seals and common seals has been reported following more recent decreases in environmental concentrations of PCBs (Helander and Bignert 1992; Olsson et al. 1994b).

Studies on the indigenous population of Beluga whales in the St. Lawrence estuary have revealed a high prevalence of various disease symptoms (> 75% in stranded animals) and high lipid concentrations of PCBs, mirex, and chlorinated pesticides such as DDT. The pathology has been interpreted as an indication of chemically induced immunosuppression (Martineau et al. 1987, 1988, 1994).

The association of effects with persistent, bioaccumulative, and toxic chlorinated compounds is further reinforced by laboratory studies. The observation of reproductive failure in ranch mink that were fed PCB-contaminated fish was followed by laboratory exposures to PCBs that resulted in reproductive failure at similar concentrations (Aulerich and Ringer 1977).

Byproducts

Byproducts and microcontaminants produced in industrial compounds have also caused effects in wildlife. As an example, a chemically induced disease was observed in the 1950s in poultry that were fed dioxin-contaminated feed. Labeled "chick edema disease," it has now been described in a variety of fish-eating birds in North America (reviewed in Gilbertson et al. 1991). The disease is similar to poisoning caused by planar chlorinated hydrocarbons such as polychlorinated dibenzo-*p*-dioxins (PCDDs), polychlorinated dibenzofurans (PCDFs), and some PCBs. Symptoms include embryonic mortality, deformities, growth retardation, edema, liver enlargement, and physiological and biochemical alterations.

The history of the discovery and characterization of the effects of polychlorinated aniline (PCA) illustrates the importance of field observations and biomonitoring programs as well as the slow response of regulatory bodies to ecotoxicological problems when cause–effect relationships are unknown. Hayes and Risebrough (1972) first attributed deformities in tern chicks from Long Island, New York to PCBs, based on their similarity to those seen in poultry. Embryo mortality and deformities in Great Lakes fish-eating birds in the early 1970s also suggested chick edema disease (Gilbertson 1983), but the causative agents were unknown. Available analytical techniques were inadequate to detect compounds such as PCDD/PCDFs in field samples (Bowes et al. 1973). In 1975, reduced hatchability and chick edema symptoms in contaminated eggs of Great Lakes herring gulls were reported to the International Joint Commission's (IJC's) Great Lakes Water Quality Board. In 1980, 2,3,7,8-TCDD was found in Lake Ontario fish. This compound was also found in gull eggs that had been stored since the early 1970s. These findings were reported by a Canadian delegation to officials at the U.S. State Department in 1981.

Based on earlier anecdotal observations, Gilbertson undertook a systematic search for chick edema disease in 2 other Great Lakes colonial bird species (Gilbertson et al. 1991). A dose–response relationship was reported between dioxin-like activity in eggs and embryo mortality of double-crested cormorants (Tillitt et al. 1992). Retrospective analysis of a sediment core from Lake Ontario indicated that 2,3,7,8-TCDD contamination of Lake Ontario occurred as early as the 1930s. Due to a lack of knowledge, a chemical first released in the 1930s was not identified as an environmental problem for many decades. This problem was not evident immediately because concentrations in birds accumulated to the threshold of toxic effects only after long periods of time.

Polychlorinated aromatic compounds also have important effects on fish. PCDDs, PCDFs, and coplanar PCBs have distinct and specific effects on embryos and larvae of lake trout (*Salvelinus namaycush*), rainbow trout (*Oncorhynchus mykiss*), and brook trout (*Salvelinus fontinalis*) (Peterson 1994). The signs of toxicity, a wasting syndrome with multiple symptoms, are analogous to those observed in mammals exposed to dioxins. Eggs hatched from both introduced and native Great Lakes salmonids develop the same signs of toxicity as those observed in eggs exposed experimentally to PCDDs, and larvae often fail to survive past swim-up.

Recently, reproductive failure of lake trout in Lake Ontario has been attributed to PCDDs, PCDFs, and PCBs (Cook 1994). Historic contamination of sediments by these compounds was strongly correlated to present concentrations in sediments, accumulation by lake trout, maternal transfer to eggs, and increased mortality rates of developing larvae, a pattern consistent with a chemical cause for the decline of lake trout in Lake Ontario after 1940. The relative risks for these same effects from nonchlorinated compounds is difficult to assess, given the apparent specific mechanism by which dioxin-like compounds interact with proteins such as the aryl hydrocarbon receptor (AhR). Interactions with the AhR lead to up-regulation of a variety of genes and induction of toxic effects (Peterson and Walker 1992). However, the specificity needs further investigation, given the small number of compounds adequately tested.

A number of other compounds have been manufactured that should have similar environmental properties and toxicities as PCBs (Table 6-1) (Giesy, Ludwig et al. 1994). Hazard assessments of PCB effects on wildlife should be extended to related compounds with similar properties, particularly when data on quantitative structure-activity relationships (QSARs) are available. For instance, on the basis of chemical structure, polychlorinated diphenyltoluenes (PCDTs) should exhibit similar environmental behavior and toxicity as PCBs and should be considered equivalent hazards (Furst 1987; Howie et al. 1990; Addison et al. 1991; Murk et al. 1991; Harper et al. 1993). Because PCDTs are used as PCB replacements in such countries as Germany and France, global distribution was expected and confirmed by analysis of bird tissues and eggs from remote areas of the Pacific Ocean (Jones et al. 1995, 1996). Currently, environmental concentrations of PCDTs are lower than those of the more widely produced PCBs. If PCDT release is stopped now, concentrations should not reach critical levels except near sources.

Aliphatic compounds

Most aliphatic OC s are not persistent (Chapter 2), do not bioaccumulate (Chapter 3), and are minimally toxic (Chapter 5). If managed appropriately, they should not pose a

Table 6-1 Compounds that should cause dioxin-like toxicity in wildlife, based on experiments or chemical structure

Polycyclic aromatic hydrocarbons	Polychlorinated anthracenes
Polychlorinated biphenyls	Polychlorinated fluorenes
Polychlorinated dibenzo-*p*-dioxins	Polychlorinated dihydro anthracenes
Polychlorinated dibenzofurans	Polychlorinated dihydro anthracenes
Polychlorinated naphthalenes	Polychlorinated dibenzothiophenes
Polychlorinated diphenylmethanes and methyl derivatives	Polychlorinated terphenyls
Polychlorinated azoanthracenes	Polychlorinated quaterphenyls
Polychlorinated anisoles	Polychlorinated quaterphenyl ethers
Polychlorinated phenoxy-anisoles	Polychlorinated biphenylenes
Polychlorinated xanthenes	Polychlorinated thioanthrenes
Polychlorinated xanthones	Polybrominated diphenyl ethers

Source: Reprinted with permission from Giesy, Ludwig et al. 1994. Copyright 1994 American Chemical Society.

risk to wildlife through biomagnification. For example, with the exception of acute toxicity due to accidental spills, no toxic effects of chlorinated alkanes or alkenes have been reported for wildlife. Hazard assessments have shown that current environmental concentrations (Chapter 7) and concentrations in tissues of wildlife are well below thresholds for known toxic effects (Delzell et al. 1994).

Nonchlorinated organics and inorganics

Nonchlorinated compounds are also hazardous to terrestrial and aquatic wildlife. Metals such as lead, selenium, and mercury can be persistent, can cause acute toxicity, and can be bioaccumulated or biomagnified (Bellrose 1959; Tejning 1967; Borg et al. 1969; Demayo et al. 1982; Hodson et al. 1984; Clark et al. 1989; Hodson 1990). Most other inorganic compounds exist as charged ionic species and are less likely to be bioaccumulated and biomagnified by aquatic organisms. However, mercury, when methylated, can be both bioaccumulated and biomagnified (Hodson 1988).

Of the nonchlorinated compounds that affect terrestrial wildlife, pyrethroid or organophosphorus insecticides are among the most acutely toxic (Anon 1984; Eisler 1986). Birds and marine mammals are also affected by polynuclear aromatic hydrocarbons (PAHs) (Albers 1991), although benthic fish may be more affected because of their continuous exposure to contaminated sediments. Brown bullhead (*Ictalurus nebulosus*) populations in the Black River, OH, have been affected by sediment PAHs downstream of a steel mill. Compared to reference populations, bullheads exposed to these sediments accumulated PAHs and showed induction of hepatic mixed-function oxidase (MFO) activity, activation of carcinogenic PAHs such as benzo(a)pyrene (BaP), increased levels of deoxyribonucleic acid (DNA) adducts, hepatocyte proliferation, elevated prevalence of liver and skin tumors, and missing-year classes due to mortality of older fish (Harshbarger et al. 1984; MacCubbin et al. 1988, 1990; Bauman et al. 1990). Effects have diminished over time as loadings decreased, further strengthening proposed cause–effect relationships.

There is also evidence at marine sites that nonchlorinated chemicals are associated with effects on natural populations of benthic fish, despite the presence of chlorinated compounds. Reproductive success of English sole was examined at several sites in Puget Sound, WA. One site was predominantly contaminated by PAHs, while others were contaminated by both PAHs and OC compounds. Fish from the site contaminated only with PAH exhibited the greatest impairment of reproduction (Johnson et al. 1988; Casillas et al. 1991; Landahl and Johnson 1993). In addition, field and laboratory studies of hepatic lesions in English sole have shown statistical associations between prevalence of lesions and sediment PAH concentrations and between concentrations of DNA adducts in livers of fish and prevalence of lesions (Myers et al. 1987; Schiewe et al. 1991; Stein et al. 1992). Lesions identical to those observed in field-captured sole were also induced by laboratory exposures of fish to BaP or to extracts of sediments from a site contaminated by PAHs but with low levels of OC compounds. These results suggest that nonchlorinated PAHs were the most likely causes of both impaired reproductive success and hepatic lesions in English sole in Puget Sound.

A similar situation exists for winter flounder from New Bedford Harbor, MA, a site contaminated with high concentrations of PCBs and moderate concentrations of PAHs. Hepatic lesions were less prevalent than in flounder from other sites contaminated with greater concentrations of PAHs and lower concentrations of PCBs (Johnson et al. 1993). These findings suggest little association between PCB exposure and hepatic lesions, although PCBs act as tumor promoters in mammals (Safe 1984, 1989).

The studies presented above are based only on associations between cause and effect. Understanding cause–effect at contaminated sites is often confounded by covariation among contaminants, both chlorinated and nonchlorinated. For example, in studies on the west coast of the United States, significant associations among concentrations of several chlorinated pesticides, PCBs, and PAHs in sediment and the prevalence of hepatic lesions in 3 fish species were observed (Myers et al. 1994). However, it was impossible to identify any group of compounds as the primary causative agents because the concentrations of all compounds strongly covaried. Although field studies can provide compelling evidence for the involvement of chemical contaminants in the etiology of effects on natural fish populations, it is often difficult to unequivocally attribute causality to any single compound or class of compounds.

Studies of the impacts of pulp mill effluents on fish have also shown no clear associations between effects on fish and discharges of complex mixtures of chlorinated compounds measured as adsorbable organic halogens (AOXs) (see Chapter 3, "Tools for assessment of exposure and bioaccumulation of organic chemicals"). Although induction of hepatic MFOs in male fish was strongly correlated with hepatic concentrations of TCDD equivalents, similar correlations existed between MFO activity and waterborne levels of chlorophenols, chloride, and effluent dilution (Kloepper-Sams and Benton 1994; Servos et al. 1994; van den Heuvel et al. 1994). Moreover, detailed field and laboratory studies have not shown that elevated tissue concentrations of PCDDs and PCDFs contributed to effects (Swanson et al. 1993). Similar physiological changes were found at sites near pulp mills with and without chlorine use (Munkittrick, Van Der Kraak et al. 1994). In other words, effects on fish were not correlated to discharges of high volumes of chlorinated organic compounds from pulp mills. Physiological changes in fish exposed to pulp mill effluent, such as MFO induction, disappeared rapidly during mill maintenance shutdowns (Munkittrick, Van Der Kraak et al. 1992) or when caged fish were removed to clean water (Munkittrick, Servos et al. 1994). Therefore, observed effects likely were due to easily excreted or easily metabolized compounds. If the effects were due to persistent, bioaccumulative OC compounds such as TCDD, such a rapid decline in biological effects would not be observed (Sijm and Opperhuizen 1988; Muir et al. 1990; Parrott et al. 1995). While MFO induction has been correlated to tissue concentrations of dioxins and furans (Hodson et al. 1992; Kloepper-Sams and Benton 1994) as well as to discharge of AOX (Ahokas et al. 1994), chlorinated compounds may simply be indicating effluent exposure and dilution.

Laboratory studies have been conducted to identify the compounds in pulp mill effluents responsible for MFO induction in fish. The association of induction with compounds extracted by methanol from particulates and separable by high pressure liquid chromatography (HPLC) indicates structures and behaviors analogous to those of PAHs (Burnison et al. 1994). The presence of inducing compounds in waste streams from nonchlorinated kraft mills and the lack of correlation of AOX to effects demonstrate that toxicity can be associated with nonchlorinated compounds and that the presence of chlorinated organic compounds near sources of contamination does not guarantee effects on fish (Swanson et al. 1993; Hodson et al. 1994; Kloepper-Sams and Benton 1994).

Recovery of Wildlife Populations Following Decline of Environmental Concentrations of DDT, PCBs, and PCDD/PCDFs

Populations of several species of birds, fish, and mammals have apparently recovered from the effects of persistent chlorinated compounds, such as PCBs and DDT, as concentrations of these compounds have decreased in the environment. However, recovery may be incomplete because some populations are maintained through replacement or immigration of individuals from unaffected areas (birds) or through stocking (some fish species).

The association of effects on wildlife with persistent, bioaccumulative, and toxic compounds has been strengthened by the recovery of populations of several species as chemical concentrations have declined (Becker and Sieg 1987; Bednarz 1990). Examples include the increase of herring gull and double-crested cormorant populations with declines in concentrations of DDT, PCBs, and PCDD/PCDFs in the Great Lakes (Gilman et al. 1979; Ludwig 1984; Mineau et al. 1984; Price and Weseloh 1986; Bauman and Whittle 1988; Anon. 1991).

As concentrations of DDT declined, there was a recovery of populations of bald eagle and peregrine falcon in most areas of North America (Morrison et al. 1978; Hoffman et al. 1990) and in birds of prey such as the white-tailed sea eagle, peregrine falcon, and eagle owl in the Baltic (Newton and Bogan 1978; Spitzer 1978; Chapdelaine et al. 1987; Lindberg et al. 1988; Posner 1991; Helander 1994). Reproductive failure and deformities in Forster's terns have also declined in frequency as concentrations of PCBs in Green Bay, WI, declined (Harris et al. 1993). Populations of fish-eating birds are also recovering in other areas. For instance, brown pelicans (Blus et al. 1974) and cattle egrets (Mora 1991) have become more abundant as concentrations of chlorinated insecticides in their eggs decreased below threshold concentrations for eggshell thinning (Lundholm 1987).

In the case of otter, there is preliminary evidence for recovery in some areas of Britain where PCB concentrations have fallen sufficiently. Similarly, some populations of marine mammals are increasing as concentrations of PCDH in their tissues decline (Hileman 1992), but recovery has been very slow for the Beluga population of the St. Lawrence estuary (Kingsley 1993). Conversely, a resurgence of effects in Saginaw Bay, MI, double-crested cormorants was observed when PCB-contaminated sediments were disturbed by a major flood (Ludwig, Giesy et al. 1993), and other adverse effects of PCBs, such as localized reproductive impairment, deformities, and developmental effects, persist. These effects are linked to PCB and/or dioxin exposure in North America and Western Europe (Gilbertson et al. 1976; Aulerich et al. 1986; Fox et al. 1991; Tillitt et al. 1992; Yamashita et al. 1993; van den Berg, Craane et al. 1994; van den Berg, DeJongh et al. 1994; Bosveld et al. 1995). Populations of lake trout in Lake Ontario are also recovering as concentrations of PCBs and dioxins decline in their tissues. However, recovery may not be due solely to decreased exposure to chemicals. For example, in lake trout, recovery may represent a complex interplay among contaminants, quality of spawning habitat, genetics of stocked lake trout, seasonal thermal regimes, and communities of foreign and native prey species (J. Casselman, Ontario Ministry of Natural Resources, personal communication). In some cases, recovery may not be complete. For eagles in regions where contamination is still close to threshold concentrations, new populations may be sustained by immigration of birds from less contaminated areas (Bowerman et al. 1995), and recovery of trout in the Great Lakes was initiated by massive stocking.

Adequacy of Wildlife Hazard Assessment Techniques

Laboratory approaches are available for characterizing most effects of chemicals on aquatic biota. However, their successful application requires that exposure methods and test endpoints be selected to suit the characteristics of the chemicals and ecosystems of concern.

The characterization of ecological effects is based on several principles and assumptions, among them that test species and endpoints in laboratory studies are appropriate and representative surrogates for assessing ecological hazards. Bioassays of survival, growth, and reproduction of surrogate species are assumed to predict ecosystem effects, and they generate dose–response data to estimate no-observed-adverse-effect concentrations (NOAECs) or exposure concentrations that represent the threshold exposures for effects in the field. Ideally, such bioassays will be modified according to the physical and chemical properties of test materials that influence their availability and accumulation and according to site-specific characteristics such as pH, temperature, organic matter, hardness, salinity, and the habitat and behavior of the organisms, which also affect chemical exposure and accumulation.

Effects characterizations should follow a tiered approach that progresses from acute to chronic tests in order to provide risk estimates with increasing certainty, as required for risk assessments. While this is a well-defined process, scientific judgment is still needed to interpret results and to decide if additional tests are necessary. In addition to its use for pure compounds, this approach can be used to characterize mixtures, effluents, and environmental samples of unknown composition. The same bioassays can be applied to identifying toxic chemicals in complex mixtures such as effluents or sediments by Toxicity Identification and Evaluation (TIE) (Mount 1989) and to demonstrating whether effects diminish when production and effluent treatment processes are modified.

Characterization of effects can involve both laboratory toxicity tests to establish dose–response relationships for extrapolation to the field and field studies to evaluate actual effects in situ (Cairns et al. 1978; Dickson et al. 1979; Mayer et al. 1981; Bergman et al. 1986; Pontasch et al. 1988; Hansen 1989; Crossland et al. 1992; USEPA 1992b; National Research Council [NRC] 1993). Effects characterizations are applicable to all chemical types and not just to specific chemical structures or groups of chemicals such as chlorinated compounds.

Effective strategies for characterizing effects must be tailored to the test chemical and ecosystem of interest. However, as described in the following section, the most commonly used laboratory toxicity tests may not adequately characterize effects on aquatic ecosystems because they do not include all relevant endpoints.

Effects on hormone regulation and biochemical pathways

The ecological risks of recently discovered endocrine modulators and disruptors are poorly characterized. To reduce uncertainty in risk assessments, there is an urgent need to evaluate the hazards of these compounds using existing tools and to develop new tools appropriate for this new class of effects.

A number of chemicals cause physiological effects during gonadal development or are transferred maternally (Miller 1993), so that toxicity is not expressed until the next generation. Responses of offspring to chemical exposure and effects in the parental generation (transgenerational effects) probably arise from multiple mechanisms. There is a key difference between indirect transgenerational effects and direct effects caused by maternal transfer of chemical residues[1].

Chemicals causing transgenerational effects exert indirect effects on offspring by acting on maternal physiology during key periods of ovarian or oocyte development. Many naturally occurring chemicals have structures similar to reproductive hormones such as estradiol and can exert deleterious effects on embryonic development. These compounds are referred to as "environmental estrogens." Possible adverse effects of environmental estrogens have recently received considerable attention (Baloer 1994), although the potential to cause congenital malformations was recognized as early as 1959. Natural steroidal alkaloids in hellebore *(Veratrum californicum)* caused cyclops deformities in sheep fetuses if they were exposed during the 14th day of gestation (Binns et al. 1959; Keeler 1984). Chlorinated compounds currently known to act as estrogen mimics are primarily persistent compounds such as *o,p'*-DDT (see Chapter 5, "Toxicants with a specific site of action"). However, because studies of endocrine effects are relatively recent, the involvement of other chlorinated compounds cannot be ruled out.

Plasma sex-steroid hormone concentrations of fish can be depressed by chronic exposure to pulp mill effluents, whether from mills using chlorine bleaching or from mills using no bleaching at all (Hodson et al. 1992; Munkittrick, McMaster et al. 1992; Gagnon et al. 1994, 1995; Munkittrick, Servos et al. 1994). Depressed steroid levels appear to result from disruptions of the hypothalamic-pituitary-gonadal (HPG) axis at the levels of gonadotropic hormone (GTH) circulation, receptor binding, steroid

[1] Since this 1994 SETAC Workshop, an intense debate has emerged concerning the possible effects of environmental contaminants on the reproduction of humans and wildlife. Multigenerational effects of persistent chlorinated compounds such as DDT and PCB and estrogenic effects of a wide variety of chlorinated and nonchlorinated compounds are of particular concern. This debate has not altered the concerns, conclusions, and research needs identified by participants in this workshop. It is obvious that considerable research is needed to evaluate the extent and nature of the problem, to develop methods for assessing reproductive hazards, and to screen candidate compounds.

synthesis, and peripheral metabolism of steroid hormones (Van Der Kraak et al. 1992). Changes in steroid concentrations are also associated with delays in sexual maturity, changes in gonad size, reductions in fecundity, and depression of secondary sexual characteristics (McMaster et al. 1991; Munkittrick, McMaster et al. 1992; Robinson 1994). Variations among mills in the extent of these effects have been observed (Kloepper-Sams et al. 1994; Munkittrick, Servos et al. 1994; Swanson et al. 1993), likely due to differences in receiving water dilution, pulping technologies, and extent of waste treatment.

β-sitosterol, a nonchlorinated plant sterol discharged by pulp mills, has been shown in laboratory studies to affect reproductive performance of goldfish (McLatchy and Van der Kraak 1995; McLatchy et al. 1997; Tremblay and Van der Kraak 1998, 1999). Whether plant sterols exert a population-level impact and are responsible for pulp mill effluent effects remains to be assessed.

Some chlorinated compounds cause receptor-mediated gene expression that affects physiological processes, including depression of vitamin levels (Mayer et al. 1978), which could have important consequences for survival of offspring. Atlantic salmon (*Salmo salar*) and lake trout in Lake Cayuga, NY, show early life-stage pathology similar (but not identical) to TCDD-induced blue sac disease, as described by Spitsbergen et al. (1991), Walker and Peterson (1991), and Walker et al. (1991). However, the increased mortality rate of salmonid larvae in Lake Cayuga and Lake Ontario may be due as much to vitamin B (thiamine) deficiency as to chlorinated compounds. Thiamine deficiency was associated with a maternal diet dominated by alewife (*Alosa pseudoharengus*), a species low in thiamine and high in thiaminase activity; consequently, maternal gonads had lower concentrations of thiamine (Fisher, Iamonte et al. 1995; Fisher, Spitsbergen et al. 1995; Fisher et al. 1996). Thiamine treatment of eggs from Lake Ontario and Lake Cayuga lake trout and Lake Cayuga salmon caused reduction of mortality to levels typical of controls (Fitzsimons 1995; Fisher et al. 1996). These findings illustrate the need for caution in assigning causality to anthropogenic chemical contaminants in natural ecosystems.

The ability of some chemicals to affect one generation after exposure of their parents challenges the assumption that short-term assays for hazard assessment will adequately predict all important effects on natural populations. Single-generation chronic toxicity tests, by definition, do not detect effects manifested in subsequent generations; more complex multigeneration protocols for reproductive and teratogenic effects are needed to assess these hazards. However, there are no standardized methods for assessing these effects in most species, and none are currently required in the regulatory process. Similarly, short-term tests of a few life stages of a test animal may also be inadequate for hazard assessments. For example, effects of pulp and paper mill effluents in standard 7-d *Ceriodaphnia* reproduction tests and 7-d fathead minnow survival and growth tests were not correlated to impacts on reproductive hormones and xenobiotic metabolism of adult white suckers from receiving waters (Munkittrick,

Servos et al. 1994; Robinson et al. 1994). This is not surprising because tests of invertebrate reproduction or larval fish growth do not include the complex physiological processes of vertebrate sexual maturation. Nevertheless, reproductive effects of pulp mill effluent were substantiated in the laboratory by complete life-cycle bioassays with fathead minnows; impaired sexual maturation typical of field-exposed fish was replicated (Robinson 1994). Clearly, the choice of the laboratory test is critical to a successful assessment of potential effects in aquatic ecosystems; chronic bioassays that involve all life stages of a species will reveal significant effects not observed in abbreviated protocols of embryo and larval stages.

Effects on immunocompetence

Immunotoxicity is poorly characterized, and suitable methods for assessing this hazard in the laboratory or in the field must be developed.

An emerging issue is the need for relevant tests of immunotoxicity to fish (see reviews by Zeeman and Brindley 1981; Anderson and Zeeman 1995). Studies have shown that immune responses of juvenile Chinook salmon (*Oncorhynchus tshawytscha*) from a contaminated urban estuary differed from those of hatchery salmon and juveniles sampled from a nonurban estuary (Arkoosh et al. 1991). Salmon from the urban estuary contained higher concentrations of both PCBs and PAHs, compared to salmon from the other locations. Exposure of juvenile chinook to PCBs or a model PAH in the laboratory induced a similar alteration in immune response (Arkoosh et al. 1994). These findings indicate that chemicals can affect immune function in fish and hence their resistance to disease and prevalence of disease (Zeeman and Brindley 1981; Norman et al. 1994; Anderson and Zeeman 1995). Despite substantial advances in methodology, the ability to assess immunocompetence of fish is in its infancy (Stolen and Fletcher 1994; Zeeman 1996), and better methods are needed to detect immune dysfunction in aquatic species exposed to chemicals.

Importance of dose

To characterize hazard, uncertainty could be reduced by expressing exposure in terms of tissue dose and the specific congeners causing toxicity.

For most compounds, tissue dose is a more accurate index of toxicity than environmental concentrations. For compounds that are readily biotransformed to toxic metabolites, the concept of tissue dose includes the concentration of the parent compound or its metabolites bound to critical cellular macromolecules. Examples include MFO-inducing compounds that bind with the AhR protein, PAH metabolites that bind to DNA (DNA adducts), or activation products of organophosphorus pesticides that inhibit acetylcholine esterase (AChE). By extension, measuring the concentration of specific PCB congeners that bind most strongly to receptors is preferable to measuring total PCBs. Therefore, to better characterize exposure, analytical methods are needed for small amounts of tissue and for compounds and metabolites associated with specific tissue receptors.

Factors that Modify Environmental Risk

Factors that contribute to the environmental risk of chlorinated compounds derive primarily from the effect chlorine may have on controlling the dose accumulated by biota, not the hazard or effects of that dose. The characteristics of chemicals that control dose are polarity, water solubility, and resistance to degradation, which in turn affect persistence and bioaccumulation. For persistent bioaccumulative compounds, the characteristics of wildlife that modify accumulated dose include lipid levels, longevity, preferred habitat, food-web structure, and food-web position.

The interactions of chemicals with a tissue receptor are not dictated solely by the presence or absence of chlorine. The precise shape of the molecule with respect to the receptor is the key to biological activity. Hence, the role of chlorine in influencing the hazard of a chemical may be to modify the kinetics of exposure and to modify (increase or decrease) the strength of interactions with receptors, but not to give chlorine unique properties by comparison to nonchlorinated compounds.

Increased hydrophobicity of a chemical shifts the primary route of exposure of aquatic species from water toward dietary uptake and maternal transfer (Miller 1993). Because the ovaries of many species of fish have a high lipid content and may constitute more than 20% of maternal body weight, eggs can represent a significant reservoir of hydrophobic chemicals (Guiney et al. 1979). Due to this maternal transfer, contaminants within the egg can be mobilized and concentrated as the yolk is consumed during embryo development. Further, the developing embryo or fetus is more sensitive to chemicals than are adults, so that effects are seen at lower exposure concentrations than expected. Maternal transfer is not restricted to chlorinated chemicals. Munkittrick and Dixon (1988, 1989) found that female white suckers at a site contaminated with mining wastes transferred significant amounts of metals to their developing eggs. Mobilization of metals during yolk utilization conferred an increased tolerance to waterborne metal exposure during embryo development.

Characteristics of species most susceptible to the effects of persistent bioaccumulative compounds include high tissue-lipid concentrations, a position high in the food web (predators), and great longevity. For example, Great Lakes species that are most affected by persistent compounds that biomagnify through food webs are lake trout and fish-eating birds such as cormorants (*Phalacrocorax carbo*) but not pelagic planktivorous fish such as smelt (*Osmerus mordax*). Simple food webs dominated by a few predatory species can lead to greater contamination of top predators than would occur in complex food webs (Chapter 4).

Species most affected by compounds that are not persistent, such as hydrophilic aliphatics (C_1–C_3) or chlorophenols, are those resident near point sources. Hydrophilic chemicals are accumulated primarily through the gills, and accumulation is not markedly influenced by the longevity, lipid content, or feeding strategy of the exposed species. Effects on aquatic biota will occur only if tissue concentrations are maintained above the threshold levels for toxicity through continuous exposure to effluents. In contrast, relatively brief exposures to persistent hydrophobic compounds can elevate tissue concentrations, although equilibria may not be attained for long periods of time. Hence, kinetics of exposure will determine whether effects are seen and over what time period.

Biotransformation usually reduces the impact of biomagnification for organisms higher in the food chain. Because new pesticides are designed to be more easily metabolized, the exposure of organisms higher in the food chain will generally be lower, and risk will be reduced compared to exposures to older persistent pesticides. However, some pathways of metabolism do not necessarily reduce toxicity. For example, some metabolites such as DDE and the methylsulfonated products of DDE and hydroxylated PCBs (Hydroxy-(OH) PCBs) are more toxic than the parent compounds (Haraguchi et al. 1992). DDE causes eggshell thinning (Ratcliffe 1970), while methylsulfonated DDE is a potent adrenocorticolytic substance (see Chapter 5, "Endocrine toxicology"). OH-PCB can bind to the trans-thyretin carrier protein (TTR) in blood plasma, thus disrupting vitamin A and thyroid hormone (T_4) dynamics (Brouwer and van den Berg 1986). Genetics play an important role in determining species sensitivity and capacity to metabolize contaminants (Tanabe et al. 1988; van den Berg, DeJongh et al. 1994;). The ability to metabolize OC compounds generally increases phylogenetically, i.e., more highly evolved species have a greater capacity than more primitive species (Chapter 3). Rates of biotransformation in mammals and birds depend largely on the number and position of the chlorine atoms (Tanabe et al. 1988; Borlakoglu et al. 1990a, 1990b, 1990c). Fish are much less capable of xenobiotic metabolism than are birds and mammals because their range of MFO enzymes is much more limited; most invertebrates have even less capacity for metabolism.

Regulatory Assessment of Chemicals

In light of the emerging awareness of previously unexpected chemical hazards such as reproductive, immunotoxic, and multigeneration effects, the data required for pesticide registration and for new chemicals notification should be re-evaluated.

The lack of an effective framework to collect data on chemical hazard contributed to the use and subsequent environmental effects of persistent chlorinated compounds such as DDT and PCBs. Based on that experience, a framework to assess chemicals was developed in the 1970s under legislation typified by the United States' Toxic Substances Control Act (TSCA 1976). There now exists a need to further improve assessment schemes to ensure adequate collection of the right type of data. For example, new industrial chemical use in the U.S. under the TSCA premanufacturing notice (PMN) process requires submission of all extant data on chemical structure and identity, production volume, byproducts, uses, environmental release, disposal practices, and human exposure (Zeeman, Nabholz, and Clements 1993). Test data on effects are not required but can be useful if they already exist. Given the new awareness of sublethal effects of chemicals on fish and wildlife, particularly the concern for endocrine effects, data requirements for industrial chemicals and pesticides should be expanded to include relevant mandatory fate-and-effects data (see Chapter 7).

Biomonitoring Programs for Wildlife

Biomonitoring of chemical effects on wildlife provides a critical check on the success of the regulation of chemical use and discharge to the environment. While laboratory methods for hazard assessment are relatively well developed, approaches for characterizing effects in natural ecosystems are much less so. The establishment of routine biomonitoring requires an improved capability to detect effects and to establish cause–effect relationships that will speed mitigation. A better integration of the measurement of chemical, biochemical, population, and ecosystem levels is required, as are sampling designs to apply the measurements.

Despite continued improvements in knowledge and the regulatory process, it is likely that effects of some chemicals on wildlife will not be predicted adequately by risk assessment techniques. Therefore, monitoring of effects and of chemical contamination is necessary to detect effects, to establish causes, and to direct mitigation in a timely manner. This monitoring should include both sensitive biochemical responses and impacts on populations and should focus on species that are top predators. The success of biomonitoring in detecting adverse effects in wildlife is well established. In countries such as Denmark, Canada, Finland, Germany, the Netherlands, Norway, Sweden, the United Kingdom, and the United States, national monitoring programs have been or are being implemented, but their effectiveness has yet to be proven.

Measures of population, community, and ecosystem structure and function are essential to a biomonitoring program (Snoo et al. 1994) because they integrate effects at lower levels of organization and are relevant to the overall condition of the ecosystem. However, these measures are currently difficult to interpret in terms of cause. To increase understanding of contaminant effects at the community and ecosystem levels, long-term monitoring to establish trends is essential, but cost dictates that a tiered approach be used to select, implement, and prioritize the various endpoints measured.

Apparent cause-and-effect field assessments of chemical hazards can also be confounded by habitat changes, nonchemical stressors, and natural gradients of factors that govern the structure and function of aquatic ecosystems. Studies by Gagnon et al. (1994, 1995) demonstrate that some effects on fish populations thought to be caused by pulp mill effluents were most likely responses to natural gradients of ecological factors in contaminated rivers. Conversely, important effects on fish population structure due to pulp mill effluents were partially obscured by nutrient enrichment of those same rivers (Gagnon et al. 1994, 1995). There is a need to identify cumulative effects of anthropogenic stressors and to distinguish such effects from variations in natural processes and effects of nonchemical stressors. However, experimental sampling designs for such assessments, and biomarkers of specific effects, are not always available for the ecosystem or chemical of interest. Each situation must be evaluated to ensure that 1) chemicals are responsible for the effects observed and 2) appropriate laboratory tests are used to verify these effects.

Despite abundant knowledge about effects of chemicals on wildlife species, establishing definitive cause–effect linkages between exposure of wildlife to chemicals and ecological effects remains extremely difficult. The requirement for "absolute proof" of environmental damage and the absence of test data can lead to unacceptable delays in regulatory action, as was the case for DDT and other persistent compounds. Adaptation of epidemiological methods to wildlife ecotoxicological research has recently improved the ability to infer cause-and-effect relationships between effects on wildlife and their exposure to specific chemicals (Fox 1991; Gilbertson and Schneider 1991, 1993; Mortensen et al. 1992; Olsson et al. 1992). These methods would be most useful if incorporated in biomonitoring programs so that regulatory authorities can make better use of this type of information in their decision-making. Regulatory systems must be sufficiently flexible to promote early action when evidence of environmental damage is observed rather than to wait for environmental evidence to become obvious and the effects serious.

Sampling design will include selection of appropriate species and sample sizes and inclusion of sites close to sources and in remote environments to assess global trends and transport. To detect the small changes that are expected in future trends of chemicals such as PCBs, it will be necessary to reduce within- and among-year variation by proper selection of sites and species, to use appropriate collection techniques and protocols, and to analyze consistent sample types (Bignert, Litzén et al. 1994; Bignert, Olsson et al. 1994). Tissue banks will be needed for retrospective analysis of both chemical and biochemical parameters when new chemical problems become evident (Olsson 1985; Odsjo 1993), but new methods must be calibrated to old to ensure the comparability of "old" and "new" data.

Summary of Improvements Needed for Hazard Assessments

This chapter has demonstrated how new concerns for hazards emerged as knowledge of subtle effects of chemicals on reproduction and immunocompetence improved. Studies in biochemistry, molecular biology, population genetics (Mortimer and Hughes 1991), and behavior (Peakall and Peakall 1973; Kubiak et al. 1989) have demonstrated significant sublethal effects of some chemicals. Increasingly, disciplines other than chemistry and classical toxicology are the sources of methods for evaluating sublethal effects in wildlife populations (Murk et al. 1994; van den Berg, Craane et al. 1994; Bosveld et al. 1995). These methods should be incorporated into hazard evaluations of new compounds before such compounds are released to commerce and the environment.

The following research needs were derived from the workshop discussions and the issues reviewed in this chapter. None relates specifically to chlorinated compounds; there are no compelling research needs exclusive to the hazard assessment of chlorinated compounds.

- To better characterize exposure, analytical methods are needed for small amounts of tissue and for compounds and metabolites associated with specific tissue receptors (DNA adducts);
- There is a need for hazard assessments to quickly incorporate new biochemical and analytical tools developed from an understanding of the mechanisms of toxic action.
- Interspecies variability in response must be evaluated to define the range of responses and the range of species sensitivities among wildlife species. Similarly, more information is needed on how exposure, reproductive effects, and transgenerational effects vary among species and critical life stages. Given that all species can never be tested under all conditions, information on variability is essential for estimating ecological risks.
- Methods are needed to assess chemicals with a spectrum of modes of action. For instance, appropriate biochemical assays could be used to screen individual chemicals and mixtures, both in pre-release evaluations and in field biomonitoring programs, to predict and confirm environmental harm. If these assays are to be used for risk assessment, their relationships to toxic effects within the species studied should be well established.
- Methods are required to distinguish between effects of chemicals and effects of more general stressors (e.g., temperature). Such distinctions will be facilitated by the continued development of chemical-specific biomarkers. However, the relative importance of other nonchemical stressors must also be measured in order to place chemical stressors in perspective.

- Knowledge of quantitative structure-activity relationships (QSARs) is needed to limit complex testing to compounds of greatest concern. QSARs should be developed for all mechanisms of toxic action.
- To understand cause–effect relationships, integrated laboratory and field experiments are needed to show how chemical exposure is related to biochemical and physiological effects, how physiological effects are related to whole-organism performance, and how whole-organism performance affects populations and ecosystems.
- Assessments in model ecosystems are needed to provide relevance to single-species laboratory testing and to improve extrapolation among species and to the field.
- Epidemiological methods must be refined for field assessments of chemical hazard. Specific requirements include sampling designs, experimental approaches, and methods to control for the effects of natural ecological gradients that cause uncertainty in measurements of exposure and effect.

In summary, there are a variety of laboratory bioassays and field monitoring tools for wildlife hazard or risk assessments. However, uncertainties about effects of chemicals on wildlife remain. Further laboratory studies are needed to identify the most relevant endpoints to characterize effects on natural populations, and further field studies are needed to detect responses not evident under laboratory conditions. Field assessment techniques cannot be standardized to the same extent as laboratory tests, nor can quality assurance always be verified in the same manner as for laboratory techniques. Linking laboratory test results to field evaluations requires a knowledge of critical tissue dose, or biologically effective dose. Nevertheless, specific case studies such as those on birds, mammals, and fish in Lake Ontario (see "Aromatic compounds," this chapter) demonstrate that the most likely causes of effects can be determined, provided enough effort is expended.

Appendix

Table 6-appx-1 Toxicity of OC and non-OC substances to fathead minnow (*Pimephales promelas*)[1]

Chemical class	Compound	MW	CAS #	Geo mean LC50 mg/L	# tests	% Cl	# Cls
Alkanes	1,2-Dichloroethane	98	107062	363	6	72	2
Alkanes	1,1-Dichhlorpropane	113	142289	23	1	63	2
Alkanes	1,2-Dichloropropane	113	78875	52	1	63	2
Alkanes	1,3-Dichloropropane	113	78875	280	1	63	2
Alkanes	Trichloromethane	119	67663	198	3	89	3
Alkanes	1,1,2-Trichloroethane	133	97005	69.9	10	80	3
Alkanes	1,1,1-Trichloroethane	133	71556	530	1	80	3
Alkanes	Tetrachloromethane	154	56235	35	1	92	4
Alkanes	1,1,1,2-Tetrachloroethane	168	630206	24	1	84	4
Alkanes	1,1,2,2-Tetrachloroethane	168	79345	41	5	84	4
Alkanes	1,1,2,2,2-Pentachloroethane	203	76017	11.7	5	88	5
Alkanes	Hexachloroethane	237	67721	0.006	1	90	6
Alkenes	1,1-Dichloroethylene	97	75354	135	2	73	2
Alkenes	Trans-1,2-dichloroethylene	97	156605	220	1	73	2
Alkenes	1,3-Dichloropropene	111	542756	6.2	1	64	2
Alkenes	1,1,2-Trichloroethylene	131	79016	51.2	7	81	3
Alkenes	Tetrachloroethene	166	127184	13.7	5	85	4
Alcohols	2-Chloroethanol	81	107073	0.003	1	44	1
Alcohols	2,2,2-Trichloroethanol	149	115208	148	1	71	3
Aro Gen	1-Chloronaphthalene	163	901311	160	1	22	1
Aro Gen	1,1-bis(p-Chlorophenyl)ethane	251	3547044	0.035	1	28	2
Benzenes	Chlorobenzene	113	108907	17	10	31	1
Benzenes	1,2-Dichlorobenzene	147	96501	0.042	2	48	2
Benzenes	1,4-Dichlorobenzene	147	106467	0.088	2	48	2
Benzenes	1,3-Dichlorobenzene	147	541731	2.07	7	48	2
Benzenes	1,2,4-Trichlorobenzene	181	120821	1.88	4	59	3
Benzenes	2,5-Dichlorobenzamide	190	2008584	858	1	37	2
Benzenes	1,2,3,5-Tetrachlorobenzene	216	634902	9.7	1	66	4
Benzenes	4-Chlorobenzenesulfonic acid, Na salt	216	5138909	7659	1	15	1
Benzenes	Pentachlorobenzene	250	308935	5.3	1	71	5
Nitrobenzenes	1-Chloro-2-nitrobenzene	158	88733	0.003	1	22	1
Nitrobenzenes	3-Chloronitrobenzene	158	121733	0.052	3	22	1
Nitrobenzenes	1-Chloro-4-nitrobenzene	158	100005	0.134	2	22	1
Nitrobenzenes	2-Chloro-6-nitrotoluene	172	83421	4.2	1	21	1
Nitrobenzenes	4-Chloro-2-nitrotoluene	172	89598	9.3	1	21	1
Nitrobenzenes	1,2-Dichloro-3-nitrobenzene	182	3209221	0.0005	1	37	2

Table 6-appx-1 *(continued)*

Chemical class	Compound	MW	CAS #	Geo mean LC50 mg/L	# tests	% Cl	# Cls
Nitrobenzenes	2,4,-Dichloro-1-nitrobenzene	102	511063	4.2	1	37	2
Nitrobenzenes	1,3-Dichloro-5-nitrobenzene	192	618322	7.5	1	37	2
Nitrobenzenes	1,4-Dichloro-2-nitrobenzene	192	89612	11	1	37	2
Nitrobenzenes	1-Chloro-2,4-dinitrobenzene	203	97007	0.8	1	17	1
Biphenyls	3-Chlorobiphenyl	189	2051618	0.43	1	19	1
Biphenyls	2-Chlorobiphenyl	189	2051607	0.71	1	19	1
Biphenyls	2,4,2',4'-Tetrachlorobiphenyl	292	2437798	0.03	1	49	4
Aniline	2-Chloroaniline	128	95512	0.015	2	28	1
Aniline	4-Chloroaniline	128	106478	0.05	1	28	1
Aniline	3-Chloroaniiine	128	108429	0.1	1	28	1
Aniline	2,4-Dichloroaniline	162	554007	0.596	2	44	2
Aniline	3,5-Dichloroaniline	162	626437	1.12	1	44	2
Aniline	3,4-Dichloroaniline	162	95761	1.13	7	44	2
Aniline	2,5-Dichloroaniline	162	95829	2.92	1	44	2
Aniline	2,3,4-Trichloroaniline	196	634673	0.73	1	54	3
Aniline	2,4,5-Trichloroaniline	196	636306	3.4	1	54	3
Aniline	2,3,4,5-Trichloroaniline	231	634833	0.64	1	61	4
Phenols	4-Chlorophenol	129	106489	4.89	4	27	1
Phenols	2-Chlorophenol	129	95578	5.24	3	27	1
Phenols	2-Methyl-4-chlorophenol	143	1570645	0.29	1	25	1
Phenols	3-Methyl-4-chlorophenol	143	59507	2	1	25	1
Phenols	2,4-Dichlorophenol	163	120832	0.26	3	43	2
Phenols	2,4-Dichloro-6-methylphenol	177	1570656	0.43	1	40	2
Phenols	3,4,5-Trichlorophenol	197	609198	0.68	1	54	3
Phenols	2,4,6-Trichlorophenol	187	88062	2.03	2	54	3
Phenols	2,4,5-Trichlorophenol	197	95954	2.7	1	54	3
Phenols	2-Methoxy-4,6,5-trichlorophenol	227	2668248	0.96	1	47	3
Phenols	2,3,4,6-Tetrachlorophenol	232	58902	0.29	1	61	4
Phenols	Tetrachlorophenol	232	25167833	0.406	1	61	4
Phenols	2,3,5,6-Tetrachlorophenol	232	935955	0.57	1	61	4
Phenols	Pentachlorophenol	266	87865	0.732	21	67	5
Phenols	Pentachlorophenol, Sodium salt	290	131522	0.4	39	61	5
Pesticides	Simazine	202	122349	94	1	18	1
Pesticides	Dichlorvos	221	62737	0.004	3	32	2
Pesticides	2,4-D	221	94757	83	6	32	2
Pesticides	Picloram	241	1918021	58.8	2	44	3
Pesticides	Trichlorfon	257	52586	0.117	2	41	3
Pesticides	Thiobencarb	258	28249776	1.2	1	14	1

Table 6-appx-1 (*continued*)

Chemical class	Compound	MW	CAS #	Geo mean LC50 mg/L	# tests	% Cl	# Cls
Pesticides	2,4-D Dimethylamine	286	2008391	184	1	27	2
Pesticides	Silvex	270	93721	140	1	39	3
Pesticides	Chloramp	281	2545600	212	1	38	3
Pesticides	Chloroxuron	291	1982474	2.95	1	12	1
Pesticides	Lindane	291	58899	0.899	3	73	6
Pesticides	Diflubenzuron	311	35367385	0.005	3	11	1
Pesticides	Chloramphenicol	323	56757	345	1	22	2
Pesticides	DDT	354	50293	0.003	2	50	5
Pesticides	Chlorfenvinphos	360	470906	0.043	2	30	3
Pesticides	Permethrin	391	52645531	0.002	7	18	2
Pesticides	Endosulfan	407	115297	0.283	17	52	6
NonCl	Benzenamine		62533	0.32	9		0
NonCl	Ethanol		64175	9509.92	3		0
NonCl	1-Propanol		71238	6415.87	7		0
NonCl	Benzene		71432	304.8	9		0
NonCl	1,1'-Biphenyl		92524	1.54	6		0
NonCl	Nitrobenzene		98053	31.48	3		0
NonCl	Phenol		108952	33.12	21		0
NonCl	Hexane		110543	2.5	1		0
NonCl	Cyclohexane		110827	36.88	5		0

[1] Source: AScI. 1994. Aqua/Info PC version of Aquire database. McLean VA: AScI Corporation.

Table 6-appx-2 Toxicity of OC and non-OC substances to water flea (*Daphnia magna*)[1]

Chemical class	Compound	MW	CAS #	Geo mean LC50 mg/L	# tests	% Cl	# Cls
Alkanes	1,2-Dichloroethane	98	107062	363	6	72	2
Alkanes	1,1-Dichhlorpropane	113	142289	23	1	63	2
Alkanes	1,2-Dichloropropane	113	78875	52	1	63	2
Alkanes	1,3-Dichloropropane	113	78875	280	1	63	2
Alkanes	Trichloromethane	119	67663	198	3	89	3
Alkanes	1,1,2-Trichloroethane	133	97005	69.9	10	80	3
Alkanes	1,1,1-Trichloroethane	133	71556	530	1	80	3
Alkanes	Tetrachloromethane	154	56235	35	1	92	4
Alkanes	1,1,1,2-Tetrachloroethane	168	630206	24	1	84	4
Alkanes	1,1,2,2-Tetrachloroethane	168	79345	41	5	84	4
Alkanes	1,1,2,2,2-Pentachloroethane	203	76017	11.7	5	88	5
Alkanes	Hexachloroethane	237	67721	0.006	1	90	6
Alkenes	1,1-Dichloroethylene	97	75354	135	2	73	2
Alkenes	Trans-1,2-dichloroethylene	97	156605	220	1	73	2
Alkenes	1,3-Dichloropropene	111	542756	6.2	1	64	2
Alkenes	1,1,2-Trichloroethylene	131	79016	51.2	7	81	3
Alkenes	Tetrachloroethene	166	127184	13.7	5	85	4
Alcohols	2-Chloroethanol	81	107073	0.003	1	44	1
Alcohols	2,2,2-Trichloroethanol	149	115208	148	1	71	3
Aro Gen	1-Chloronaphthalene	163	901311	160	1	22	1
Aro Gen	1,1-bis(p-Chlorophenyl)ethane	251	3547044	0.035	1	28	2
Benzenes	Chlorobenzene	113	108907	17	10	31	1
Benzenes	1,2-Dichlorobenzene	147	96501	0.042	2	48	2
Benzenes	1,4-Dichlorobenzene	147	106467	0.088	2	48	2
Benzenes	1,3-Dichlorobenzene	147	541731	2.07	7	48	2
Benzenes	1,2,4-Trichlorobenzene	181	120821	1.88	4	59	3
Benzenes	2,5-Dichlorobenzamide	190	2008584	858	1	37	2
Benzenes	1,2,3,5-Tetrachlorobenzene	216	634902	9.7	1	66	4
Benzenes	4-Chlorobenzenesulfonic acid, Na salt	216	5138909	7659	1	15	1
Benzenes	Pentachlorobenzene	250	308935	5.3	1	71	5
Nitrobenzenes	1-Chloro-2-nitrobenzene	158	88733	0.003	1	22	1
Nitrobenzenes	3-Chloronitrobenzene	158	121733	0.052	3	22	1
Nitrobenzenes	1-Chloro-4-nitrobenzene	158	100005	0.134	2	22	1
Nitrobenzenes	2-Chloro-6-nitrotoluene	172	83421	4.2	1	21	1
Nitrobenzenes	4-Chloro-2-nitrotoluene	172	89598	9.3	1	21	1

Table 6-appx-2 (*continued*)

Chemical class	Compound	MW	CAS #	Geo mean LC50 mg/L	# tests	% Cl	# Cls
Nitrobenzenes	1,2-Dichloro-3-nitrobenzene	182	3209221	0.0005	1	37	2
Nitrobenzenes	2,4,-Dichloro-1-nitrobenzene	102	511063	4.2	1	37	2
Nitrobenzenes	1,3-Dichloro-5-nitrobenzene	192	618322	7.5	1	37	2
Nitrobenzenes	1,4-Dichloro-2-nitrobenzene	192	89612	11	1	37	2
Nitrobenzenes	1-Chloro-2,4-dinitrobenzene	203	97007	0.8	1	17	1
Biphenyls	3-Chlorobiphenyl	189	2051618	0.43	1	19	1
Biphenyls	2-Chlorobiphenyl	189	2051607	0.71	1	19	1
Biphenyls	2,4,2',4'-Tetrachlorobiphenyl	292	2437798	0.03	1	49	4
Aniline	2-Chloroaniline	128	95512	0.015	2	28	1
Aniline	4-Chloroaniline	128	106478	0.05	1	28	1
Aniline	3-Chloroaniiine	128	108429	0.1	1	28	1
Aniline	2,4-Dichloroaniline	162	554007	0.596	2	44	2
Aniline	3,5-Dichloroaniline	162	626437	1.12	1	44	2
Aniline	3,4-Dichloroaniline	162	95761	1.13	7	44	2
Aniline	2,5-Dichloroaniline	162	95829	2.92	1	44	2
Aniline	2,3,4-Trichloroaniline	196	634673	0.73	1	54	3
Aniline	2,4,5-Trichloroaniline	196	636306	3.4	1	54	3
Aniline	2,3,4,5-Trichloroaniline	231	634833	0.64	1	61	4
Phenols	4-Chlorophenol	129	106489	4.89	4	27	1
Phenols	2-Chlorophenol	129	95578	5.24	3	27	1
Phenols	2-Methyl-4-chlorophenol	143	1570645	0.29	1	25	1
Phenols	3-Methyl-4-chlorophenol	143	59507	2	1	25	1
Phenols	2,4-Dichlorophenol	163	120832	0.26	3	43	2
Phenols	2,4-Dichloro-6-methylphenol	177	1570656	0.43	1	40	2
Phenols	3,4,5-Trichlorophenol	197	609198	0.68	1	54	3
Phenols	2,4,6-Trichlorophenol	187	88062	2.03	2	54	3
Phenols	2,4,5-Trichlorophenol	197	95954	2.7	1	54	3
Phenols	2-Methoxy-4,6,5-trichlorophenol	227	2668248	0.96	1	47	3
Phenols	2,3,4,6-Tetrachlorophenol	232	58902	0.29	1	61	4
Phenols	Tetrachlorophenol	232	25167833	0.406	1	61	4
Phenols	2,3,5,6-Tetrachlorophenol	232	935955	0.57	1	61	4
Phenols	Pentachlorophenol	266	87865	0.732	21	67	5
Phenols	Pentachlorophenol, Sodium salt	290	131522	0.4	39	61	5
Pesticides	Simazine	202	122349	94	1	18	1
Pesticides	Dichlorvos	221	62737	0.004	3	32	2
Pesticides	2,4-D	221	94757	83	6	32	2
Pesticides	Picloram	241	1918021	58.8	2	44	3

Table 6-appx-2 *(continued)*

Chemical class	Compound	MW	CAS #	Geo mean LC50 mg/L	# tests	% Cl	# Cls
Pesticides	Trichlorfon	257	52586	0.117	2	41	3
Pesticides	Thiobencarb	258	28249776	1.2	1	14	1
Pesticides	2,4-D Dimethylamine	286	2008391	184	1	27	2
Pesticides	Silvex	270	93721	140	1	39	3
Pesticides	Chloramp	281	2545600	212	1	38	3
Pesticides	Chloroxuron	291	1982474	2.95	1	12	1
Pesticides	Lindane	291	58899	0.899	3	73	6
Pesticides	Diflubenzuron	311	35367385	0.005	3	11	1
Pesticides	Chloramphenicol	323	56757	345	1	22	2
Pesticides	DDT	354	50293	0.003	2	50	5
Pesticides	Chlorfenvinphos	360	470906	0.043	2	30	3
Pesticides	Permethrin	391	52645531	0.002	7	18	2
Pesticides	Endosulfan	407	115297	0.283	17	52	6
NonCl	Benzenamine		62533	0.32	9		0
NonCl	Ethanol		64175	9509.92	3		0
NonCl	1-Propanol		71238	6415.87	7		0
NonCl	Benzene		71432	304.8	9		0
NonCl	1,1'-Biphenyl		92524	1.54	6		0
NonCl	Nitrobenzene		98053	31.48	3		0
NonCl	Phenol		108952	19.21	30		0
NonCl	Hexane		110543	3.87	1		0
NonCl	Cyclohexane		110827	3.78	1		0

[1] Source: AScI. 1994. Aqua/Info PC version of Aquire database. McLean VA: AScI Corporation.

Chapter 7

Ecotoxicological Risk Assessment of Organochlorine Chemicals

Risk assessment has evolved in the last decade from a limited focus on the role of single chemicals in human carcinogenesis toward a holistic assessment to protect ecosystems, including effects from exposure to chemical mixtures. The basic scientific procedures for ecological risk assessment have been developed and incorporated into regulatory frameworks in North America and Europe. Detailed technical guidelines for ecological risk assessment are currently being developed. These guidelines employ a tiered assessment process in which chemical, biological, and production volume or end-use criteria trigger data requirements and related assessment tiers. Ecological risk assessment is broadly applicable and capable of estimating risk from both chlorinated and nonchlorinated chemicals.

A significant challenge for ecological risk assessment is to address the complex chemical and biological behavior associated with persistent and bioaccumulative chemicals. The data requirements are large, intensive, and detailed. Progress in understanding the behaviors of these types of chemicals has been rapid in the last decade. The recent models for polychlorinated biphenyls (PCBs) in Green Bay and Lake Michigan and for tetrachlorodibenzo-*p*-dioxin (TCDD) and related chemicals in Lake Ontario indicate that persistent and bioaccumulative chemicals can be successfully addressed by ecological risk assessment.

Despite significant progress, further improvements in the capabilities and accuracy of ecological risk assessment should be made. First, uncertainties in risk assessment are generally due to inadequate and imperfect data. A minimum chemical and biological dataset for individual chemicals has now been incorporated in a European risk assessment framework. The requirements and the availability of this dataset should be harmonized internationally. Second, ecological risk assessments should address endpoints beyond survival, growth, and reproduction, such as neurotoxicity, immunotoxicity, and the potential for disruption of hormonal and developmental processes. Third, ecological risk assessment should incorporate the potential interaction of chemical and nonchemical stressors.

Chapter Preview

Risk Assessment

Risk assessment is a process that involves comprehensive review and integration of 3 factors that interact to determine risk: 1) the toxicity (effects) of a chemical substance or mixture, 2) the intensity of exposure to the substance, and 3) the probability of exposure to the substance. Risk assessment considers both the potential effects of chemical substances or mixtures and the data gaps or uncertainties in a transparent manner, so that risk managers may make clear and consistent decisions about human and environmental safety. Risk assessment applications may be classified as retrospective (historical), current, or prospective (future) evaluations. Uncertainties in risk assessment result from a lack of knowledge regarding potential effects, imprecise measurements of effects or exposure, and the inherent variability in effects at each ecological level of organization, from individual organisms to whole ecosystems. Most uncertainties are inversely proportional to the richness and quality of data describing the chemistry of the substance, the toxic effects, the exposure conditions, and the ecosystem conditions.

The public usually is not familiar with risk assessment or risk management concepts. Often, the public mistakenly believes that zero risk is the goal of risk managers, whereas in reality a state of zero risk is never achievable. Further, risk assessors often have difficulty communicating the uncertainties in risk assessments. This makes risk communication about chlorinated compounds difficult for scientists, regulators, legislators, and nongovernmental organizations.

As a result of the technical issues and public perceptions, doubts have been expressed about the capabilities of risk assessment to deal with organochlorines (OCs), especially persistent and bioaccumulative chemicals. For example, the International Joint Commission (IJC 1994) has concluded "that there is no acceptable assimilative capacity for persistent, bioaccumulative toxic substances." Therefore, the IJC concluded that risk assessment is not directly relevant to the basic commitments of Canada and the United States to "virtually eliminate" persistent toxic substances from the Great Lakes ecosystem, pursuant to the Great Lakes Water Quality Agreement (IJC 1994). Many scientists, the public, and nongovernmental organizations concerned with environmental protection concur that there are enough data on some of the persistent and toxic substances to show that they do present a risk to the environment and that it is now time for action. They do not believe that further risk assessments are needed and prefer the alternative of phasing-out the production, use, and release of such chemicals. While this is clearly appropriate for some of the persistent and bioaccumulative substances, there are others for which there is less knowledge and for which risk assessment is essential to deciding the appropriate strategy for mitigation.

This chapter includes a brief history of the development of risk assessment, the structure of the risk assessment procedure, and the current status of environmental risk assessment for OC chemicals, including those chemicals that are persistent and bioaccumulative, and it suggests improvements for the risk assessment process. The judgment of most scientists participating in the Workshop on Environmental Risk Assessment for Organochlorine Compounds was that

1) the risk assessment procedure is scientifically sound,
2) the risk assessment procedure works well when adequate data are available, and
3) risk assessment can address persistent and bioaccumulative chemicals, chlorinated and nonchlorinated, when the resources are applied to obtain the necessary information.

Brief History of Risk Assessment

Ecological risk assessment has an established scientific and regulatory framework in North America and Europe. As the regulatory framework and the risk assessment procedures have been developed, OC chemicals, including several persistent and bioaccumulative chemicals, have been included. This has resulted in a variety of strategies to regulate and control these substances, depending upon their individual characteristics and properties.

Bulk production and release of persistent and bioaccumulating OCs such as dichlorodiphenyltrichloroethane (DDT), polychlorinated biphenyls (PCBs), and tetrachlorodibenzo-*p*-dioxin (TCDD) in Agent Orange in the 1940s, 1950s, and 1960s preceded the scientific understanding of environmental fate, transport, and bioaccumulation, the development and implementation of risk assessment procedures, and the construction of a regulatory framework to evaluate chemicals prior to their production. With regulations in place since the mid 1970s, a series of important steps has already been taken. Most industrialized countries have ceased deliberate production of these particular chlorinated chemicals and have made steady progress in disposing of stocks and remediating contaminated sites. The inadvertent production of these chemicals has also been reduced. As a result, concentrations in the environment have decreased, the body burdens of biota are generally decreasing, and many previously affected populations, such as the bald eagle, are recovering. However, production and use of some persistent and bioaccumulative chemicals, such as DDT, continue in a number of countries because of public health benefits, agricultural needs, and lower economic costs, compared to the alternatives.

Persistent, bioaccumulative chemicals such as DDT, PCBs, and TCDD have been the catalysts for the development of ecological risk assessment over the past 30 years. These substances and chemicals with related properties have prompted the study of numerous important environmental processes, such as atmospheric transport, partitioning among environmental compartments, bioaccumulation and biomagnification, trophic food-web interactions, and reproductive and other toxicological endpoints. As a result, the basic environmental properties of these chemicals are relatively well characterized in multiple environmental settings and conditions.

United States

The history and development of risk assessment are intertwined with the concern for chemical carcinogenicity that began in the late 1950s and 1960s (Armitage and Doll 1957). In the late 1970s and 1980s, linear extrapolation of rodent tumor dose-responses determined in life-time bioassays at high doses was used to estimate the cancer risks of the low chronic doses encountered in the environment (Crump et al. 1977; Crump and Watson 1979; Moolgavkar and Venzon 1979; Moolgavkar and Knudson 1981; U.S. Environmental Protection Agency [USEPA] 1986). There is now detailed toxicological information for over 400 chemicals and 100 OC chemicals, such as pesticides and solvents that have been primary candidates for bioassay testing in the National Toxicology Program (Selkirk and Soward 1993). These data have been incorporated with additional use-and-exposure data to estimate human risk. The numerical value that became accepted to represent insignificant risk was that of a theoretical occurrence of 1 additional cancer case in 1 million people, or a risk of 10^{-6}. This process became known as "quantitative risk assessment."

In the mid to late 1970s, the need for consistent use and application of human risk assessment fostered the active cooperation of U.S. federal agencies to develop a common and robust procedure. Ultimately, this was achieved with the National Academy of Sciences (NAS) report, "Risk Assessment in the Federal Government: Managing the Process" (NAS 1983). The steps and roles of conducting risk assessment were clearly defined and clearly separated from risk management. The human health risk assessment steps are briefly described in "The Risk Assessment Process" box below. Since the 1983 report, efforts to identify and remedy shortcomings in the NAS work have been undertaken, both inside the government (the Committee on Risk Assessment Methodology) and outside (the Society for Risk Analysis).

The Risk Assessment Process

Risk assessment is a process leading to an estimation of the probability that an adverse effect will result from exposure to a substance, usually within a defined set of circumstances. Risk assessment is distinct from risk management. The latter uses the results of a risk assessment to develop policies to identify and reduce risks. Risk assessment in its classical human-health context consists of 4 separate steps, each with a defined purpose: 1) hazard identification, 2) dose–response assessment, 3) exposure assessment, and 4) risk characterization.

1) *Hazard identification* identifies the link between exposure to a chemical and specific adverse effects. Several endpoints may be involved with a single compound. Epidemiologic, field, animal toxicity, and in-vitro toxicity data are all used to establish the link. Findings derived from quantitative structure-activity relationships (QSARs) and computer modeling sometimes are used.

2) *Dose–response assessment* describes the quantitative relationship between exposure to various concentrations of a compound and the nature, incidence, or severity of a toxic effect. Data are usually drawn from experimental animal studies, but field data may also be used. Separate dose–response assessments are needed for each separate endpoint of toxicity. Various extrapolations and assumptions may be used to address significant gaps in the knowledge base, which can be contentious in the risk assessment process.

3) *Exposure assessment* identifies and characterizes the populations exposed and also determines the magnitude and duration of the exposure. This step is critical because without sufficient exposure, no effect will occur. Exposure assessment relies on environmental monitoring or modeling.

4) *Risk characterization* integrates the outcome of the first 3 steps to estimate the likelihood that an adverse health effect will occur, the frequency and severity of the outcome, the expected population in which the effect will occur, and the specified conditions (scenario) under which the effect will occur. The latter is especially important for risk management decisions.

Parallel to the development of risk assessment, the need to apply these techniques in legislative and regulatory situations outside the U.S. Food and Drug Administration began growing in the 1970s. New regulatory agencies, e.g., the USEPA, the Occupational Health and Safety Administration (OSHA), and the Consumer Products Safety Commission, were established by such legislation as the Clean Air Act Amendments (CAAA 1955); the Water Pollution Control Act (1948); the Toxic Substances Control Act (TSCA 1976); the Resource Conservation and Recovery Act (1976); the Food Quality Protection Act (FQPA 1996); and others. A notable feature of the Federal Insecticide, Fungicide and Rodenticide Act (FIFRA 1947) is the defined, categorical requirement for extensive test procedures for pesticides (over 120 individual studies, including toxicology, dietary exposure, worker exposure, environmental fate, and ecotoxicology). FIFRA mandates a periodic review and updating of these initial studies for chemicals registered prior to the FIFRA legislation. This permits continuous development of a substantial scientific database to oversee the use of these chemicals. In contrast, TSCA does not specifically require the performance of particular procedures or the submission of certain risk-oriented test data prior to manufacture of a new chemical. TSCA does require a review resulting in the conclusion that a chemical does not present an unreasonable risk, allowing a flexible approach by regulators to the characteristics of a specific chemical. As a result, the USEPA has latitude to evaluate submitted chemicals and to request data and specific testing of a chemical's individual properties. TSCA calls for a periodic review, termed the "Inventory Use Review" (IUR) to monitor changes in production volumes and use patterns that could trigger risks different from those posed in the original filing. In addition, TSCA's Section 4 provides for regulatory procedures to demand environmental data and testing for chemicals manufactured and used in commerce prior to the TSCA legislation. Section 4 is triggered by recommendations from the Interagency Testing Committee (ITC) to the USEPA for specific tests of certain existing chemicals. Walker (1993a) has summarized the existing ITC chemical review process as of 1992, noting that 38,000 chemicals had been reviewed up to that time. Walker also describes the coordination of the ITC testing recommendations with other U.S. and international agencies and the similarity of the list to other international recommendations, especially for persistent, bioaccumulative, and toxic chemicals. Historically, the need for further data to assess ecological effects and the need to assess bioconcentration or bioaccumulation for greater-volume chemicals have been primary reasons for the ITC to request additional data on chemicals (Walker 1993a, 1993b). Requests have been filed for several OCs, including chlorinated paraffins, 2-chlorotoluene, trichlorobenzenes, and dichloropropene (Walker 1993b).

The assessment data used by the USEPA under TSCA are derived from field and laboratory measurements of chemical properties, production volumes, tiered testing of aquatic and terrestrial toxicity, food chains, contamination, QSARs, bioconcentration, and ecological effects. These procedures recently have been described by USEPA guidance and *Federal Register* publications referenced therein (Nabholz et al. 1993; Smrchek et al. 1993; Zeeman and Gilford 1993). It should also be noted that the TSCA legislation and its precursors were the basis for prohibiting the manufacture of certain persistent and bioaccumulative OC chemicals, such as PCBs, DDT, mirex, and dieldrin in the 1970s and toxaphene in 1983.

Ecological risk assessment is fundamentally more complex than human risk assessment because it includes multiple species and characteristics unique to individuals, populations, communities, and ecosystems. A less-mature scientific discipline than human health risk assessment, ecological risk assessment is still developing and evolving. Ecological risk assessment faces a fundamental question: What should be protected? The diversity of species and the multiple possible toxicological endpoints make it impossible to test all potential effects, but ecological risk assessment has begun by protecting populations, not individuals, and by addressing integrative endpoints such as survival, growth, and reproduction. Ecological risk assessment also may have several diverse purposes: the assessment of old chemicals, new chemicals, manufacturing processes, physical stressors, in-place problems, and remedial options at contaminated sites.

The USEPA published its basic risk assessment procedure in "Framework for Ecological Risk Assessment" (USEPA 1992b) and in the proposed refinements (USEPA 1996) (Figure 7-1) after a series of workshops in the early 1990s. Within the framework, risk analysis consists of 2 fundamental steps: effects characterization and exposure characterization. Effects characterization quantifies the potential for adverse effects and integrates hazard identification and dose–response assessments, as conducted for human health risk assessment. Exposure characterization involves the determination of a chemical's emissions, paths and rates of transport or movement in the environment, transformation and degradation, and accumulation and metabolism by biota (food chains). The framework notes the iterative nature of risk assessment, which occurs as additional data lead to a refined estimate.

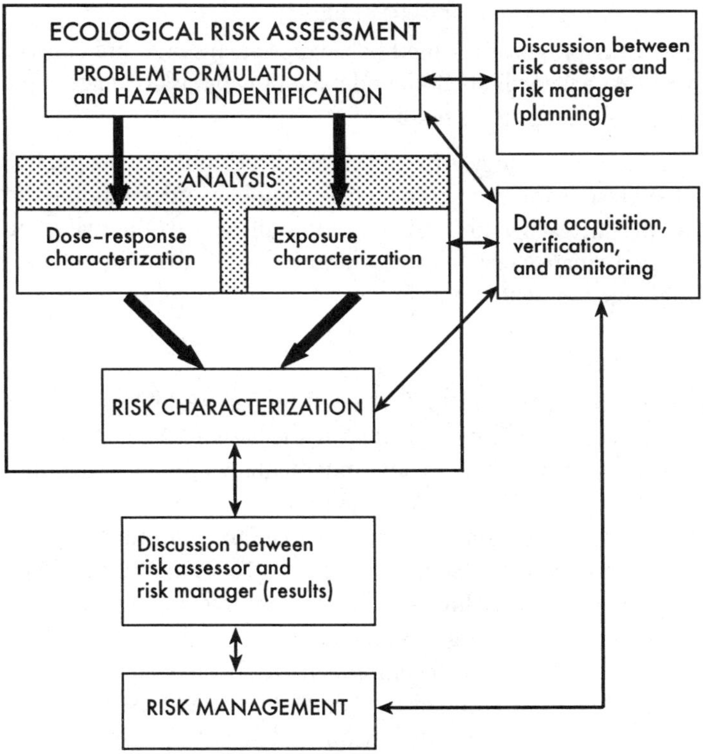

Figure 7-1 Framework for USEPA risk assessment procedure

The USEPA has begun to prioritize its efforts based on the development of relative risk assessment procedures with "Unfinished Business: A Comparative Assessment of Environmental Problems" (USEPA 1987) and "Reducing Risks: Setting Priorities and Strategies for Environmental Protection" (USEPA 1990b). In "Reducing Risks," USEPA managers ranked high, medium, and low risks. Habitat alteration, biodiversity loss, ozone depletion, and climate change were ranked as higher overall risks; acid deposition, water quality problems (including toxic chemicals, herbicides, and pesticides), and airborne toxic industrial chemicals were ranked as medium overall risks; groundwater pollution, oil spills, and radionuclides were ranked as low risks (USEPA 1987).

The USEPA's risk assessment framework recognizes that society benefits from risk assessments only when they are effectively communicated to risk managers, who then use risk information as a basis for policy decisions. Although the risk assessment process is subject to the possibility of errors resulting from incorrect models and assumptions and lack of knowledge, the public must trust that the scientific process is the best procedure for evaluating risks to complex ecological systems. While risk assessment is conducted prior to risk management actions, there is a distinct need to

- ensure that the problem formulation supports an understanding of the risks of concern to risk managers and the public, and
- develop descriptions of the methods, data, conclusions, and uncertainty analyses that will facilitate both understanding by risk managers and a communication of findings to the public.

If risk assessment methodology is to gain greater acceptance by the public, there should be a long-term commitment to increasing public understanding through education, public outreach, and consistent communication. Otherwise, the public may not be able to participate effectively in the prioritization of environmental risks and environmental improvement strategies.

Canada and Europe

In Canada, the development and application of human health risk assessment also preceded ecological risk assessment, including the basic scientific process to achieve a quantitative estimate of risk. This is not fundamentally different from the U.S. experience, except that in Canada a safety margin approach is used for the cancer endpoint. For ecological risk assessment, the Canadian Environmental Assessment Act (CEAA 1992) and the Canadian Environmental Protection Act (CEPA 1997) sets out a stepwise risk assessment procedure, yielding a screening or comprehensive report, a review by an expert panel, and the design and implementation of a follow-up program. Public comment is required for any comprehensive risk assessment report, and provisions are made for mediation of legal disputes. Canada also performs environmental impact assessments that include scoping, forecasting, risk evaluation, and follow-up steps. The latter includes monitoring to validate the outcome of any mitigation steps.

In Europe, development of human health risk assessment again preceded ecological efforts. In most European procedures, a quantitative estimate of risk is not used; rather a "weight-of-evidence" approach is employed. European models also use a safety margin approach for a cancer endpoint. For ecological risk assessments, recent European Union (EU 1993a, 1993b) legislation requires extensive data and testing and gives specific detailed technical guidance. This testing is based on

1) a direct linkage between the chemical tonnage produced or to be produced (amounts greater than 10 metric tons);
2) a tiered approach to chemicals testing, based upon chemical and biological characteristics and production volume triggers; and
3) inclusion of several trophic levels in aquatic and terrestrial environments (a minimum of 3 in the base or minimum set of data).

Finally, a quantitative model is generated whereby the predicted environmental concentration in a given medium is divided by the lowest-observed-adverse-effect level (LOAEL) in the most sensitive species in that medium or the predicted no-effect concentration (EU 1993c, 1994). This ratio, which must be less than 1, is considered a "margin of safety." The size of this ratio is a determinant for additional tests, with an emphasis on chronic testing. The potential for bioaccumulation includes the consideration of body burdens in top predators, particularly fish-eating predators. In addition, there are provisions whereby the data can be refined with environmental monitoring, both for chemical residues and for biological effects. These procedures target the protection of a population of a species, not the individual within that population.

Overview of OCs and Risk Assessment

Human and ecological risk assessment procedures have been applied to OC chemicals on numerous occasions. The subset of persistent and bioaccumulative chemicals presents special challenges to risk assessment because of their complexity and large data requirements. For these persistent and bioaccumulative chemicals, recent successful risk assessments include PCBs in the case of Green Bay and Lake Michigan and the mixture of polychlorinated dibenzo-*p*-dioxins (PCDDs), polychlorinated dibenzofurans (PCDFs), and some PCBs expressed as TCDD equivalents in Lake Ontario lake trout. These examples demonstrate increasing capability of risk assessment procedures to address the full spectrum of OC chemicals.

Organochlorines such as PCBs, DDT, dieldrin, and PCDD/PCDFs have been one focus for the development and application of risk assessment since the 1970s (Delzell et al. 1994). Other OC chemicals have been subjects for human cancer risk assessments, beginning with industrial feedstocks such as vinyl chloride, solvents such as methylene chloride, pesticides such as DDT, and the byproducts of drinking water disinfection such as chloroform. The USEPA has produced national water quality criteria for over 60 OCs, including major solvents such as methylene chloride, tetrachloroethylene, and chloroform; intermediates such as 1,2-dichloroethane, vinyl chloride, and chlorobenzenes; and persistent, bioaccumulative substances such as DDT, dichlorodiphenyldichloroethylene (DDE), TCDD, and PCBs (USEPA 1992a). According to the FIFRA statute and its regulatory promulgation, all new pesticides must be thoroughly tested for human and ecological safety and submitted to the USEPA for assessment, as part of the registration process prior to marketing. In addition, use of the linear low-dose extrapolation for human cancer risk assessment led to the development of detailed metabolic and physiological data for OC solvents. In the 1980s, this fundamental understanding provided the foundation for the application of pharmacokinetics to refine risk assessment, including chlorinated chemicals (Andersen et al. 1987; Reitz et al. 1988; Corely et al. 1990; Nichols et al. 1990; Connolly and Anderson 1991). There have been numerous site-specific risk assessments for OC chemicals, usually involving operating plant sites; siting of disposal facilities; remediation of groundwater, sediments, waste sites, and soils; and fisheries advisories. As a result, considerable environmental data have been generated for OCs under a variety of environmental settings. These data have led, in turn, to a growing understanding of the diverse properties and behavior of these chemicals and to models to predict their individual behavior. In an iterative pattern, this allows more refined monitoring, test designs, and models.

Ecological risk assessments have ambitious goals in seeking to estimate which species, which stressors (biological, chemical, or physical), and which life stages should be used to determine no-observed-adverse-effect levels (NOAELs) suitable for the protection of populations, community structure, and ecosystem functions. The ecological risk assessment of persistent, bioaccumulative chemicals, including certain OCs, presents special technical challenges. These challenges include

- bioaccumulation from water concentrations at or below the analytical detection limit, resulting in detectable amounts in some biota through the food chain;
- accumulation in long-lived species that are sometimes distant from the point of entry into the environment; and
- occurrence of sediment and body burdens exceeding the NOAEL, for years and even decades after production and direct discharges have ceased.

Despite these challenges, successful exposure and risk assessments have been achieved for persistent chemicals, such as PCBs in the Green Bay ecosystem, and the mixture of PCDDs, PCDFs, and some PCBs expressed as TCDD equivalents in the Lake Ontario ecosystem.

Case studies on risk assessment of organochlorine chemicals

Polychlorinated biphenyls in Green Bay

Polychlorinated biphenyls are a complex mixture of 209 structurally similar chemical congeners, but each congener has different chemical and biological properties that depend on the number and arrangement of chlorine atoms. For example, the log K_{OW} of PCBs, which predicts the tendency to bioaccumulate, ranges from 4.3 to > 8. Volatilization, water solubility, sediment partitioning, microbial degradability, metabolism by vertebrates, and toxicological properties differ, sometimes significantly, among congeners (see Chapters 2 through 6). Environmental persistence ranges from a few weeks to many years as these materials partition to sediments. These environmental data for PCBs have been summarized in Mackay et al. 1992a. The risk assessment of PCBs is complex. PCBs found in sediments can become available to organisms with biomagnification up the food chain. Not only must the partitioning of each congener from sediment organic carbon be considered, but other factors must be evaluated, such as the physical movement of sediment, as affected by sediment particle size and the flow velocity of the overlying water, and the amount and the nature of colloidal organic matter in the water column. These interactions all influence the transport of PCBs and the movement of PCBs through aquatic food webs. Biological characteristics of organisms add further to this complexity. Predator–prey relationships within food webs may change with season and life stage of each organism. The rate at which chemicals are absorbed in the gut, held in the body in association with lipids, and metabolized and excreted differ among species. Finally, the endpoints affected by PCBs, such as survival and reproduction, and their relationship to concentrations of PCBs in tissues must be understood. Thus, complex mixtures such as PCBs exemplify the significant amounts of data required for risk assessment.

The most comprehensive effort to date to assess ecosystem exposure to persistent and bioaccumulative chemicals (i.e., PCBs) has been in the Green Bay system of Lake Michigan in Wisconsin. PCBs in the sediments of the Fox River at the head of Green Bay are a substantial reservoir and nonpoint source. Modeling distributions of PCBs in Green Bay indicates that the use of basic chemical properties of PCB congeners, such as K_{OW} and K_{OC}, and a fundamental understanding of organic carbon dynamics in the system are critical parameters for risk assessment (DePinto et al. 1994). PCBs are transported into the Green Bay system primarily through adsorption onto particulates, and they enter the food web by partitioning to water, absorption to plankton, and passage through fish to fish-eating birds. Reproductive and developmental effects were associated with PCB exposure in fish-eating birds in the 1980s and included a 0.5 to 1% rate of bill defects in hatched double-crested cormorant chicks (Fox et al. 1991), significant rates of reproductive failure, and major deformities in Forster's terns. These effects have decreased since the late 1980s, as the concentrations of PCBs in birds have decreased with time (Harris et al. 1993).

A major PCB mass-balance study in the Green Bay ecosystem has recently been completed (see Chapter 6). It addressed the physical movement of PCBs in the river and bay among sediment, suspended sediments, and the water column; the food-chain transfer of individual PCB congeners; the exposure of various organisms to individual congeners; the potential for remediation efforts; and modeling of future behavior and effects of PCBs. Importantly, the chemical-specific behavior of 6 groups of PCB congeners across a range of persistence and bioaccumulation values was successfully modeled. The result is a detailed, working exposure model for biota in Green Bay that predicts a slow decrease in body burdens, the periodic but transient increases in PCBs in biota due to sediment disturbance by floods, and the difficulty in removing contaminated sediments without causing additional exposure of the ecosystem (Bierman et al. 1992; Great Lakes National Program Office [GLNPO] 1992; Endicott, Richardson et al. 1992; DePinto et al. 1993, 1994; Connolly and Parkerton 1994; Steuer et al. 1994; Velleux and Endicott 1994). Predictions from this model and the knowledge of chemical and biological characteristics, especially the relationship between PCB exposures and concentrations of PCBs in fish tissues, can be applied in risk assessments to protect sensitive species at the apex of the food chain in larger ecosystems such as the Great Lakes (Ludwig, Giesy et al. 1993).

PCDDs, PCDFs, and TCDD-like PCBs in Lake Ontario

Ecological risk assessment must apply the results of laboratory toxicity tests and NOAELs to the same or to other species under field conditions. For the Great Lakes lake trout (*Salvelinus namaycush*), laboratory NOAELs for 2,3,7,8-TCDD for larval mortality at swim-up have been established using water transfer to eggs (Spitsbergen et al. 1991), egg injection (Walker et al. 1991), and material exposure and transfer to eggs (Walker et al. 1994). In Lake Ontario, lake trout were beset by major stressors in the 1930s; commercial over-fishing and sea lamprey predation led to a rapid population decline. By the 1940s, chemical production introduced substantial quantities of TCDD and other chemicals to Lake Ontario. Lake trout disappeared from the lake by 1960. Efforts to restock Lake Ontario began in the 1970s, but there was no evidence for successful reproduction. The onset of some reproduction in the late 1980s gives hope for the restoration of natural lake trout recruitment.

Major studies have described the fate and distribution of TCDD in the Lake Ontario ecosystem (Endicott and Cook 1994) and the effects of TCDD-like chemicals (PCDDs, PCDFs, and certain PCBs) on lake trout (Cook 1994). Knowledge of the sediment, water column, and food chain properties of a persistent, bioaccumulative OC were again successfully integrated into a model. A key finding was that measuring TCDD distribution between sediments and fish and in the food chain was sufficient to predict lake trout body burdens; the technical problems of measuring TCDD and related chemicals in the water column were no longer an issue. This finding contradicted assertions that a lack of sufficiently low analytical detection limits for water prevented accurate risk assessment. More importantly, the body burden of TCDD and related chemicals in lake trout was modeled for retrospective, current, and future conditions to improve the information available to risk managers. The model was consistent with sediment and tissue concentrations of lake trout over time and was also consistent with concentrations in herring gulls and their eggs. The model output was consistent with field observations that TCDD levels were sufficient in the 1970s to prevent successful reproduction of lake trout. TCDD levels in lake trout have now declined to levels where successful lake trout reproduction should not be prevented, but survival of the young may still be at risk (Cook 1994). The gradual improvement in conditions for lake trout recruitment in Lake Ontario and the continual decrease in TCDD-like chemical concentrations suggest that a threshold for survival of all life stages of lake trout is near or has been exceeded. These findings suggest that an assimilative capacity exists for chemicals such as TCDD.

Efforts to improve risk assessments of persistent, bioaccumulative OCs continues. In the Great Lakes, investigators are attempting to estimate the potential threshold levels of PCBs in fish-eating birds (Harris et al. 1993; Giesy, Bowerman et al. 1994; Giesy, Ludwig et al. 1994; Giesy, Verbrugge et al. 1994). These studies have recognized the plausible interaction of several chemicals based on a shared mechanism of toxic action: aryl hydrocarbon receptor (AhR)-mediated toxicity of PCDDs, PCDFs, and coplanar PCBs (USEPA 1989; Cook 1994). These studies represent the growing effort to address chemical mixtures that have common mechanisms of action.

In summary, ecological risk assessment procedures are applicable to the complete spectrum of OC chemicals, from volatile, labile chemicals such as vinyl chloride to persistent, bioaccumulative chemicals such as PCBs and TCDD, when adequate data are available.

Further Development and Improvement of Ecological Risk Assessment

Despite the progress in developing ecological risk assessment methods, further improvements in risk assessments are needed. Risk assessment procedures do not stand alone but rely on the availability of accurate scientific data to estimate risks. Risk assessment must cover all endpoints that threaten species, and current risk assessments do not address all relevant endpoints. Therefore, substantial opportunities for improving risk assessment processes include 1) the improved generation and collection of data on chemical behavior and toxicity, 2) the inclusion of additional important endpoints, and 3) the improvement of toxicological testing procedures. One step in this data improvement process is the development of minimum or base data requirements for all chemicals. Among the relevant endpoints to consider for inclusion are critical developmental processes. The toxicological dataset can be improved with protocols for chronic exposures to better address noncancer endpoints such as reproduction and development.

The success of some current ecological risk assessment procedures does not preclude the need to improve the quantity and quality of data available for use in risk assessment. There is a continual need to improve the methodology by which the data are collected and to explore the need to assess additional effects or endpoints.

Uncertainty in risk assessment

The uncertainty of a risk assessment is directly related to the quality and completeness of available data, and minimum datasets should be generated for all chemicals. Ecological risk assessments of fish and wildlife rely on sufficient data from field and laboratory studies and on chemical properties, environmental fate and transport, exposure, and biological effects. For new chemicals, field data are not available. The current data requirements to market a pesticide can easily exceed $20 million (U.S.) and take over 6 years to complete, primarily because pesticides are designed to have effects on pest organisms and are released deliberately to the environment in large quantities. The entire process for pesticides, with its significant use of resources, is not necessarily appropriate and cannot be expected for every chemical. There is a common misperception that all chemicals on the TSCA inventory are in commercial use and are produced in significant quantities. Actually, of the more than 70,000 chemicals on the TSCA inventory, the most recent IUR survey shows that about 14,000 are produced in quantities exceeding 5000 kg/y. For OCs, 821 chemicals are produced in the U.S. in quantities over 5 metric tons/y; production of 354 is less than 45 metric tons, 223 range from 45 to 450 metric tons, 141 range from 450 to 4500 metric tons, 71 range from 4500 to 45,000 metric tons, and 32 exceed 45,000 metric tons (Figure 7-2). Thus, only a limited number are produced in very large quantities.

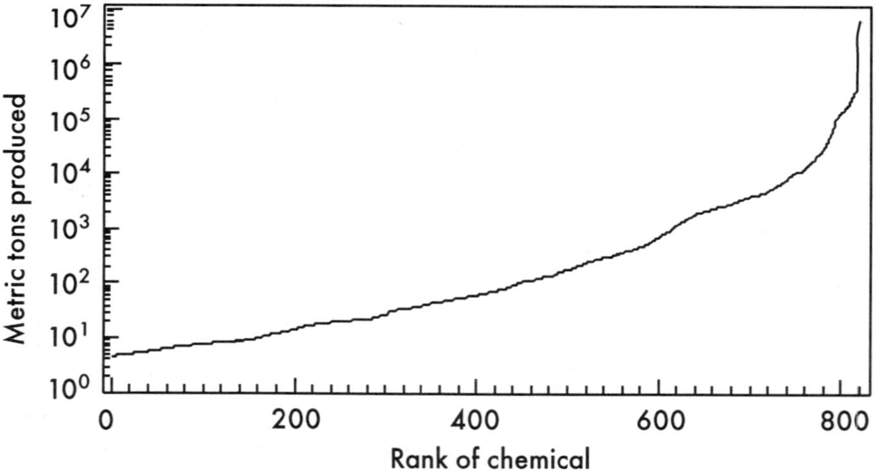

Figure 7-2 Current TSCA-reported production volumes of OCs in the U.S., according to USEPA (1990a). A total of 821 individual chemicals are ranked from smallest to greatest production.

A tiered approach, which generates data for risk assessment in sequential steps, is the paradigm that is practiced in both North America and Europe, especially for evaluation of new chemicals. In the initial tier, the objective is to generate information on basic chemical and toxicological properties as well as information on the amount of material produced and how it will be used. These data will allow estimates of environmental fate and transport, characterize the potential for toxicity, and predict the magnitude of concentrations in environmental compartments. Only those chemicals that 1) trigger either chemical or biological criteria, 2) have significant production volumes, or 3) have a greater probability of release to the environment would require additional testing steps. This proposal is a balance between the current TSCA and the present EU requirements for extensive data to make chemical-specific evaluations (EU 1993a, 1993b). Methods currently exist for tiered assessment of chemical risks (Cairns et al. 1978; CEAA 1992; EU 1993c, 1994; Nabholz et al. 1993; National Research Council [NRC] 1993; Smrchek et al. 1993; Suter 1993; Zeeman and Gilford 1993; USEPA 1992b, 1996; CEPA 1997).

The base set of information builds a framework for further assessments. Information on production volumes and intended uses and applications allows an understanding of amounts and conditions of release into the environment. Basic chemical properties, including molecular formula and weight, aqueous solubility, vapor pressure, K_{OW}, Henry's Law constant (HLC), and K_{OC}, provide a reasonable estimate of environmental partitioning and an indication of which fate properties to consider. QSAR estimates of hydroxyl radical reactions, hydrolysis, and photolysis are reasonable to estimate the fate of most chemicals. For effects, basic toxicity protocols for mammalian, fish, invertebrate, and algal or plant species are reasonable expectations for large-volume chemicals and, where QSAR estimates cannot be performed, also for smaller-volume chemicals (< 1000 metric tons/y). This information, when assembled, has a second benefit: the more comprehensive and accurate the available data on chemical and biological properties of numerous chemicals, the better the capability to perform refined QSAR analyses on new compounds. The minimum dataset can be complemented by new techniques in QSAR that have been developed to estimate acute toxicity and to predict chronic effects (Mayer et al. 1994; Sun et al. 1994; Lee et al. 1995) as well as chemical fate (Kaiser 1986; Hermens and Opperhuizen 1991). International harmonization of testing protocols and exchange of test data are encouraged, both to preserve scarce testing resources and to provide rapid development of the datasets.

Expansion of risk assessment procedures

Current risk assessment endpoints can be expanded in both laboratory and field studies. Survival, growth, development, and reproduction are appropriate endpoints for assessing risk and are included in current assessments. As previously noted, one challenge to ecological risk assessment is the myriad of potential endpoints; all possible endpoints can never be practically measured, but several deserve consideration. Current concerns about potential developmental, endocrine, immune, and nervous system endpoints from exposure to chemicals, both chlorinated and nonchlorinated, are candidates. These endpoints have not been ignored. For example, the testing of teratogenic and developmental effects for human safety have historically included in-utero exposures of laboratory animals, where development processes are known to be susceptible to effects only during specific times during prenatal development. Some studies have included multigeneration exposures as well.

In fact, procedures and protocols to evaluate estrogenic and reproductive effects in laboratory animals have been developed and validated, and they are currently available in a range of sensitivities (Lamb et al. 1998). For example, the textbook responses in female laboratory rodents to acute or chronic estrogen exposure are increased uterine growth and vaginal cornification. These responses occur in adult and juvenile females. Measurement of uterine weight and observation of uterine size (pathology) and histology are standard in most reproductive and chronic protocols. Most of these protocols now also include evaluation of vaginal histology and/or estrous cyclicity daily, by vaginal cytology, during a 2-to-4 week period. This noninvasive procedure detects the vaginal cornification that is characteristic of estrogen responsiveness. Persistent vaginal cornification distinguishes the abnormal response from the cyclic onset of estrus elicited by the animal's endogenous estrogens. Other parameters that have been evaluated in females in reproduction and fertility studies and that also are affected by estrogenic chemicals include fertility, mating efficiency (time to mating), maintenance of pregnancy, duration of gestation, and offspring development and survival. Estrogenic effects observed in males following chronic exposure include altered spermatogenesis, impaired fertility, sterility, testicular atrophy, altered secondary sex-organ weight and function, and the acquisition of female secondary sex characteristics. Alterations in spermatogenesis and testicular atrophy are detectable in reproduction, fertility, and chronic studies by measurement of testicular weight and testicular histological examination. Alterations in secondary sex-organ function and size can be detected by gross and histological examination and by organ weight. Decreases in fertility are detected in reproduction studies by mating trials and histological examination of the testes and epididymis. More subtle effects on sperm production (and therefore, on fertility) are detected by evaluation of parameters of sperm production and quality, including epididymal sperm concentration, motility and morphology, and testicular spermatid concentration. These techniques, along with evaluation of secondary sex-organ weight, are being incorporated more frequently into chronic studies. In both sexes, circulating hormone levels can also be measured, although they may not always provide additional information on hormone modulators.

It should also be noted that antiestrogenic and antiandrogenic effects would be detected in modern reproduction and chronic toxicity tests through evaluation of the same parameters described above. Antiestrogens should be identified by decreased uterine growth, altered estrous cycle, and decreased fertility; antiandrogens should affect sperm production, testicular weight, and secondary sex-organ weight and function. The potential for in-utero effects can be evaluated by continuous exposure of the dam prior to mating and during pregnancy with subsequent evaluation of the offspring, including reproductive performance after reaching sexual maturity in multigeneration studies. Estrogenic agents have a number of adverse chronic health effects, particularly cancers in estrogen-responsive organs. These effects may occur in the reproductive tract, mammary tissue, and liver in rodents. These types of tumors are easily detected in standard chronic toxicity studies. Other aspects of adverse estrogenic effects that are produced in shorter-term studies, such as testicular atrophy, can also be observed in chronic studies.

The need to evaluate chemical mixtures has been recognized since the mid 1980s, when specific studies were begun to address these needs in the U.S. National Toxicology Program (NTP). For example, the abovementioned reproductive and estrogenic protocols have been included in studies to evaluate exposure to environmentally relevant concentrations of a 25-chemical mixture of agrochemicals, including putative estrogenic chemicals such as atrazine. The program included a full battery of studies: 26-week exposures of adults, teratology studies for effects on offspring from in-utero exposure, continuous breeding and second-generation evaluations, evidence for genetic toxicity that might affect subsequent generations, and neurobehavioral evaluations. No exposure-related effects were observed, even in sensitive estrous-cycle and sperm-count or morphology endpoints (NTP 1993).

While good in-vivo assay tools are available for identifying endocrine responses in laboratory animals, these often require long exposures and are therefore expensive and not useful for screening large numbers of substances quickly. Although some new in-vitro assays have been developed to assess hormone-receptor agonism and antagonism, the relationship of these responses to in-vivo effects is not yet well understood. Assays similar to those used in laboratory animals are not generally available for most wildlife and fish species. The development of similar procedures and protocols for other organisms, such as invertebrates and plants, where significant exposures may occur, could be important for assessing potential effects of chemicals produced in large quantities and with significant distribution in the environment.

Detection of subtle developmental and endocrine effects will still require the refinement of old methods and the development of new testing procedures. To include endocrine and other endpoints in these reproductive and developmental protocols will also significantly increase the numbers of test animals. Immunotoxicological and neurological laboratory protocols are currently under development. In field monitoring, reproductive success and developmental effects have received increased emphasis in the past decade. Studies of Great Lakes bird populations have associated contamination with developmental effects (Fox et al. 1991; Tillitt et al. 1992; Harris et al. 1993; Giesy, Ludwig et al. 1994; Giesy, Verbrugge et al. 1994). These complex analyses should be reserved for the chemical uses, volumes, and releases that require these tests, and a tiered system of screening should be developed and applied to conserve testing resources.

Research Needs

Development of parallel laboratory and field methods to increase understanding and reduce uncertainty, thus improving risk assessments, should receive broad encouragement and financial support. For legislators, it is also important to recognize that laboratory testing protocols are periodically revised at about 5-y intervals in the U.S. for effluent toxicity tests, and improved or additional protocols are developed, such as for saltwater species and sediment toxicity. Therefore, mandated testing protocols would be unwise and would eventually result in testing procedures that become outdated and difficult, if not impossible, to improve.

Full life-cycle and multigeneration tests

Full life-cycle and multigeneration tests may be useful to assess specific chemicals. Reproduction and development of progeny are processes that are potentially susceptible to disruption by cumulative exposure until the reproductive portion of the life cycle is completed. In the case of potentially bioaccumulating chemicals, partial and even full life-cycle exposures are sometimes required (Walker 1993a, 1993b). Reproductive tests for certain chemicals currently involve multigenerational tests, usually in mice or rats. The current *Ceriodaphnia* and *Selenastrum* aquatic toxicity tests are in fact multigenerational. Instances where multigeneration tests may be needed for such vertebrates as fish or birds include bioaccumulative chemicals, both chlorinated and nonchlorinated. These chemicals may potentially be passed from parents to offspring in egg lipids, in-utero, or during lactation. For chemicals that are hormonally active, exposure and testing during development within the egg or in-utero is essential at concentrations expected in the environment. Toxicity tests should always be selected with consideration of expected modes of action, exposures, and vulnerable species.

Interactions and mixed exposures

Interaction of mixtures and nonchemical stressors can be considered in risk assessment, but this is not feasible under all circumstances. Sufficient resources and time are simply not available to test all potential impacts of all combinations of chemicals and other stressors. It is prudent, based upon known properties, to consider the potential interaction of one chemical with others under the specific conditions or concentrations that may occur in a contaminated environment. Stressors in the natural environment, such as climate, habitat alteration, availability of food, disease, and predation in site-specific risk assessments, should also be considered.

Field validation of risk predictions

Risk assessments should be validated in the field to verify the accuracy of predictions and to identify where procedures need refinement. Field monitoring of environmental fate and effects should be encouraged and is the basis for some current recommended practice (USEPA 1992b; EU 1993a, 1993b).

There should be a continued effort to develop and apply innovative instrumental and biological techniques, to address variability and fluctuations in field conditions, and to compile the fundamental understanding of natural systems that forms the basis for environmental risk assessment. It is also worthwhile to note that the seminal data leading to the recognition of PCBs as an environmental issue in the mid 1960s was the analytical detection of body burdens in field organisms. Therefore, chemical monitoring for unique or previously unobserved chemicals should be included in field work.

Assessment in the global context

Global modeling of persistent, bioaccumulative chemicals is necessary. For persistent, bioaccumulative, globally transported chemicals, a goal of ecological risk assessment should be to predict concentrations on a global scale. Environmental scientists have begun to model global dispersion of chemicals, including lindane, DDT, and PCBs (Wania and Mackay 1993b, 1994). The greatest shortcoming faced by current global models appears to be the absence of accurate data on environmental loadings of these chemicals. The actual amounts released each year from specific regions can only be estimated because of poor production and release records in developing countries. Thus, a global inventory system is needed to understand and ultimately to manage persistent, bioaccumulative chemicals.

Life-cycle concepts

Life-cycle concepts for chemicals should be included in risk assessment, with the recognition that both are separate but potentially complementary tools. The life cycle of a chemical involves resource extraction, energy utilization, manufacture of feedstock and byproducts, use and possible further conversion, transportation, and ultimate disposal. This inherent complexity will lead both to exposure of different chemicals of different toxicities at various life-cycle stages and to varying exposure of the same chemical. Risk assessments should be conducted with awareness of chemical life cycles in order to appropriately review the total potential risks resulting from use of the chemical.

Speeding up the process

Some have argued that chemical-by-chemical or site-by-site risk assessments are too slow, when rapid action is needed to control persistent and bioaccumulative substances before large releases to the environment can occur. The need for rapid action is recognized in the tiered approaches to risk assessment described above. Those substances, chlorinated or not, that are most likely to cause adverse effects in the environment will be filtered out by early tiers of assessment and can therefore be addressed in a timely manner. This presupposes that appropriate methods are available for prioritization and that these are not circumvented by public or other pressures that derive from perceived rather than actual risks.

Conclusions

- The basic assessment framework for evaluating the risk of chlorinated and other chemicals is sound. This framework is largely in place and is similar in design in both North America and Europe. The framework has been steadily enhanced to further address the risks from the diverse spectrum of both OC chemicals and non OC chemicals, including persistent, bioaccumulative chemicals.
- While the basic risk assessment procedures are technically sound, current uncertainties for environmental risk associated with use of chlorinated and other chemicals can be reduced. Specific recommendations that further reduce these uncertainties are identified in Chapter 6, "Summary of improvement needed for hazard assessment."
- Persistent and bioaccumulative chemicals present additional data requirement challenges to risk assessment for environmental behavior and food-chain bioaccumulation and biomagnification. The increased data requirements for risk assessment posed by these chemicals currently have been addressed with some success, such as PCBs in Green Bay and PCDDs, PCDFs, and TCDD-like PCBs in Lake Ontario.
- Endpoints beyond survival, growth, and reproduction should be developed for ecological risk assessment. The ultimate choice of these relevant and meaningful endpoints will require a better understanding of organisms, populations, and ecosystems.

Recommendations

To improve the power of risk assessment procedures, and to reduce uncertainty, it is recommended that the following actions be taken:

- Establish a globally acceptable minimum dataset for use on all chemicals, utilizing a tiered approach to generate additional data when necessary, based on triggers for either chemical or biological criteria and production volumes. Similarly, certain testing procedures do not automatically need to be performed when physicochemical properties and fate data indicate that they are unnecessary. The objective is to allow comprehensive, rapid, and fundamental estimations of environmental fate and transport, the potential for toxicity, and the magnitude of likely concentrations in environmental compartments. The goal is a cost-effective system to identify at an early stage the problematic properties of persistence, bioaccumulation, and greater toxicity.
- Refine existing test protocols to generate toxicological data for risk assessment, and develop new protocols to reduce uncertainties that exist for some endpoints, such as immunotoxicity and endocrine modulation.

- Modify the design and methods of present studies, where needed, to include more realistic test duration, appropriate exposure conditions (water chemistry and light), critical timing of exposure such as in utero, and routes of exposure such as sediment or food.
- Provide for the omission of tests that are not applicable to particular circumstances. For example, the requirements to conduct chronic exposure assays for chemicals that bioaccumulate will not usually provide adequate effects data on the reproduction process unless these are specifically included in the exposure period.
- Include multigenerational effects in risk assessment procedures, and include interactions between toxicity and environmental stressors in ecological risk assessment.
- Evaluate chemicals with consideration of other stressors and environmental conditions that may add to, or otherwise ameliorate, effects attributable to the chemicals.
- Develop an understanding of mechanisms of toxicity, and measure these mechanisms in risk assessments.
- Update, in a continuous, periodic manner, both tier criteria to select appropriate tests and the protocols themselves.
- Avoid legislative fiat of test protocols and procedures that will inhibit further evolution of the risk assessment process and the development of improved test procedures.
- Design and implement effective field monitoring programs to validate the outcomes and the accuracy of ecological risk assessments.
- Incorporate life-cycle analysis concepts to complement risk assessment procedures so that the complete "cradle-to-grave" generation and release of chemicals can be considered.
- Improve risk communication to the public, and educate the public on the risk assessment procedure.
- Address the global nature of many persistent and bioaccumulative chemicals by improving and developing global transport models and by compiling an accurate and updated inventory of global environmental loadings of these chemicals.

Chapter 8

Risk Management for Organochlorine Chemicals

Risk management focuses on the reduction of risks to humans and the environment by combining the information and the perspective provided by risk assessment with a broad set of technical, economic, societal, and policy inputs. The approach to the risk management of organochlorine (OC) compounds is not inherently different from that of nonchlorinated compounds. The standard approach of emission limitations, which uses emission criteria and improved technology, applies in almost all cases. Additional nonstandard scrutiny should be applied to potential byproducts during the life cycle of chlorinated materials to monitor the formation of persistent and bioaccumulative compounds. Use restrictions and bans for OCs that are persistent, bioaccumulative, and highly toxic are already in place at the national level in many countries.

Several factors combine to make risk management for chlorinated compounds more difficult than for nonchlorinated compounds. First, the public often appears to associate all chlorinated compounds with a few persistent and bioaccumulative chemicals such as dichlorodiphenyltrichloroethane (DDT), polychlorinated biphenyls (PCBs), polychlorinated dibenzo-*p*-dioxins (PCDDs), and polychlorinated dibenzofurans (PCDFs). Second, environmental problems with these compounds continue, and they remain in the public agenda. The long depuration times for contaminated sites, extraordinary cleanup costs, disposal and landfill issues, and the residual effects in wildlife at several previously contaminated sites all act in concert to reinforce the public misperceptions that all OCs are problems. These factors make risk management and, especially, risk communication more difficult tasks when dealing with OC chemicals.

Risk managers may continue to rely on the basic and most commonly used regulatory option: improved technology to control emissions and to reduce risks from chemicals. More flexible strategies may be considered, including voluntary agreements and economic instruments. The more severe options, such as bans, replacements, and phaseouts, are limited to a few cases of persistent compounds.

Chapter Preview

The goal of risk management is to develop and implement policy that reduces the risk or hazard of an environmental problem, where needed, both to humans and to the environment. Risk management is important for all chemicals, organochlorine (OC) and non-OC alike. Risk management uses the results of risk assessment in concert with a broader set of factors, including technical and engineering capabilities (best available technology [BAT]), economic analysis, societal issues and concerns, and policy issues (pollution prevention), to determine the management of health and environmental issues. Risk assessment estimates whether there is a risk and, if so, attempts to identify the type of risk and to quantify the magnitude of the risk and the uncertainty in the assessment. Then, using results of the risk assessment, the risk manager incorporates into the decision-making process the following factors: who or what may be affected, the severity of the impact, the technologies available to mitigate, what voluntary measures are available, the costs of various options to reduce the risk, the administrative and regulatory framework for addressing the particular risk, the public perception of the risk, and whether there are political concerns. Risk management, therefore, not only must consider the risk assessment but also must integrate numerous additional issues and act within the established legal and regulatory framework. The perception of risk is an important factor, and opinions differ on whether a given risk is unacceptable, both between nations and within different segments of a society. In this situation, it is especially important to have well-defined, public criteria for what constitutes unacceptable risk and to clearly communicate these criteria.

The technical and economic considerations regarding risk management of chlorine are extensive, both in developed and developing countries. In industrial settings, several hundred chlorinated compounds are used as chemical feedstocks and intermediates in a very wide range of production volumes (see Chapters 1 and 7). From the original manufacturing sites, many compounds are, in turn, used in hundreds and possibly thousands of different manufacturing situations. The volumes and chemicals are not static but are undergoing continuous change. Many of the "old generation" of chlorinated chemicals were persistent and bioaccumulative, but they are no longer used in most industrialized and many nonindustrialized nations. In these nations, production has stopped and recovery or disposal of existing stocks is underway. These chemicals include chlorinated insecticides such as DDT, chlordane, endrin, aldrin, dieldrin, toxaphene, kepone, mirex, heptachlor, hexachlorobenzene (HCB), pentachlorophenol (PCP), and industrial chemicals such as PCBs. As a result, these chemicals have been replaced by either nonchlorinated compounds or "newer generation" OC chemicals that pose significantly lesser risks. However, production bans and recovery or disposal are not universal. This leads to continued dispersal of these compounds into the global environment in several instances.

As noted in Chapters 1 and 3, the formation of byproducts during industrial processes and incineration is a specific concern for chlorine compounds. Risk managers should employ techniques such as substance-flow analysis and life-cycle analysis to ensure scrutiny of byproducts from direct production, recovery and recycling of process streams, off-site recycling and recovery operations, disposal, and incineration. Risk managers should strive to reduce the environmental release and dispersal of persistent and bioaccumulative compounds. Risk managers should employ standard assessment techniques to reduce the formation of byproducts directly and improved technology to recover and eliminate byproducts directly at the site of formation.

Management issues are involved in balancing competing risks. The threat to human life from malaria and other diseases in developed nations is not as severe as in some developing countries. Some countries have decided that the benefits of older OC insecticides, such as DDT for controlling the spread of these diseases, outweigh the potential environmental risks. Other countries have opted for alternative control mechanisms for disease vectors in dealing with these risks. Several African nations control malaria through chemotherapy and chemoprophylaxis programs in the most vulnerable groups. Decisions such as these must carefully balance the potential effects of the pesticides and the benefits that are gained from its use. The situations meant to be resolved by these management strategies are not static in either case. As societal resistance to the use of OC insecticides has developed, the malarial parasite and others have also developed resistance to chemotherapy and prophylactic drugs (several of which are OCs themselves). Therefore, the cost–benefit analysis inherent in most risk management is not straightforward. The real and potential environmental risks of DDT and the resultant costs often are not as easy to quantify as the loss of human lives or crop production. Furthermore, the environmental "costs" are generally displaced to some other place and to a future time, whereas the public health crisis is immediate; environmental costs thus are not easily quantified by "local" risk managers.

Societal Considerations

The public is very aware that certain persistent and bioaccumulative OC compounds, such as DDT, PCBs, PCDDs, and PCDFs, are associated with adverse effects in the environment. The public does not appear to be fully aware that production of these compounds is banned in most countries. In addition, residual contamination and effects from these compounds continues to maintain a public awareness, leading to fears that risks may be increasing rather than decreasing. This awareness also generates a mistaken association of the harmful properties of these compounds with all OC chemicals. This public misperception makes risk management and risk communication problematic when dealing with OC chemicals.

Societal and political concerns are also important considerations in risk management decisions. Legislation, such as the Delaney clause (now replaced) in the United States, which forbade any detectable anthropogenic carcinogen in food, or the guideline of

0.1 µg/L of total pesticides in ground water established by the European Union (EU), may sometimes supplant scientific risk management. In the case of OC compounds, societal concerns have been and continue to be important factors. PCBs, DDT, and other well-known persistent compounds were an important impetus to the Toxic Substances Control Act (TSCA 1976); the Federal Insecticide, Fungicide, and Rodenticide Act (FIFRA 1947); and Superfund legislation (SARA 1986) in the U.S. and similar legislation elsewhere. The public is also aware of the very high toxicity of tetrachlorodibenzo-*p*-dioxin (TCDD), which is associated with the spill of TCDD at Seveso, Italy in 1976, the human health effects of Agent Orange on military personnel and civilians during the Vietnam War in the 1960s and 1970s, and the use of PCB- and dioxin-contaminated waste oil at Times Beach, MO, in 1970. These compounds are still mistakenly associated in the public mind with unrelated chlorine-containing chemicals. Although the bald eagle is now removed from the endangered species list, the public is reminded of continuing evidence of low rates of subtle developmental problems (largely associated with PCBs) such as deformed beaks in some fish-eating bird colonies in the Great Lakes. The public often is unaware that most of the persistent and bioaccumulative OCs that cause these continuing effects are now banned or severely restricted in developed countries.

The public continues to be reminded about these persistent and bioaccumulative compounds, as society attempts to remediate the mistakes of the past. Many members of the public do not understand (and may react with frustration and disbelief) when risk managers indicate that the decontamination of such areas as Green Bay is not feasible and that it will necessary to wait several decades for PCB levels to decline to levels believed to be acceptable. There is also considerable anger and frustration with the failure of current legislation, such as Superfund, to provide for the rapid decontamination of identified harbors, landfills, and old manufacturing sites. Simultaneously, evidence continues to emerge for additional reproductive, hormonal, and development concerns from PCDDs. This continues to reinforce the public concern with persistent and bioaccumulative chemicals. The overall result is that the public may sometimes equate the properties of these few compounds with those of all OC chemicals. This generates a very difficult communication job for risk managers, who will need to deal with the continuing societal perceptions and misperceptions well into the future.

The public needs direct communication with both regulatory agencies and those being regulated in the private sector. Environmental reports need to demonstrate efforts to reduce pollution, especially those involving persistent and bioaccumulative compounds. Recent examples include the thorough USEPA and Environment Canada review of all pulp and paper sites for the formation of PCDDs and PCDFs, followed by yearly reports from U.S. and Canadian industry that demonstrate the progress of their efforts.

Regulatory Strategy Considerations

The basic and most commonly used regulatory option to reduce risks from chemicals relies on improved technology to control emissions. Different regulatory frameworks are in place to manage risks through improved technology. Some frameworks provide flexibility to choose technologies as long as specific emission limits are met, while others require the use of a specific BAT to meet emission limits. More flexible strategies are being tested, including voluntary agreements and economic instruments. The more severe options, such as bans, replacements, and phaseouts, are limited to a few cases of persistent compounds that clearly pose unacceptable risks.

The basic options for risk managers, in increasing order of severity, are

1) to provide better information and communication to prevent uninformed misuse, such as safety data sheets on use and disposal and labels on hazard or transportation;
2) to control discharges and releases of chemicals and their byproducts to particular levels that are judged to be safe at various points in the life cycle, such as air or water criteria or voluntary emission reduction programs;
3) to restrict chemical use and applications, e.g., lindane and PCP; or
4) to impose a total manufacturing ban, such as on DDT and PCBs.

The information option (1) will not be discussed further.

Restrictions on use and release

The implementation of discharge and release controls is the normal regulatory pattern for most chemicals, including OCs. A significant number of OC and nonchlorinated chemicals have targeted, environmental-concentration limits such as air and water criteria developed by regulatory agencies (USEPA 1992c). The legislative and regulatory framework is in place to restrict use and release of additional OC and other compounds in most nations and including such multinational organizations as the EU. Within nations, additional regulatory frameworks may exist (e.g., in the U.S., the California Proposition 65 regulations) that require risk assessments for occupational and public exposures for listed compounds. Some regulatory agencies have enlisted public concerns to motivate emission reductions outside the formal regulatory framework. For example, in the U.S., public disclosure of environmental releases for some compounds has resulted in further declines in releases. Often, the current release levels are already well below those identified to be of concern by the risk assessment process.

Technologies that reduce discharges and emissions are continuously changing and evolving, and new technologies are emerging. As a result, the legal framework for most regulations requires periodic review of industrial emissions and of the technology available to reduce these emissions. In theory, this ensures a steady decline in emissions. However, there are practical implementation issues. Many regulatory agencies find themselves understaffed and may not make these reviews on an appropriate time schedule. As a result, public pressure and even lawsuits are often necessary to initiate reviews by regulatory agencies. This approach is inflexible and has the disadvantage of taking control of priorities and resources away from the policy and risk managers. Further, an inflexible approach may hinder the development of comprehensive, rational policies, based on both credible science and economic or risk benefit analyses, by circumventing their efforts to apply prioritized or relative risk strategies for the use of limited agency resources. Where technologies have substantial costs, short time intervals for installation of new technologies as required in some statutes may result in either reluctance by the agencies to demand the technology or resistance to the regulatory process by the affected parties.

There is no uniform system for selecting remediation and control technologies. In some cases, criteria are established and the choice of technology is left to the discharger, based on economics and process considerations at particular sites. In other cases, the framework provides that regulatory agencies determine and require BAT. Therefore, the technology may be predetermined by the regulatory agency, with little input from industry. Because BAT changes, regulations set by this standard must be periodically reviewed and assessed. Normally, BAT is determined in a context of economic feasibility and costs versus the environmental and health benefits to be achieved.

A restrictive exception is the U.S. Clean Air Act Amendments of 1990 [CAAA 1990]. For instance, a number of hazardous air pollutants (HAPs), including both chlorinated and nonchlorinated chemicals, are specifically identified. These HAPs must be addressed by the maximum available control technology (MACT), without provision for economic considerations. After the application of MACT, the residual risk of the identified chemicals is determined. If the risk is unacceptable, then processes that use or generate the HAP may be restricted or banned. Most chlorinated solvents are special cases because they are emitted into the atmosphere during use. In most of these applications, almost complete solvent recovery and recycling are possible. One example is tetrachloroethylene, which is used in dry cleaning. Recovery of this solvent and emissions restrictions at both dry-cleaning and recovery operations have increased substantially in recent years. Further reductions may be needed as environmental monitoring shows ambient concentrations that exceed safe values for long-term exposure. One risk management option is to replace one technology or substance with another. Alternatives to chlorinated solvents include water or other solvents. However, the energy (and, therefore, emissions from fossil fuels) required to recover HAPs and the danger presented by the flammability of replacement solvents indicate that economic and safety trade-offs must be considered and assessed.

A new and evolving regulatory strategy is to seek proactive pollution prevention efforts from the regulated parties. The objective is to move from capture and pre-release treatment of pollution toward modification of old processes or development of new processes that reduce or eliminate pollutant production. This strategy may also seek to reduce overall risks (e.g., the development of substitute solvents to reduce risks from emitted solvent) or, more radically, the development of nonsolvent processes. Initial, limited experience with pollution prevention shows that risk–risk trade-offs may be encountered in some cases (Graham and Wiener 1997). However, long-term research and development efforts in industry now often include pollution prevention criteria and targets.

Making alternative choices means carefully comparing relative risks by studying the life cycles of both existing and alternative processes. Comparisons are technically difficult when both processes are well known; they may be impossible without extensive experience in the replacement process. In addition, much less may be known about new processes, so it is important that sufficient information about replacement processes be available. Therefore, there is a danger that ignorance of the hazards for new processes will be mistaken for an indication that risks of the process are acceptable. Because risk managers have little experience with assessing relative risks and identifying trade-offs between old and new processes and chemicals, agencies are reluctant to undertake this exercise.

One possible strategy is to include in the risk assessment and risk management process a comprehensive review of chemical production, use, and disposal. Identifying chemicals released and locating where these releases are generated assists in developing appropriate risk reduction strategies. In these situations, a life-cycle assessment (LCA) — which tracks material and energy use and environmental release during the manufacture, use, and disposal of a product — may be a valuable analysis tool. These analyses should be applied to substitute chemicals as well, as they may present greater risks of explosion or worker injury. The release of compounds associated with almost all of these processes could be managed by controlling manufacture and disposal practices. For example, polyvinylchloride (PVC) is a relatively inert polymer during its use, and its primary risks involve possible releases during production and disposal; PVC is produced at a limited number of major chemical complexes, where air emissions have been increasingly controlled since the 1970s through process engineering. Improved on-site disposal of tars and solid wastes that contain chlorinated chemicals has been introduced at several sites and should be considered at all sites. PVC incineration, possibly resulting in combustion byproducts such as PCDD/PCDFs, has been a concern. Control strategies include recycling PVC scrap from construction sites and automobiles as well as using landfill disposal rather than incineration. Another successful approach is modification of incinerators to minimize PCDD/PCDF emissions, as is legally required in Austria, the Netherlands, and Germany and recommended in Sweden. Note that these measures will never entirely eliminate the releases of OCs associated with PVC.

Bans and phaseouts

Bans and restrictions have most often occurred for high-volume, persistent, bioaccumulative, hazardous compounds. While these restrictions and bans have been effective in halting production within jurisdictional boundaries, production and use can continue in other jurisdictions, which is the case with DDT and other compounds. These restrictions and bans are obviously not retroactively effective in decontaminating manufacturing sites and landfills or other sites.

Where bans are imposed, their implementation may not be immediate but may take place over a given time period. This option is referred to as a "phaseout." An example, in which significant risks were identified and the time needed to convert the manufacturing process was determined, is the development of the Montreal Protocol and the London amendments for chlorofluorocarbons (CFCs) (United Nations Environmental Program [UNEP] 1987, 1990, 1992). In this case, replacement chemicals (e.g., hydrofluorocarbons) were identified that were acceptable from environmental, human health, and performance perspectives. The original protocol provided for a decade to convert manufacturing processes in industrialized countries and did not set specific deadlines for less developed countries.

Much of the recent debate about OC chemicals and about the future use of chlorine has been initiated by a recommendation of the International Joint Commission (IJC) to the United States and Canadian governments. The Commission's proposed approach envisions governmental consultation with industry and other interests to restrict, phase out, and eventually ban all uses of chlorine and chlorine-containing compounds from industrial feedstocks. This approach explicitly recognizes the need to take socioeconomic and other consequences, including those of the available alternatives, into account in determining timetables (IJC 1994). This recommendation is a particular challenge for risk managers and agencies, as it totally departs from previous practice. In fact, most of the subject individual chlorinated chemicals are now judged by the regulatory agencies to be largely produced and used safely under the in-place legislation.

In considering a widespread ban on such a broad and diverse group of chemicals, the primary question is whether current legislation is adequate 1) to prevent the future production of compounds such as PCBs and 2) to deal with risks from currently produced chemicals. The framework of current legislation in most developed countries does appear adequate. These national controls responded to and targeted the widespread production of DDT, dieldrin, PCBs, and other chemicals prior to the mid 1970s. The current framework addresses recognized risks and continues to improve. New institutional arrangements are being devised in some jurisdictions to address 1) the data gaps for existing chemicals, e.g., as in the EU (1993a, 1994), and 2) those unintentional sources and byproduct emissions, e.g., PCDD/PCDFs. The current framework is also sufficiently flexible to consider and deal with new risks such as the emerging issue of endocrine disruption during sensitive developmental stages. This flexibility can take into account the timing of exposures, such as in utero, and the development of sensitive protocols to assess molecular indices such as receptor binding and hormonal levels in later life stages. In addition, countries in the Organization for Economic Cooperation and Development (OECD) have begun to collaborate on the international harmonization of assessment methodologies. From a national standpoint, governments have been enacting appropriate legislation to prevent future, deliberate production of highly persistent, bioaccumulative, and toxic chemicals such as PCBs. Institutional arrangements at the international level should be further developed to address continued production of certain persistent and bioaccumulative chemicals in large quantities and the potential stocks and production sites that may exist in some developing countries. The current United Nations Economic Commission for Europe draft (UN-ECE 1994) that describes persistent chemicals is a step toward an international protocol. No current system however, can adequately address unknown byproducts from future processes.

Voluntary agreements

A newly emerging approach to controlling emissions and risks is the development of agreements between industry and regulators to achieve certain target reductions or other actions. Because these agreements exist outside the force of existing regulations, they are referred to as "voluntary." One example of a successful, voluntary emission reduction is the case of the pulp and paper industry and the U.S. regulatory agencies. When dioxin was discovered in pulp mill effluents, products, and sludges in 1988, a voluntary joint survey of all mills using chlorine was arranged between industry and regulatory agencies, the first complete multimedia survey of an industry. In conjunction with the survey, a joint process study helped to identify means of controlling dioxin formation. Simultaneously, analyses of paper products and risk assessment exercises were conducted by industry with those several agencies having various jurisdictions. The end results included voluntary agreements on limits ranging from food packaging to sludges used for soil amendment, including dates for implementation. The primary advantages to this approach were the very rapid implementation of the process changes and the rapid reduction in risks versus prolonged regulatory proceedings. A second advantage was that the approach allowed industry to identify cost-effective approaches on both interim and permanent bases.

Voluntary agreements are not exclusive to the U.S. In 1991, a voluntary program to reduce the risks from residual epichlorohydrin in various industrial and consumer products began in Europe. Due to the demonstrated reduction in regulatory resources and to the benefits of rapid implementation, there is growing interest by regulatory agencies in the use of voluntary agreements to set emission reduction targets and specific deadlines.

Economic incentives and disincentives

Several potential economic methods can be used as strategies to reduce exposure from OC chemicals. Basically, government actions impose or raise monetary charges for activities that incur greater risks or attempt to reduce charges for those activities with lower risks. However, experience with these models is limited, but it is hoped that these methods would add to the growing environmental consciousness of both businesses and consumers that purchase such materials and products.

Taxes

A "use tax" can be levied on the producer or user of a chemical when that chemical is deemed by policy makers to be harmful to the environment. Obviously, to be effective, the tax must be equal to or exceed the cost of implementing normal technological emission controls. However, whether the economic disincentive provides better overall reductions in emissions than regulatory strategies remains to be tested. The use tax results in a hidden tax to the consumer because these costs must be recovered by the producer.

Disposal fees

When the chemical of concern is a waste that cannot be recovered and when emission reductions are desired, disposal fees could encourage waste reduction or prevention processes. However, large manufacturing sites increasingly use on-site disposal or destruction. For small manufacturing sites and small users, a disposal tax could become an incentive for improper disposal and dumping, an unfortunate situation that has already become an issue for hazardous wastes.

Trade sanctions

It has been proposed that trade sanctions could be applied to countries that continue to produce clearly harmful OC chemicals. However, given the increasing complexity of international trade agreements like the General Agreement on Tariffs and Trade (GATT), it is unlikely that such sanctions could be implemented without extensive negotiations. National laws preventing use and importation of certain chemicals are already in place for a number of persistent and bioaccumulative compounds like PCBs. Environmental treaties and protocols such as those for CFCs (UNEP 1987, 1990, 1992) would be a more likely avenue to approach.

German approach

Primary objections to the current regulatory approach are 1) the rate at which some regulations are initiated or produce the desired results and 2) the tendency of environmental legislation and regulation to become enmeshed in political agendas and litigation. Germany has attempted to provide a process that addresses both complaints. The process incorporates the basic elements and advantages of the voluntary agreements mentioned previously.

The German chemical law (*Chemikaliengesetz*) was formulated to provide comprehensive environmental protection from many chemical releases. However, this law does not cover pharmaceuticals, pesticides, inorganics, or polymers. The law defines various hazards, from minor toxicity to carcinogenicity, and methods to determine their extent. The administrators of the law are entitled to act on their own when cause is demonstrated. However, industries can challenge such actions in court. The objective, however, is to avoid extensive negotiations and court challenges by achieving mutual industry-administration agreement based on sound scientific evidence. Negotiations occur through an advisory committee (*Beratergreminum umweltrelevante Altstoffe* [BUA]) that reports to the government. The BUA has been established to develop recommendations on which actions should be taken. The BUA consists of 5 members from the academic community, 5 members from the chemical industry, and 5 members from the various federal agencies involved in environmental affairs. The scientists act independently and represent their own disciplines as well as the public. The scientists are elected to 4-year terms by the board of the German Chemical Society (*Gesellschaft Deutscher Chemiker*) and cannot be replaced during a term. The chairman and the co-chairman of the BUA must be scientists. The scientist group cannot be overruled in committee decisions; that is, they have essentially complete veto power in their areas of expertise, as individual members and as a group.

The BUA performs risk assessments for which the chemical industry provides basic information in the form of draft reports on the compound under discussion. The governmental agencies review these drafts and provide additional information as needed. The final draft report to the BUA is a joint report by industry and federal agencies. This paper is peer reviewed by the BUA scientists and their staff. Information about adverse effects or hazards is not subordinate to any other data. The BUA's risk assessment leads to a report, and recommendations are adopted by the committee. Reports are made public in printed form, and all information is kept as an original source by the BUA.

If industries and federal agencies cannot agree, the BUA scientists arbitrate a final decision or leave the question open for further experimentation to be done by industry. In practice, industry acts during the negotiations to begin implementation of necessary changes, as is currently practiced in other jurisdictions. To date, more than 150 chemicals have been evaluated, and the full reports are available. However, this process can be seen to circumvent the opportunity for public participation that occurs in most formal regulatory processes.

Additional Risk Management Considerations

Several noteworthy concerns and circumstances lead to conflict over risk assessment and pose difficult decisions for risk managers. These include the inability to effectively remediate some sites, the costs and the degree of remediation at other sites, situations where data gaps lead to a public mistrust of the process, source apportionment, and the existence of differences in national approaches to both regulatory limits and restrictions. Further, no international strategies exist for dealing with the transboundary movement of persistent and bioaccumulating compounds. These circumstances will continue to raise questions in the minds of some about the effectiveness and the fairness of risk assessment and risk management.

Risk management should be focused on the reduction of environmental and human health risks, identified on the basis of risk assessments and not on the class of chemical or the mere presence of chlorine or any other element in the molecule. The current methods of risk characterization of OC chemicals as well as other nonchlorinated chemicals, which include a tiered assessment approach, are appropriate. As noted in Chapter 7, establishing a minimum dataset for all chemicals would considerably improve risk assessment; the same is true for risk management decisions. The need for data grows as the quantity of a chemical produced increases, where there are significant releases, and as the likelihood of exposure rises. In particular, the effects of chemical persistence and bioaccumulation on the long-term exposure of organisms must be carefully reviewed. This review should include environmental partitioning, fate, and steady-state concentrations. The confidence of risk managers increases when risk assessment procedures incorporate information derived from field studies for existing chemicals. This is true for both chlorinated and nonchlorinated compounds. These methods and decisions normally lead to establishing specific limits on the release of compounds. Risk managers must realize that data gaps may often be perceived with suspicion and fear, which undermines confidence in the process.

Risk management decisions are required by a myriad of situations other than the release of manufactured compounds. For example, there are existing stocks of PCBs. These stocks lead to fugitive emissions from manufactured products (electrical transformers and capacitors) and to accidents with significant releases. These PCB stocks should be collected and destroyed. The process of collection, transportation, and destruction should be assessed to ensure that risks from these operations are minimized.

Conflicts on the appropriate course of risk management also occurs at and near disposal facilities, contaminated manufacturing sites, and landfills. Intense debate centers on the degree of remediation effort or the residual level targeted at such sites. As the clean-up time and remediation costs accelerate rapidly with increasing remediation efforts, there is growing support to target the risks for the intended future uses of such sites and to remediate on that basis. Elsewhere, contaminated soils and sediment–water exchange must be recognized as reservoirs for continuing environmental releases, as exemplified by extensive atmospheric recycling and sediments of PCBs in the Great Lakes. Risk managers are thus continually challenged to balance the level of risk and the enormous costs of its mitigation. These difficulties emphasize and reinforce the need to prevent the manufacture and the release of chemicals with properties and characteristics similar to the PCBs and DDT through specific assessments of persistence and bioaccumulative potential.

For risk assessment and risk management, source apportionment issues are growing. A single source for a compound should be not assessed in isolation but in the context of all relevant inputs or sources of the compound. There is also a need to understand environmental transformations that may mask sources. Often, uncertainty exists regarding the actual sources responsible for observed environmental concentrations. The sources of chlorinated acetic acids in air, precipitation, and surface water have not been fully elucidated. The observations of extractable organic chlorine in precipitation and sediments stress the need for chemical characterization to identify these compounds and their sources.

Risk assessment combined with risk management must clearly define those unacceptable risks that lead to consideration of bans. The obvious examples are the particular combination of chemical and biological properties with the right exposure conditions: environmentally persistent, highly bioaccumulative, and highly toxic chemicals with sufficient production to have effects on fish and wildlife, particularly those species at the apex of the food chain. For unintentionally produced compounds that have a similar combination of properties, such as PCDD/PCDFs, regulatory efforts have been made to identify sources, such as in Germany, Sweden, and the Netherlands, and to minimize emissions, ranging from BAT for incineration to banning particular production processes generating these compounds, such as 2,4,5-T and PCP. Efforts should be taken to prevent the production of new chemicals with these properties, to halt the deliberate production and release of these types of compounds, and to rigorously address the sources of unintentional byproducts.

Controls on the production and use of chemicals are not harmonized internationally, and this is especially true for persistent, bioaccumulative, and toxic compounds. Currently, countries that have ceased to produce compounds such as DDT and lindane are still subject to long-range transport and exposure (Chapter 2). Depending upon the chemical's properties, dispersion may occur primarily within a localized region or may result in global dispersion to remote regions. For certain persistent, semivolatile compounds such as toxaphene and lindane, global models indicate that the greater local concentrations may occur far from the regions where the compounds are released (Wania and Mackay 1993b, 1994). This may be especially true for cold Arctic and Antarctic regions. Consequently, long-term risk management solutions for persistent and bioaccumulative compounds must be global in perspective.

Strategies for Assessing Risk Management Results

The effectiveness of risk management strategies and decisions should be continually assessed in order to improve the process and the results. Monitoring the environmental chemical concentrations and the health of previously affected and sentinel populations is paramount to ensure that remediation is effective and that potential, unrecognized effects are quickly discovered. This process requires considerable effort to design and execute proper sampling and analysis programs, given spatial and temporal considerations. It is especially important that the health of biota be properly assessed. The monitoring outcomes must be clearly communicated to the public in order to build their understanding of complex issues and to maintain their confidence in risk management.

Risk management decisions taken to reduce the risks from hazardous compounds must be followed by assessments of the decisions' effectiveness. Principal environments to consider are 1) environments where the production of the chemical takes place; 2) environments where the products are used, recycled, and disposed of; and 3) environments where the products may appear before they are degraded. For example, all are specifically considered in the EU assessment process for new and existing chemicals (EU 1993a, 1993b, 1993c, 1994) and are considered as part of the USEPA TSCA process for new chemicals (Nabholz et al. 1993; Smrchek et al. 1993; Walker 1993a; Zeeman and Gilford 1993).

Monitoring activities are needed to support the accuracy of assessments in receiving media such as water and air, in the vicinity of industrial sites, and in specific sinks such as sediments and biota. It is important to identify the specific purpose of such monitoring and to decide whether to study long-range transport, environmental persistence, bioaccumulation, or cause-and-effect linkages. The available data are normally included in regulatory reviews for development of water-quality criteria or industrial-effluent or air-emission regulations. This process normally results in several parties, including industry, academia, and regulatory agencies, gathering additional information to address identified gaps. However, there are few arrangements in place to ensure that sampling and analytical procedures are comparable or that there is sufficient spatial and temporal coordination of sampling either among parties or among sites. This is a significant area for data improvement, particularly for long-term trends (Bignert, Litzén et al. 1994). These chemical data do not address the need for biological observations on the health of receiving ecosystems, particularly in view of the growing public concerns about long-term, subtle biological effects like potential developmental and hormonal effects of chemicals. More comprehensive ambient monitoring programs in various potential emission sites and in "pristine" areas subject to long-range transport would be valuable in providing early detection of environmental contamination.

The measures for evaluation of risk management outcomes should be carefully chosen and prioritized. There is a wide diversity of opinion about the measures that should be reported to demonstrate that environmental chemicals have been successfully controlled. The USEPA has systematized these reporting measures as follows:

1) number of regulations implemented,
2) amount of money spent by industry on environmental controls,
3) effluent concentration of chemicals,
4) total loadings at a particular location or to a particular medium,
5) ambient environmental concentration of the chemical,
6) concentration in organisms, and
7) the incidence of effects in organisms.

These are different kinds of measures of success. Ultimately, an accurate assessment of the health of biota in their natural environment is the most important evidence for the success of risk assessment and risk management.

It is important in a long-term perspective to continuously inform the public about the outcome of monitoring studies. Dissemination of information on improvements such as the recovery of the bald eagle, the diminished effects in Great Lakes fish and fish-eating birds, or the apparent recovery in Baltic seals build and sustain public confidence in risk assessment and risk management. Otherwise, there may be public despair, confusion, and loss of confidence. Periodic reports by governmental agencies, local as well as federal, such as the USEPA's biennial report on national air and water quality (USEPA 1994d, 1994e), address the need for information.

Legislative Issues

Some countries have promulgated legislation to provide a comprehensive framework that addresses chemical risks. Such legislation has generated significant progress in reducing risks in industrialized countries. Developing countries need assistance in creating regulatory frameworks that do not drain scarce resources or hamper economic development. For persistent, bioaccumulative, and toxic chemicals, full international coordination is currently lacking. These chemicals pose particular cost and management challenges for developing countries to achieve control of production and release, substitution of other chemicals, remediation of contaminated sites, and disposal of existing stocks.

Many countries have passed legislation to obtain information on chemical production, use, and disposal; to gather available data on a chemical and its biological characteristics and to provide for testing to generate necessary data; to enable use restrictions and even prohibitions on all activities involving substances that pose an unreasonable risk; and to monitor the results of regulatory decisions. These regulations have resulted in significant progress, particularly in industrialized countries. However, these individual actions beg the need for international coordination and harmonization for the subset of persistent and bioaccumulative compounds. In addition, the substantial scientific and administrative resources that these laws require must be recognized. Attempts by developing countries to implement their own independent laws and administrations will consume scarce resources and will encourage the proliferation of test protocols that may be incompatible. The latter will result in significant economic costs in many countries, will raise potential de facto barriers to both trade and manufacturing, and may hamper economic development in countries where this is most needed.

One major development toward international harmonization is the establishment of the Chemicals Group under the auspices of the OECD. Based on findings in individual member states and the decision of the primary manufacturer in the U.S. to restrict production, the OECD member countries passed an agreement in 1973 prohibiting the use of PCBs in all open systems. Subsequently, this group has initiated harmonization of the testing guidelines for new and existing chemicals. In recent years, member countries have been involved in collaborative projects to assess the risks posed by existing chemicals.

Currently lacking, however, is full international coordination of legislation that controls the production, use, and release of the identified group of persistent, bioaccumulative, and toxic chemicals. Although a growing number of developing nations are joining the bans on chemicals such as DDT and restrictions on chemicals such as hexachlorocyclohexane (HCH), monitoring indicates that substantial production and use is still occurring. It appears that only environmental protocols analogous to the Montreal Protocol for CFCs (UNEP 1987, 1990, 1992) will be sufficient to halt production and to both recover and destroy existing stocks. Such protocols may be neither simple to develop nor easy to implement, as they must require either systematic inspection of production facilities or monitoring of new releases in receiving environments in order to trace releases to production facilities. One challenge will be that many developing nations face considerable increases in costs for new chemicals, including the expenditure of precious foreign exchange for chemical imports, and some form of financial assistance for emission control must be considered. In addition, there will be contaminated manufacturing and disposal sites for pesticides that must be remediated, and there will be electrical equipment with PCBs that require disposal. These will act as source reservoirs for future releases unless remediated. Because many nations have neither the financial nor the technical resources for remediation and disposal, international protocols should contain provisions for some assistance.

Policy Recommendations

Established risk assessment procedures are sufficient to identify and quantify most environmental hazards of deliberately produced chemicals. Special attention should be given to the formation of byproducts whose identity and properties may be difficult to ascertain. Once these byproducts are identified, the established procedures of risk assessment appear sufficient. As noted in Chapter 7, improvements can be instituted in the information requirements for chemicals, the further development of appropriate test protocols, and the investigation of effects of concern, such as developmental and hormonal disruption.

The present state of knowledge indicates that chemicals with the combined properties of persistence, bioaccumulation, and high toxicity represent a substantial environmental risk. Policy makers and regulators should be especially cognizant of the implications for environmental risk that persistence and bioaccumulation confer on a chemical. Risk-control programs for new and existing chemicals should require vigilant and thorough review of chemicals for persistence and bioaccumulation.

Those chemicals that are recognized to be highly persistent, bioaccumulating, and highly toxic, such as PCBs, DDT, mirex, chlordane, and toxaphene, require international initiatives to halt production, to minimize releases, and to encourage substitution, modeled on the successful Montreal Protocol for CFCs. These initiatives should recognize that certain OC chemicals such as lindane (γ-HCH) and DDT are still widely and beneficially used in countries that cannot afford more expensive and less persistent compounds or that lack the infrastructure to promote the use of these compounds. For unintentionally produced compounds, there should be a determination of the significant sources of PCDD/PCDFs, resulting in a timetable of pragmatic steps to virtually eliminate their formation or emission.

The basic operating principle for risk managers is to ensure that risk assessment yields sufficient and robust data on the fundamental properties of environmental persistence and bioaccumulation. For almost all chemicals, an understanding of environmental fate and biodegradation is needed to assess persistence and bioaccumulation. In the case of chemicals produced in substantial volumes or deliberately released to the environment, full life-cycle and development testing may be required to assess core toxicological concerns. Given the expense and the time required to conduct these tests, international harmonization of testing protocols is important to prevent the inhibition of technical innovation and economic growth that would otherwise occur with multiple and conflicting protocols.

The current regulatory process in developed countries and the response of the chemical manufacturing industry have resulted in the development of practices to safely synthesize, store, and transport most chemicals. There is no apparent need to replace these regulatory processes with broad phaseouts or bans, but there is a need to periodically review and, where needed, to improve these regulatory processes and approaches. It is desirable that these basic practices be adopted globally. It is to the mutual advantage of regulators and industry to cooperatively and proactively scrutinize operations in the life cycle of compounds for the formation of byproducts that may be persistent and bioaccumulative.

Risk managers must ensure that risk assessment procedures are continually improved and are not based on generalized approaches such as the content of one or more atoms of a particular element. Environmental monitoring for contamination and adverse effects is a necessary step to further verify and refine risk assessment procedures. Because hazard assessment involves the review of known effects, laboratory investigations may need to be supplemented by field monitoring for possible unknown effects. In addition, the effectiveness of regulatory controls and remediation efforts must be assessed with data on past, current, and future concentrations of chemicals and on the health of previously affected populations. Finally, communicating the results of regulatory action is necessary to maintain public confidence.

Risk management options such as voluntary agreements, economic incentives or disincentives, and the model of the German chemical law should be carefully explored. The objective is to speed the reduction of chemical emissions, where necessary. Identified barriers include the inability of adversarial systems to develop consensus, the resource limitations of regulatory bodies to address all chemicals in detailed studies, and the litigation of regulatory procedures and decisions by both public-interest groups and industry.

APPENDIX 1

Application of Risk Assessment in Policy and Legislation in North America

Randall S. Wentsel, U.S. Environmental Protection Agency, Washington DC
Adapted with permission from Handbook of Environmental Risk Assessment and Management by Peter Calow, Blackwell Science, Oxford UK, 1998

Advances of science and technology give us new tools to address environmental issues. As we face up to indoor air pollution issues like radon, regional issues like nonpoint sources of water pollution, or global issues like ozone depletion, biodiversity, and climate change, we must use risk analysis and other new tools more frequently. However, risk assessment cannot be applied indiscriminately to all environmental issues.

Policy-makers who support the use of risk analysis believe that federal programs to protect public health and safety and the environment should be carefully targeted to address the worst risks first, and that the risk reduction achieved should be worth the cost. Politicians also hear complaints about unfunded federal mandates to state and local governments and about the growing cost of compliance of environmental requirements by industries.

Opponents of this use of risk assessment argue that exclusive reliance on risk assessment and costs to define problems and evaluate solutions ignores other equally important facets of policy decisions, such as timeliness, fairness to all segments of society, and practicality. They also state that risk assessment and cost–benefit analysis undervalue environmental and health benefits, exaggerate costs, and focus on relatively small costs and risks spread throughout the entire population.

History

Application of risk to environmental losses

The history of the field of risk analysis is the history of the development and use of various techniques for gathering and analyzing information about hazards. Many different analytical techniques have been employed, both qualitative and quantitative. Most are borrowed from other disciplines, including actuarial accounting, economics, biology, geology, geography, and engineering.

Agreement on terminology in the field of risk analysis has often been difficult to do. However, there is general agreement on the following terms.

> "Risk analysis": a term that encompasses various risk activities such as risk assessment, risk management, risk communication, and comparative risk.
>
> "Risk assessment": "the use of the factual base to define the health effects of exposure of individuals or populations to hazardous materials and situa-

tions" (National Research Council [NRC] 1983). Risk assessment is a process that evaluates scientific data. It has 4 basic steps of analysis: hazard identification, dose–response relationships, exposure analysis, and risk characterization.

"Risk management": an approach that uses the results of the risk assessment along with economic analysis, societal issues and concerns, policy issues, and technology and engineering input to determine the management of the health or environmental issue of concern.

"Relative risk": relative (quantitative) magnitudes of the risk estimates.

"Comparative risk": relative (subjective) magnitudes of the concerns after the risk management factors have been included.

Impact of federal statutes

Rapid development of environmental risk analysis methods was stimulated in the last 2 decades, in part by the enactment of major federal environmental laws, such as the Clean Air Act Amendments (CAAA) of 1970; the Federal Water Pollution Control Act Amendments of 1972; the Toxic Substances Control Act (TSCA) in 1976; the Resource Conservation and Recovery Act (RCRA) of 1976; and the Comprehensive Environmental Response, Compensation, and Liability Act (CERCLA) of 1980. Implementation of these laws required the development of criteria, standards, limits, and other numerical descriptors of environmental quality. Some also required agencies to weigh social and economic impacts of regulations against the benefit of improved safety that the regulations were intended to achieve. Risk analysis was adapted to satisfy both requirements.

These statutes to protect human health or the environment vary in specificity; some grant federal agencies broad discretionary power, while others limit the flexibility of the regulators. Factors such as costs, benefits to use, social impacts, or technical feasibility are not used consistently in environmental statutes. Risk analysis methods were developed to measure the risk avoided, the major benefit of most environmental regulations. The U.S. Environmental Protection Agency (USEPA) began in the late 1970s to estimate both costs and benefits under 2 environmental statutes: the TSCA and the Federal Insecticide, Fungicide, and Rodenticide Act (FIFRA) (Schierow 1994).

Initial activities of federal agencies

Federal agencies began to use chemical risk assessment in the 1970s to estimate the cancer-causing potential of chemicals in commerce. Their efforts in the field of risk analysis were encouraged when, in 1980, the U.S. Supreme Court issued an opinion that required the Occupational Safety and Health Administration (OSHA) to perform risk assessments of toxic chemicals as a basis for regulating occupational exposures. The use of risk assessment in the development of regulations continued to be controversial, in part because authorizing legislation often provided little guidance about how risks should be balanced against other factors. Thus, for example, in 1987, a federal appeals court nullified a USEPA rule restricting air emissions of vinyl chloride because the agency had not used risk assessment properly (Rosenthal et al. 1992). Other risk assessment decisions also have been overturned by federal courts because the assessment was judged to be of insufficient technical quality.

By the mid 1970s, agencies had begun efforts to improve coordination among programs and to ensure consistent use of uniform risk assessment procedures within and across agencies of the federal government. Formal procedures for extrapolating research results to estimate human health effects—i.e., chemical risk assessment—were adopted in the late 1970s. In 1976, the USEPA established an internal working group, the Carcinogen Assessment Group (CAG), which published the first internal guidelines for assessing risks of suspected carcinogens. In 1977, the USEPA, the Consumer Product Safety Commission (CPSC), the Food and Drug Administration (FDA), and OSHA formed the Interagency Regulatory Liaison Group (IRLG). The IRLG met as a forum for voluntary coordination and information exchange until 1980. One of its products was a "cancer policy" that attempted to present the scientific basis for determining the carcinogenicity of a substance.

National Academy of Sciences framework

The agencies' efforts to systematize risk analysis were criticized by some scientists and industrial representatives who were concerned that policy matters were overly influencing scientific judgments, and thus the risk assessment process. In response, Congress requested a study by the National Academy of Sciences (NAS) of institutional arrangements to improve agencies' use of risk assessment. This lead to the landmark report entitled "Risk Assessment in the Federal Government: Managing the Process" (NRC 1983). This report presented the framework for conducting risk assessment. It discussed the need to separate risk assessment from risk management and recommended that uniform risk assessment guidelines be adopted by the federal government. While a uniform guide for the federal government has not been adopted, in 1985 the Office of Science and Technology Policy (OSTP) in the Office of the President produced a report on chemical carcinogens and proposed a method to assess their hazards. This report described the state of the science upon which decisions could be made.

In the Clear Air Act Amendments of 1990, Congress asked the NAS National Research Council to assess the current state of the USEPA risk analysis. The 1994 NAS report "Science and Judgment in Risk Assessment" (NRC 1994) listed their concerns and identified major themes for their review. The concerns have included the following:

- the lack of scientific data quantitatively relating chemical exposure to health risks,

- the divergence of opinion within the scientific community on the merits of the underlying scientific evidence,
- the lack of conformity among reported research studies needed for risk characterization, and
- the uncertainty of results produced by theoretical modeling, which is used in absence of measurements.

In response to its mandates, the USEPA has adopted risk assessments that, for the most part, incorporate conservative default options (i.e., those that are more likely to overstate than to understate human risk). Better data and increased understanding of biological mechanisms should enable risk assessments that are less dependent on conservative default assumptions and are more accurate as predictions of human risk.

Major questions addressed in the report were as follows:

Default options: Is there a set of clear and consistent principles for modifying and departing from default options?

Data needs: Is enough information available to the USEPA to generate risk assessments that are protective of public health and scientifically plausible?

Validation: Has the USEPA made a sufficient case that its methods and models for carrying out risk assessments are consistent with current scientific information available?

Uncertainty: Has the USEPA taken sufficient account of the need to consider, describe, and make decisions in light of the inevitable uncertainty in risk assessment?

Variability: Has the USEPA sufficiently considered the extensive variation among individuals in their exposures to toxic substances and in their susceptibilities to cancer and other health effects?

Aggregation: Is the USEPA appropriately addressing the possibility of interactions among pollutants in their effects on human health and addressing the consideration of multiple exposure pathways and multiple adverse health effects? (NRC 1994).

The report contains 70 recommendations for improving risk assessment (NRC 1994). The major recommendations from the 1994 report are as follows:

- The USEPA should generally retain its conservative approach to risk assessment in the initial phase of setting standards, but the USEPA should more clearly state its principles.
- The USEPA should develop and use an iterative approach to risk assessment, beginning with relatively inexpensive screening techniques and moving on to more resource-intensive levels of data gathering, model construction, and model application as each situation warrants.

- The USEPA should work to continually improve the models and data used in risk assessments and develop a standard procedure for deviating from its conservative approach to risk assessment when warranted by scientific considerations.
- In its reports to decision-makers and the public, the USEPA should present information about the resources and magnitudes of uncertainty as well as point estimates of risk.
- Risk assessment is a set of tools, not an end in itself. The limited resources available should be spent to generate information that helps risk managers to choose the best possible course of action among the available options.

Cancer and noncancer guidelines

Following the 1983 NAS framework, agencies such as the USEPA have promulgated their own risk assessment guidelines. The USEPA guidelines for cancer risk assessment were finalized in 1986, and updated draft guidelines were released for comment in late 1995. In a related report, the USEPA in 1986 published a classification guide for carcinogens. Revised proposed guidelines for carcinogen risk assessment were released for comment in August 1995 (USEPA 1995a).

The risks of noncancer effects of chemical exposure to humans also have been recognized and analyzed. As early as 1980, just before it was disbanded, the IRLG was developing guidelines for the risk assessment of effects on reproduction and human development. To date, the USEPA has promulgated final guidelines for risk assessment of genetic mutations and adverse effects on human development. Draft guidelines for neurotoxicity risk assessment were recently released for comment (USEPA 1995b). Guidelines have yet to be established for assessing risks of reproductive failure, respiratory effects, or damage to the immune system.

The USEPA has also provided more guidance on risk characterization and the risk assessor/risk manager interface (Habicht 1992). Concerning discussions between the risk assessor and risk management, the risk assessment information must be clearly presented, separate from any nonscientific risk management considerations. Discussion of risk management options should follow, based on consideration of all relevant factors, both scientific and nonscientific. Key scientific information on data and methods in both risk characterization (e.g., use of animal or human data for extrapolating from high to low doses, use of pharmacokinetics data) must be highlighted. The assessor should write a statement of confidence in the assessment that identifies all major uncertainties, along with comment on their influence on the assessment. The USEPA established policy to present information on the range of exposures derived from exposure scenarios and on the use of multiple risk descriptors (i.e., central tendency, high end of individual risk, population risk, important subgroups, if known) consistent with terminology in agency guidelines.

Guidelines are necessary to ensure that risk assessments are conducted consistently and are, therefore, more easily evaluated by independent experts. However, guidelines do not ensure that scientists will agree with resulting risk estimates. In fact, some criticize the USEPA's risk estimates for carcinogens because they do not like the guidelines or think different rules should apply to certain chemicals. Controversy surrounds risk assessment, partly because the field is so young that its methods have not been studied thoroughly and validated.

Council, committee, and societal actions

Other groups are also active in risk assessment and risk management. The National Research Council (NRC) of NAS has a permanent Committee on Risk Assessment Methodology (CRAM). They are working to consider changes in the scientific foundation of risk assessment that have occurred since the 1983 report. They issued a report entitled "Issues in Risk Assessment" in 1993 (NRC 1993).

A Commission on Risk Assessment and Management was also authorized by the Clean Air Act Amendments. The Commission became active in 1994. They are mandated to do the following:

- Consider issues related to how human health risk assessments are performed and, most importantly, how the results of human health risk assessments are used in regulatory decision-making.
- Focus on the 5 aspects of their mandate:
    1) uses and limitations of risk assessment in decision-making,
    2) appropriate exposure scenarios,
    3) uncertainty and risk communication,
    4) risk management policy issues, and
    5) consistency across agencies.
- Consider the 1994 NRC report "Science and Judgment in Risk Assessment."
- Report to Congress in March and September 1996.

Within the White House National Science and Technology Council, a Committee on Environment and Natural Resources (CENR) was formed. The mission of CENR is to develop a strategic guidance for environmental research and development activities across the federal government. Each of 10 subcommittees under the CENR is charged with defining critical policy questions or issues relevant to their environmental issue area (e.g., air quality, toxic substances, water resources) or cross-cutting methodological (e.g., risk assessment, social and economic sciences) area (Figure 1). The scientific knowledge and corresponding research necessary to meet those policy challenges are then to be identified. Within CENR, a subcommittee on risk assessment has been established (OSTP 1995).

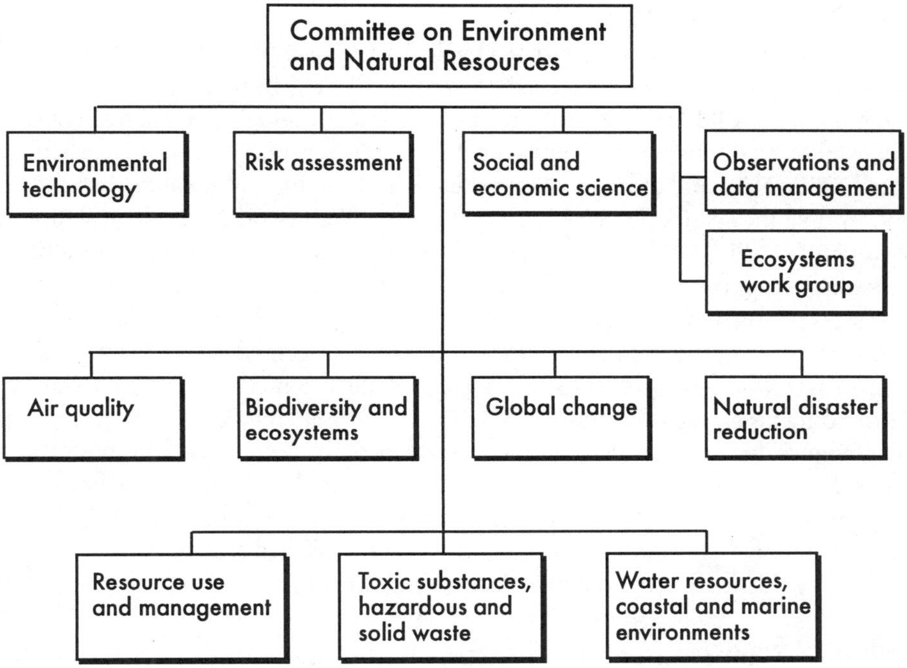

Figure 1 Structure of the U.S. Committee on Environment and Natural Resources

The subcommittee on risk assessment focuses on methodological research topics and upon the development of means for effective translation and transmittal of scientific knowledge to policy-makers. They have a dual role: to understand how key data gaps in each issue area influence (and perhaps inhibit) risk assessment and to identify high-priority needs for methodological research for assessment, characterization, and communication of risks. The subcommittee was established to review federal risk assessment research programs and to develop ways to improve the enhancement and coordination of federal activities. In examining the current state of risk assessment research across the federal agencies and proposing research priorities that might enhance the performance of risk assessments, the subcommittee is considering issues such as the following:

- gaps in scientific knowledge that influence assessments for different types of risks and endpoints,
- data needs for the performance of risk assessments and the potential for sharing data within and across agencies, and
- alternative approaches to risk assessment across agencies, across federal statutes, and internationally (CENR 1995).

Ecological risk assessment

Ecological risk assessments were initiated by the USEPA in order to develop water quality criteria required under the Clean Water Act (CWA). The first ecological risk assessment was a 1981 Synfuels assessment (Suter 1993). Ecological risk assessments initially followed the 1983 NAS framework, but in 1991 several workshops were held by the USEPA, NAS, and others to reassess the procedures for ecological risk assessment. In 1992 the USEPA published a report entitled "Framework for Ecological Risk Assessment" (USEPA 1992). This new framework, while similar to the human health framework, recognizes the difference between ecological and human processes. Differences between them include the various endpoints in ecological risk assessments, such as protection of bald eagles or maintaining species diversity in an aquatic system. This framework appears to have been generally accepted as the most appropriate procedure for ecological assessments. In December 1995 the USEPA released a draft guidance document on ecological risk assessment for peer review and comment (USEPA 1995c). The Departments of Energy and Defense also have guidance documents providing additional details on ecological risk assessment (USDOE 1994; Wentsel et al. 1994). In addition, the Canadians have published ecological risk assessment guidance for remediation efforts (Gaudet 1994).

Key USEPA reports

In 1987 the USEPA published "Unfinished Business: A Comparative Assessment of Environmental Problems" (USEPA 1987). This effort was the first time that environmental problems were compared to each other without regard to programmatic context. Much of the Congressional activity and interest in comparative risk or risk ranking was generated from this report. In this report, 31 environmental problems were ranked by USEPA scientists based on the relative risk that each posed in 4 categories: cancer risks, noncancer health risks, ecological effects, and welfare effects (which included such effects as impairment of visibility and materials damage). It is important to note that the risks were ranked assuming that current regulatory controls were in place. So the low relative risk of hazardous waste sites was, in part, due to the existence of regulations and funds to treat that problem. Other problem areas (e.g., indoor air pollution) were ranked as relatively high risks because they have never been regulated by the USEPA or any other federal agency. In addition, the report emphasized that the rankings were based on information available at the time, which, according to the scientists, contained "substantial gaps." The results indicated that no environmental problem ranked relatively high or relatively low in all 4 categories of risk. Problems that were ranked as relatively high or moderate in 3 or more categories included air pollutants, stratospheric ozone depletion, pesticide residues on food, and other pesticide risks. Other problems that were ranked as relatively high risks to human health included hazardous air pollutants, indoor air pollution, indoor radon, pesticide applications, exposure to consumer products, and worker exposures to chemicals. The other problems that ranked high in risk to ecology or human welfare included global warming; surface water pollution; physical alteration of wetlands, estuaries, and other aquatic habitats; and mining wastes. One of the conclusions of the report was that the scientists' ranking of risks to human health differed from the ranking implied by national polls of public concerns. The public reported greatest concern about chemical waste disposal, water pollution, chemical plant accidents, and air pollution. They had moderate concern about oil spills, worker exposure, pesticides, and drinking water. The public had relatively low concern about indoor air pollution, consumer products, radiation (other than nuclear power), and global warming. The USEPA budget appeared to correlate better with the priorities assigned by the public than with the risks thought to be greatest by the USEPA scientists. Interpretation of these results requires caution. For example, the low relative risk of hazardous waste sites (as indicated by data available in 1987) was due, in part, to the existence of regulations and the availability of funds to treat the problem. Problems such as indoor air pollution were characterized as relatively risky, at least in part because they were not regulated by any federal agency. Additional concerns of the report have centered on the methods used to rank the environmental problems. Ranks were assigned using qualitative professional judgments, incomplete datasets, and other unvalidated methods. The feasibility of controlling risks, the nature of the activities that posed risks, and the distribution of risks and benefits were not considered.

In a follow-up study, the USEPA Science Advisory Board produced a well-regarded report entitled "Reducing Risk: Setting Priorities and Strategies for Environmental Protection" (USEPA 1990). The purpose of the report was to determine strategies for reducing risks and to recommend approaches for ranking and reducing risk in the future. Two committees were formed, one to study ecological and welfare risks and a second to address human health risks. The committees were to rank risks within their areas of concern. However, the report pointed out that the data gaps, uncertainty, and judgment issues made the process of ranking risk tenuous and unreliable. The authors of the "Reducing Risk" report also recommended 10 actions for the USEPA to take to reduce risk and improve the risk assessment process. While these recommendations have been well received, they have not been implemented by the USEPA through a top-down approach. The report also addressed problems with current cost–benefit analysis of environmental resources. Current methods assume that the future value of an ecological resource must be less than its present value. This policy leads to depletion of irreplaceable natural resources. The "willingness to pay" technique is also misused in economic analysis. The public may not care about wetlands, but the contribution that wetlands make to the larger ecosystem is of value to the public now and in the future. Techniques need to be developed to assess the real long-term value of ecosystems.

Comparative Risk Assessment

As a result of the 2 reports on risk by the USEPA Science Advisory Board and their members, comparative risk assessment, a procedure for ranking environmental problems by their relative risk, has received support from risk managers and policy-makers. It is used to prioritize by risk alone, risk and cost–benefit, or using all of the risk management components. Due to the federal regulatory mandates, which impact the budgets of state and local governments, they are in favor of comparative risk assessment. No standard methods are available for the ranking of risks. The USEPA Office of Policy, Planning, and Evaluation (OPPE) has, however, funded comparative risk projects for states and has supported human health risk rankings and ecological risk rankings for the USEPA regions (USEPA 1993). These studies have had a mixed success. However, the involvement of various environmental offices, policy-makers, and the public in priority setting of environmental issues is very important. Van Houtven and Cropper (1993) reported on the use of risk ranking within an environmental category. Benefit–cost analysis was used to evaluate 3 USEPA regulatory actions. One example on the cost–benefit of banning asbestos gave results in cost per cancer case avoided, and it was a good example of how relative risk assessment, within a category, could be used. Comparative risk assessment is an attractive use of risk assessment, cost–benefit analysis, and involving the public. The unfunded federal regulatory mandates to state and local governments and the limits in the federal budget will push decision-makers toward this process. However, data gaps, subjective issues, and uncertainty in any ranking of risks must be recognized. The USEPA Science Advisory Board report on reducing risk ranked ecological risks in 3 broad categories. Groupings of environmental problems in broad categories may minimize the imprecision of the process. The OPPE efforts in this area should be evaluated for use as a framework for ranking environmental issues. Improvements in methods for comparative risk assessment will increase its use in budgeting and prioritization of environmental problems.

Some groups argue against priority setting based on risk because it focuses on the annual probability of death or disease. They believe that the process ignores other equally important factors, such as the degree of force used to implement the risk reduction strategy and the fairness of the result to all segments of society (Schierow 1995).

The California Comparative Risk Project (CCRP 1994) differed from previous comparative risk projects by developing 3 lists of environmental issues instead of just 1 list of environmental problems for analysis. The CCRP developed "environmental topic lists," which compared and ranked environmental hazards from 3 distinct and relatively consistent subsets of environmental issues.

- List I included the traditional statutory division of environmental problems into the media (air, water, and land) that are impacted by the release of toxic substances from different sources.
- List II addressed major environmental stressors.
- List III considered overarching categories of human activity, which impact the environment.

Slovic et al. (1995) presented the results of a comparative risk assessment study in Canada. The results showed that Canadian toxicologists had a lower perception of risk and were more supportive of the use of chemicals than the Canadian public. The attitude of the public was quite negative and showed the same lack of dose–response sensitivity found in the earlier U.S. studies. Both of the groups studied lacked confidence in the value of animal studies for predicting human health risks. Technical judgments of toxicologists were found to be associated with factors such as affiliation, gender, and world view.

Uses of risk assessment

Many functions of the federal government are intended, either directly or indirectly, to reduce risks to public health and safety and the environment. In this function, risk analysis is used as a tool for evaluating what is known about things that cannot be known with certainty due to effects of hazards that are unpredictable, lack of scientific understanding of data, or uncertainty with models or other extrapolations. Risk analysis produces an estimate, and it varies due to the quality of information used by the assessor. Risk analysts can only discuss the likelihood of various outcomes, discuss the role of scientific judgment, and present risks as statistical probabilities. Agencies and departments evaluate risks, formally or informally, in most of their activities. Uses of risk assessment include the following (CENR 1995):

- defining problems and predicting risks;
- selecting risk avoidance or mitigation strategies and developing management programs;
- setting standards to protect human or ecological health, and evaluating ongoing risk reduction activities; and
- determining management and policy priorities.

Policy or nonscientific decisions in risk assessment include choices of assumptions and methods to address uncertainty in available scientific data. These decisions become significant when 2 federal agencies evaluate the same data and calculate different levels of risk. Risk assessment techniques typically rely on multiple assumptions that may be untested, or even untestable. In human health risk assessment, assumptions include the relevance of animal data to humans, the choice of a dose–response model to extrapolate from high (experimental exposures) to lower environmental doses, exposure pathways, and protections levels, i.e., 95% of the affected population. The NRC report (1994) identified at least 50 decisions required in conducting a cancer risk assessment that cannot be made on a scientific basis, and many of these decisions have strong implications for public policy.

To improve risk assessment, the NAS National Research Council (NAS 1994) panel concluded that the most effective way would be by improving the quality and comprehensiveness of knowledge. The NAS recently evaluated the availability of data for risk analyses of 189 hazardous air pollutants and concluded that the USEPA did not have "sufficient data to assess fully the health risks—within the time permitted by the Clean Air Amendments of 1990."

Risk assessment has a major role in California's Proposition 65—the Safe Drinking Water and Toxic Enforcement Act of 1986—which became effective on 1 January 1997 (Risk Policy Report [RPR] 1995). The governor must publish a list, updated each year, of chemicals known by the state to cause cancer or reproductive toxicity. Businesses cannot "knowingly and intentionally" expose any individual to the chemicals, 12 months after they are listed, without first giving "clear and reasonable warning." Exceptions are that no warning is required if exposures would result in a risk lower than "no significant risk," defined as "one excess case of cancer per 100,000 individuals exposed over a 70-year lifetime" for carcinogens and as "less than one-thousandth of the no-observed-effect-level" for reproductive toxicants. Other exceptions also apply that are not risk-related (e.g., businesses with fewer than 10 employees), and the law prohibits discharges of chemicals into state drinking water sources 20 months after listing. Proposition 65 now lists approximately 500 chemicals, including some with no commercial use and 75 to 100 that are drugs whose potential health risks are told to users when they are prescribed. For commercially useful chemicals, more than 250 numerical standards have been set, and none of these standards has been challenged in court.

There may be a greater role for risk assessment in the Endangered Species Act (ESA). A recent NAS National Research Council report (NRC 1995) stated, "Despite the major advances that have been made in models for predicting mean extinction times, the existing models still have substantial limitations. Most of the models deal with only one risk factor at a time and fail to incorporate the interactive effects of multiple factors on reducing the time to extinction. This might result in a tendency for such models to underestimate the risk of extinction."

The NRC offered several recommendations for strengthening the implementation of the ESA, which deal with the use of risk assessment (NRC 1995):

1) Extinction models should play a more central role in ESA decisions, especially as guides to research.
2) Extinction models should be used to categorize species according to relative risk of extinction.

3) Levels of risk to trigger ESA decisions must be framed as a probability of extinction during a specified time period, and it will be necessary to view extinction over longer time periods (i.e., on the order of hundreds of years).
4) An estimate of the number of individuals needed for long-term survival must be based on the biology of the organism.
5) To the degree that they can be quantified, the levels of risk associated with endangered status should be higher than those for threatened status.
6) The procedures used to make ESA decisions should be explicit and well documented.

The NRC Committee identified 2 main categories of risk: the risk of extinction and the risk associated with unnecessary expenditures in the face of substantial uncertainties about future events. They noted that policy decisions or public input had played a role in setting different levels of risk for different taxonomic groups. Since implementation of the ESA, numerous models have been developed for estimating the risk of extinction that provide valuable insights into potential impacts of management activities and recovery plans. However, despite major advances, most of these models still have substantial limitations, such as not taking into account multiple risks and the interaction of all the factors promoting extinction. They also noted that many ESA listing decisions and recovery plans have not met the guidelines suggested by current scientific thinking (i.e., population viability and risk assessment). Those favoring the use of extinction models and other risk assessment tools for listing and recovery planning state that they are scientifically sound and should lead to better rates of species survival. Their weakness is that extinction models are still relatively crude and have substantial limitations. In addition, such modeling is expensive and time consuming, especially when combined with the field research needed to produce the data needed for model input.

Application of Risk Assessment

Public policy and risk management

When establishing public policy in environmental protection, the decision maker must consider the various risk management parameters. These include cost–benefit, engineering, societal issues, political issues, and risk assessment. The goal of risk management is to reduce the risk or hazard of an environmental problem to humans and the environment. Risk assessment estimates whether there is a risk and, if so, the uncertainty bounds. The product of a risk assessment is judged by the risk manager as to how serious the risk is, availability of mitigation measures, cost to mitigate, and other management concerns. Examples of the management concerns are: Who or what is impacted? Are there technologies available to mitigate? Does the cost reduce the risk significantly? Does the public care about this specific issue? Are there political concerns?

A White House OSTP paper (1995) listed several pertinent questions:

1) How should a risk assessment be used and presented? How can risk assessment best be used to inform environmental policy and management decisions?
2) As a means to ensure accountability and transparency, what is the most effective way to characterize and communicate information about risks, uncertainty in assessments, and limits to assessments?
3) How might uncertainty be weighed into policy decisions?

Risk assessment as a tool

Risk assessment is appropriate as an analytical tool to help identify problems, set regulatory priorities, compare effectiveness of risk management options, communicate to the public, and identify research needs. Since the purpose of environmental regulations is to protect human health and/or the environment, risk assessment will estimate needed protection levels either quantitatively or qualitatively. Risk assessment is often involved in the generation of health or environmental criteria used in the regulations. Typically, risk assessment alone will not provide a hard and fast number for regulation. Other questions asked in the OSTP report were as follows:

1) When should a risk assessment be undertaken? When will it enhance policy decisions?
2) What is the appropriate level of effort and precision to dedicate to risk assessment?
3) What is the estimated incremental value of obtaining additional information through increased research, versus the incremental cost of delaying a decision?
4) How should a risk assessment be conducted? How should risk assessments be framed and bounded to inform policy decisions appropriately and most effectively?
5) How should environmental justice and other social, cultural, and ecological concerns be integrated into framing a risk assessment and defining relevant data needs?

Implementation in environmental statutes

In the implementation of environmental statutes by the USEPA and other agencies, there are variations in how risk is addressed in the regulations. There is also variation between USEPA offices in the parameters incorporated into risk assessment. Both of these variables affect the determination of the appropriate use of risk assessment in environmental statutes. A uniform approach would enhance the usefulness of risk assessment as a regulatory tool. The previous USEPA administration formed a Risk

Assessment Council and initiated action toward a unified approach to risk assessment (Habicht 1992).

The environmental statutes are written as either narrative or numerical directives. The USEPA implements narrative directives primarily through 3 different approaches: health-based standards, technology-based standards, and no unreasonable (balanced) risk standards (Table 1). The primary environmental laws use one or more of these approaches (Rosenthal et al. 1992).

Health-based standards

Health-based standards regulate on the protection of human health or the environment without regard to technology or cost factors. The primary health-based standard is the Delaney Clause. The Delaney Clause is a zero-risk standard and prohibits the approval of any food additive that has been found to induce cancer. The FDA has made few rule-making decisions that might trigger Delaney's prohibitions and largely avoided the strictness of the Delaney Clause covering food additives. However, the USEPA, which must consider the carcinogenicity of pesticide residues on foods when making registration and re-registration decisions about pesticide products, has had to address the clause. While the USEPA argued that the cancer risks posed by these pesticides were negligible, a federal appeals court ruled in July 1992 that the Delaney Clause prohibits food additives that are carcinogenic, regardless of whether or not the magnitude of the cancer risk is negligible. It has been used only twice by the USEPA to refuse a new food use of a carcinogenic pesticide. The ESA currently requires the protection of listed biota without regard to technology or cost factors. However, this Act is not a zero-risk standard.

Section 112 under the Clean Air Act (CAA) requires values that protect the public health allowing an ample margin of safety. This is defined, by the USEPA, as an increase in risk by no more than 1 in a million. Initially, this Act was designed to regulate as a health-based standard. However, between 1970 and 1990, extensive litigation occurred over the role of risk assessment under Section 112. The wording "ample margin of safety" was found to be too vague, and it is a good example of where risk assessment cannot be used to establish a single number for regulation. In 20 years, only 7 air toxicants were listed and regulated. Congress turned to technology-based standards in 1990 in the hope that the USEPA would be able to reduce the delay in standard setting. The technology-based standards are the initial phase of risk reduction. If, after 8 years, the residual risk to the maximum exposed individual (MEI) is greater than 1 in a million, then the USEPA must apply health-based standards. During the Congressional debate, there was much dissension over risk assessment. This is reflected in the directive to NAS to evaluate USEPA methods. Also, Congress

Table 1 Example of risk-related provisions in selected federal laws (OSTP 1995)

Statute	Regulatory authority	Effects of concern	Approach to risk
Consumer Product Safety Act	Consumer products	"An unreasonable risk of injury"	Balance risks against product utility, cost, and availability
Occupational Safety and Health Act	Risks in the workplace	"Material impairment of health or functional capacity"; what is reasonably necessary or appropriate to provide safe and healthful employment?	Attain highest degree of health and safety protection; best available evidence; technical and economic feasibility
Federal Water Pollution Control Act (Clean Water Act)	Waste water discharges	"The toxicity of the pollutant, its persistence, degradability..."	Best available technology that is economically achievable
Clean Air Act Section 109	National ambient air quality standards	Protect public health	Set standards to provide ample margin of safety
Section 112	Emissions standards for hazardous air pollutants	Adverse effects to health and the environment	Reduce emissions using maximum achievable control technology and later address "residual risk"
Section 202	Emissions standards for new motor vehicles	Unreasonable risk to health, welfare, or safety	Greatest degree of emission reduction achievable through technology available, taking into consideration cost, energy, and safety factors
Federal Insecticide, Fungicide, and Rodenticide Act Section 3	Pesticides	Unreasonable risks to health and the environment	Balance risks against economic benefits to pesticide users and society
Toxic Substances Control Act Section 6	Existing chemicals in commerce	Unreasonable risks to health and the environment	Balance risks against economic benefits, considering alternative technologies
Comprehensive Environmental Response, Compensation, and Liability Act Section 313	Toxic release inventory	Hazards to human health or the environment	Reporting is based largely on hazard and quantity used
Section 9621	Hazardous waste site remediation	Persistence, toxicity, mobility, and propensity to bioaccumulate; short- and long-term health effects	Protect human health and the environment in a cost-effective manner
Safe Drinking Water Act Section 300g-1(b)	Drinking water	Known or anticipated adverse effects on human health	Set a goal (maximum containment level goal [MCLG]) with an adequate margin of safety, and define a maximum contaminant level (MCL) as close as feasible to the goal

created the Commission on Risk Assessment and Management to study the proper use of risk assessment and propose legislation.

Technology-based standards

Technology-based standards require best practicable control technology, best available technology, and other controls for pollution reduction or treatment. Parameters evaluated include the determination of the effectiveness of methodologies on concentration reduction and often costs, rather than reduction of risk. The SWDA sets drinking-water standards using the best technology, treatment technique, and other methods. When maximum contaminant level goals (MCLGs) are established, risk assessment provides the scientific basis for the value. The enforceable standards are called maximum contaminant levels (MCLs) They are set as close to the MCLG as is "feasible with the use of the best technology, treatment techniques, and other means... which are available (taking cost into consideration)." The enforceable standards are set using technology criteria such as the affordability analysis of analytical technology for detecting contaminants (Rosenthal et al. 1992). While MCLGs have limited impact under SDWA, they are used in CERCLA to establish cleanup levels. The radon issue in the proposed SDWA reauthorization is addressed with a unique multi-media risk reduction approach. Radon exposure from drinking water does not have to be controlled if the locality can regulate the reduction of radon in indoor air. The CWA requires technology-based controls to treat water pollution. Two sets of water quality criteria are used to protect aquatic life and to protect human health. These water quality criteria are based on risk assessment input. More recently, criteria to protect aquatic life have been set based on effluent bioassays or in-stream criteria. The criteria are the goals that the technology-based standards seek to achieve. The use of risk assessment to support technology-based standards is an appropriate use of risk assessment.

Balanced risk-based standards

The use of "no unreasonable risk" in several statutes requires the balancing of risk assessment results against cost–benefit to determine the risk management approach. Risk assessment is used under the Toxic Substances Control Act (TSCA), and it is balanced against the benefits of the substance, the availability of substitutes, and the economic cost of the action to limit or ban. The TSCA uses a balanced cost–benefit approach to limit unreasonable risk of injury to health and the environment. In the Fifth Circuit's ruling on the asbestos ban, "...the court did not forbid the agency from controlling toxic chemicals. Rather, it ruled that the USEPA had to follow every step in the analysis of risks and costs required under the law's Section 6—which authorizes permanent controls up to bans—before issuing a rule" (Rosenberg and Wheeler 1993). The court refused to let the USEPA ban asbestos, under the Act, because the USEPA did not show that the ban on asbestos represented the least burdensome process for reducing risk. The data on risk reduction, in terms of cancer cases avoided, were not significant when compared with the costs. The court required the USEPA to use the balanced risk management that the Act required. Rosenberg and Wheeler (1993) stated that the TSCA is effective because it requires balancing, it has the voluntary cooperation by Premanufacture Notice submitters, and it has the unseen benefits of the chemicals that were never produced because of the scrutiny of the TSCA.

However, few regulatory decisions have been made and the TSCA is regarded by many as ineffective legislation. This is due to several factors. The term "unreasonable risk" has an unclear meaning and is open to wide interpretation. The fear of judicial oversight has led to inaction. There are few instances where the TSCA has primary authority. The discretionary nature of the narrative standard has made clear implementation difficult. Also, the focus on individual chemicals combined with the expensive and time-consuming regulatory process leads to a limited number of regulatory decisions (Shapiro 1990; Rosenthal et al. 1992).

FIFRA also balances the health and environmental impact of a chemical versus economic cost–benefit and societal concerns. This statute, instead of setting a level of acceptable risk, provides the USEPA with flexibility to make determinations on a case-by-case basis using a balance of diverse factors. However, FIFRA has been criticized for being a slow process, with industry using the risk assessment process as a delaying technique.

Under CERCLA, sites are initially evaluated for their hazard to humans or the environment by the hazard ranking system (HRS). This system uses a health assessment along with exposure and persistence data on the chemicals to rank sites for inclusion as Superfund sites. Those HRS values above 28.5 require the site to be listed on the National Priority List (NPL). At an NPL site, risk assessment is used in 2 ways. First, a baseline risk assessment of health and ecological concerns is conducted to establish whether the risk justifies mitigation. Second, in the remedial investigation/feasibility study (RI/FS), risk assessment is used to establish risk levels for different areas of the Superfund site. In CERCLA, health and ecological risk are relatively equal in importance. Applicable, relevant, and appropriate requirements (ARAR) are standards from other federal or state environmental statutes. These are used, when available, to set cleanup criteria. The CERCLA also stresses a preference toward permanence in rendition decisions. Risk assessment can identify areas of the site that have elevated risk. These risks can be related to chemical levels for remediation. An example of where risk assessment is used would be to support the ESA. The risk assessment could establish the chemical concentration (with uncertainty bounds) to protect a given organism. The results from risk assessment are used with the other risk management tools to determine the mitigation of a site. Risk assessment is also used to evaluate alternative remedial actions.

Section 307 of the CWA requires the USEPA to set effluent limitations for discharges of toxic pollutants to surface waters achievable by applying the best available technology that is economically achievable and "will result in reasonable further progress toward the national goal of eliminating the discharge of all pollutants" (paragraph 301[b][2][A]). In addition, the CWA requires effluent standards to provide an ample margin of safety, taking into account "the toxicity of the pollutant, its persistence, degradability, the usual or potential presence of the affected organisms and the nature and extent of the effect of the toxic pollutant on such organisms, and the extent to which effective control is being or may be achieved under other regulatory." The CWA does not instruct the agency in how it should balance these considerations relative to one another.

In summary, each environmental statute approaches the problem of controlling risk from a different vantage point and authorizes consideration of different factors by the USEPA. Some statutes authorize several different approaches for controlling different kinds of risk. One statute, CERCLA, incorporates all of the other statutory approaches to risk, at least in effect. These diverse statutes, however, seem to conform to a few general rules. They generally allow consideration of the costs of regulation at some stage of risk management, either explicitly or by reference to feasible, practical, or available technology (the Delaney Clause is an exception to this rule); they tend to exclude costs from consideration in the development of scientific documents (e.g., water quality criteria), safety goals (e.g., safe drinking-water goals), or health-based standards of ambient environmental quality (e.g., primary air quality standards), all of which clearly are meant to be protective of health and the environment; and they require consideration of costs when the USEPA directly regulates commerce, i.e., the production, distribution, and use of commercial products.

Importance of flexibility in risk management

Flexibility in the management of environmental risk reduction is an important consideration in the establishment of public policy. Portney (1990) discussed how decisions on risk reduction are eventually made on the costs of protection versus the added benefits. At some point, society will decide that the reduction in risk is too small to justify the cost. The health-based approach is inflexible to make these tradeoffs. Technology-based standards may not be flexible enough to recognize that densely populated or pristine areas may need special regulation. These tradeoffs may be difficult to do under technology-based standards. The balanced risk-based approach has the potential to make these tradeoffs for effective management of risk reduction. The tradeoffs of the multimedia approach and pollution prevention will work best under a balance-based regulatory framework. Ruckelshaus (1985) also supported flexibility in environmental regulations. He called for agencies to be given greater flexibility to address environmental problems at the local level and to confront economic and social tradeoffs required for effective risk management. A well-known example where Ruckelshaus had the affected public decide the issue was in Tacoma, WA. Arsenic was released into the air from a copper smelter. After controls were in place, it appeared impossible to eliminate the carcinogenic risk from arsenic. To eliminate the risk would mean shutting down the plant. Ruckelshaus turned the decision over to the Tacoma citizens. The public considered the risk management tradeoffs and came up with ideas to minimize the risk. Their rational approach to risk and the use of tradeoffs enabled the plant to stay open.

Hoskins et al. (1994) calculated the occupational safety and health risks associated with hazardous waste site remediation. The 2 occupations with the highest death rates, truck driver and laborer, contributed most to total exposure hours for the alternatives considered. The calculated expected number of fatalities was converted, using the Poisson distribution, to the risk of experiencing at least 1 fatality, as follows: 0.149 for excavation and landfill, 0.012 for capping, and 0.014 for capping plus slurry wall. They stressed the need for a more scientific, quantitative approach to remediation decisions involving risks to workers.

Breyer (1993) (recently appointed to the U.S. Supreme Court) describes a "vicious circle" that inhibits proper prioritization of environmental risks. The circle is caused by the public's irrational reaction to risk, Congress' attempt to regulate risk through detailed legislation, and uncertain and irrational methods in technical regulation. He proposes to address the problem through restructuring risk regulation in the executive branch agencies. Through his various experiences, Breyer has observed 3 main problems in the regulation of risk: tunnel vision, random agenda selection, and inconsistency. An example of tunnel vision ("getting the last 10%") is where regulation of a single medium requires strict treatment of an environmental problem. The strict treatment imposes high cost without significant additional environmental benefits. The random agenda problem refers to the lack of prioritization of environmental problems. This can be minimized through the ranking of risks. The inconsistency problem refers to agencies and offices using different methods to estimate risk and cost–benefit. This results in varied and confusing information on the impact of an environmental problem.

Canada

Lind (1995) reported on the development and use of policy goals for health and safety in Canada. In 1990, the Royal Society of Canada and the Canadian Academy of Engineering formed a joint Committee on Health and Safety. The committee proposed 3 guiding principles for economic scrutiny of health management and safety regulation. The first was that accountability, which was defined as decisions for the public, must be open and apply across the complete range of hazards. Balancing all important benefits, hazards and costs must be demonstrated and made available openly for comparison with other potential or realized alternative uses of the resources. Next was net benefit, which states that risks shall be managed to maximize the total expected net benefit to society. The net benefit of an option is the excess of the total of all benefits over the total of all costs and hazards. Since benefits and detriments are usually different in kind and accrue to different degrees to individuals, it is necessary to arrive at a common basis of measurement. The last principle was the measurement value. The committee recommended the safety benefit to be promoted as quality-adjusted life-expectancy (QALE). The committee proposed to maximize the net benefit to all members of society at all ages, and QALE aggregates net benefit over all individuals, health states, and ages in a society or group. They discussed and acknowledged that the net benefit of any particular intervention is not evenly distributed. Lind (1995) suggested that 2 other issues be considered: procedure (processes for public health and safety must be quantified and follow accepted protocol) and sovereignty (management of risk to the public should be subject to public consent).

The Government of Canada is introducing a Toxic Substances Management Policy (TSMP), which is based on science (Environment Canada 1995). The policy calls for virtual elimination from the environment of toxic substances that result from human activity and are persistent and bioaccumulative. The policy also calls for cradle-to-grave management of all other substances of concern that are released to the environment. It establishes a preventative and precautionary approach to deal with substances that enter the environment and could cause harm to human health or the environment.

The key objectives are as follows:

1) virtual elimination from the environment of toxic substances that result predominantly from human activity and are persistent and bioaccumulative (i.e., Track 1 substances), and
2) the management of other toxic substances and substances of concern, throughout their entire life cycles, to prevent or minimize their release into the environment (i.e., Track 2 substances).

A substance that meets all 4 criteria outlined in Table 2, i.e., persistent, bioaccumulative, toxic, and primarily the result of human activity, will be targeted for virtual elimination from the environment (Track 1 substance). This objective will be achieved by addressing sources of release to the environment or by removing or managing the substance if it is already in the environment. Pollution prevention strategies will be used to prevent the measurable release of a Track 1 sub-

Table 2 Criteria for selection of substances for Track 1 for Canadian toxic substances[1]

Persistence				
Medium	Half-life	Bioaccumulation[2]	Toxicity	Predominantly anthropogenic
Air	≥ 2 d	BAF ≥ 5000	CEPA-toxic	Concentration in
Water	≥ 182 d	or	or	environment largely
Sediment	≥ 365 d	BCF ≥ 5000	CEPA-toxic	resulting from
Soil	≥ 182 d	or	equivalent	human activity
		log K_{OW} ≥ 5.0		

[1] Source: Environment Canada 1995.
[2] BAF=bioaccumulation factor, BCF=bioconcentration factor, log K_{OW}=octanol/water coefficient

stance from domestic sources. A Track 1 substance that cannot be managed successfully throughout its life cycle will be targeted for phaseout of generation and uses. Through bilateral or multilateral agreements, the federal government will work to eliminate Track 1 substances that originate from sources outside the country. Remediation may be undertaken when a Track 1 substance is already in the environment. Where the benefits to the ecosystem or to human health of removing the substance outweigh cleanup costs, remediation will be considered. Otherwise, management strategies will focus on minimizing exposure and the site's potential risks.

The federal government will identify toxic substances and other substances of concern subject to management under Track 2 through a variety of existing programs. Substances that do not satisfy all 4 criteria are candidates for life-cycle management to prevent or minimize their release into the environment. Management strategies, including pollution prevention, pollution control, and remediation, will be based on a life-cycle approach. Risk assessment and risk management approaches will be used to identify Track 2 substances and management options. Socioeconomic factors will be considered when determining long-term environmental goals, targets, strategies, and time lines for Track 2 substances.

Criteria to determine Track 1 or Track 2 have been documented. A substance is considered toxic if, after rigorous scientific assessment and based on decisions taken under federal programs, it either conforms or is equivalent to "toxic" as defined in Section 11 of the Canadian Environmental Protection Act (CEPA). The document entitled "Toxic Substances Management Policy—Persistence and Bioaccumulation Criteria" provides details about these criteria, including their numerical values, the process and rationale used in establishing them, and information about how they are applied. Many Track 1 and Track 2 substances are already subject to federal statutory management strategies that are consistent with this policy—CEPA, the Pest Control Products Act, and the Food and Drugs Act. No new action will be required for a substance that is adequately managed under existing programs.

Summary

In summary, risk assessment is used in 3 basic ways by environmental regulations: health-based, technology-based, and balanced risk-based approaches. In health-based standards, where risk is used without regard to cost or technology considerations, risk assessment alone has been found to be too restrictive, inflexible, and has resulted in reduced enforcement. The use of risk assessment in technology-based standards is primarily to set a treatment goal. Technology treatments are enacted to approach the goal with the best or most appropriate technology available. This use of risk assessment has worked well. Standards can be implemented and enforced. However, technology-based standards may not be flexible enough to consider multimedia approaches. A balanced risk-based approach, where risk assessment is used along with economic, technological, and societal concerns, provides an encompassing risk management approach to environmental problems. The balanced approach may take more time for implementation. The TSCA, FIFRA, and CERCLA each have had problems with the implementation of this approach. Enforcement will not be as clear. There are case-to-case variations depending on the regulator or site-specific information. However, the inclusion of the public in the decision-making process is important. The flexibility of the risk management can provide an avenue for novel mitigation of environmental issues.

References

Breyer S. 1993. Breaking the vicious circle toward effective risk regulation. Boston MA: Harvard Univ Pr.

[CCRP] California Comparative Risk Project. 1994. Toward the 21st century: planning for the protection of California's environment. Sacramento CA: CCRP.

[CENR] Committee on Environmental and Natural Resources. 1995. Draft research strategy and implementation plan: risk assessment research in the federal government. Washington DC: CENR, National Science and Technology Council.

Environment Canada. 1995. Toxic substance management policy. Message from Minister of the Environment. June 1995. Ottawa: Environment Canada.

Executive Order 12866 of 30 September 1993. *Federal Register* 58(190):51735–51744.

Gaudet C. 1994. A framework for ecological risk assessment at contaminated sites in Canada; review and recommendations. Ottawa: Environment Canada.

Habicht F. 1992. Guidance on risk characterization for risk managers. U.S. Environmental Protection Agency Memorandum; 26 February 1992. Washington DC.

Hoskins AP, Leight JP, Planck TW. 1994. Estimated risk of occupational fatalities hazard waste site remediation. *Risk Analysis* 14(6):1011–1017.

Lind NC. 1995. Policy goals for health and safety. *Risk Analysis* 15(6):639–644.

[NRC] National Research Council. 1983. Risk assessment in the federal government: managing the process. Washington DC: National Academy Pr.

[NRC] National Research Council. 1993. Issues in risk assessment. Washington DC: National Academy Pr.

[NRC] National Research Council. 1994. Science and judgment in risk assessment. Washington DC: National Academy Pr.

[NRC] National Research Council. 1995. Science and the Endangered Species Act. Washington DC: National Academy Pr.

[OSTP] Office of Science and Technology Policy. Science, risk, and public policy. Washington DC: OSTP.

Portney P. 1990. The evolution of federal regulation. In: Portney P, editor. Public policies for environmental protection. Washington DC: Resources for the Future. p 7–26.

[RPR] Risk Policy Report. 1995. Lessons for the national risk debate? Risk Policy Report, California's Prop 65. 20 January 1995. Washington DC. p 40–41.

[RPR] Risk Policy Report. 1996. White House issues benefit cost guidance. Ending months of debate. Risk Policy Report 31, January 1996, Special Report. Washington DC.

Rosenberg F, Wheeler J. 1993. Unreasonably at risk. *Environ Forum* 10:18–22.

Rosenthal A, Gray G, Graham J. 1992. Legislating acceptable cancer risk from exposure to toxic chemicals. *Ecol Law Quarterly* 19:269–362.

Ruckelshaus W. 1985. Risk, science, and democracy. *Issues in Science and Technology* 1:19–38.

Schierow L. 1993. Environmental risk and public policy. Part I: Cost–benefit risk analysis of environmental regulations. Washington DC: Congressional Research Service, Library of Congress.

Schierow L. 1994. Risk analysis and cost–benefit risk analysis of environmental regulations. Washington DC: Congressional Research Service, Library of Congress. 94-961 ENR.

Schierow L. 1995. The role of risk analysis and risk management in environmental protection. Washington DC: Congressional Research Service, Library of Congress. IB94036.

Shapiro M. 1990. Toxic substances policy. In: Portney P, editor. Public policies for environmental protection. Washington DC: Resources for the Future. p 195–242.

Slovic P, Malmfors T, Krewski D. Mertz CK, Neil N, Bartlett S. 1995. Intuitive technology II. Expert and lay judgments of chemical risks in Canada. *Risk Analysis* 15(6):661–675.

Suter II GW. Ecological risk assessment. Chelsea MI: Lewis. 538 p.

[USDOE] U.S. Department of Energy. Incorporating ecological risk assessment into remedial investigation/feasibility study work plans. Washington DC: USDOE, Office of Environmental Guidance. DOE/EH-0391.

[USEPA] U.S. Environmental Protection Agency. 1987. Unfinished business: a comparative assessment of environmental problems. Washington DC: USEPA, Office of Policy Analysis.

[USEPA] U.S. Environmental Protection Agency. 1990. Reducing risk: setting priorities and strategies for environmental protection. Washington DC: USEPA, Science Advisory Board. SAB-EC-90-021.

[USEPA] U.S. Environmental Protection Agency. 1992. Framework for ecological risk assessment. Washington DC: USEPA, Risk Assessment Forum. EPA/630/R-92/001.

[USEPA] U.S. Environmental Protection Agency. 1993. A guidebook to comparing risks and setting environmental priorities. Washington DC: USEPA Office of Policy, Planning, and Evaluation. EPA 230-B-93-003.

[USEPA] U.S. Environmental Protection Agency. 1995a. Draft proposed and interim guidelines for carcinogen risk assessment. Washington DC: USEPA, National Center for Environmental Assessment, Office of Research and Development.

[USEPA] U.S. Environmental Protection Agency. 1995b. Draft proposed guidelines for neurotoxicity risk assessment. Washington DC: USEPA Risk Assessment Forum.

[USEPA] U.S. Environmental Protection Agency. 1995c. Draft proposed guidelines for ecological risk assessment. Washington DC: USEPA Risk Assessment Forum. EPA/630/R-95/002.

Van Houtven G, Cropper M. 1993. When is a life too costly to save? Washington DC: Resources for the Future. CRM93-02.

Wentsel RS. 1994. Environmental policy and risk assessment. *Environ Toxicol Chem* 13(9):1381.

Wentsel RS, La Point TW, Simini M, Ludwig D, Brewer LW, Checkai RT. 1994. Procedural guidelines for ecological risk assessment at U.S. Army sites, Volume 1. Aberdeen Proving Ground MD: U.S. Army. ADA-297968, EEDECTR-221.

Appendix 2

Canada and Sweden—
Contrasting Regulations for Chlorine Discharge from Pulp and Paper Industries

Peter V. Hodson and John H. Carey
National Water Research Institute, Environment Canada
Burlington, Ontario, Canada

Kelly R. Munkittrick and Mark R. Servos
Great Lakes Laboratory for Fisheries and Aquatic Sciences
Burlington, Ontario, Canada

Sweden and Canada have adopted markedly different regulations for controlling the liquid effluents from pulp mills: Sweden has placed limits on the discharge of adsorbable organic halogen (AOX), whereas Canada has not. Both countries have based these decisions on scientific evidence derived largely from field studies of effluent effects on fish. A landmark study in Sweden demonstrated that fish in the Baltic Sea exposed to effluent from bleached kraft pulp mills showed chemical contamination, increased activity of mixed-function oxygenase (MFO) enzymes, pathology, altered responses to stress, changes in energy use and storage, enhanced growth, and impaired reproduction. The extent of effects was proportional to exposure along a dilution gradient, and differential species sensitivity caused a change in fish community structure. Effects were less intense following improved effluent treatment and substitution of chlorine dioxide (ClO_2) for molecular chlorine (Cl_2) in bleaching. Swedish scientists concluded that these effects were due to the effluent, specifically to the presence of chlorinated compounds as measured by AOX, an index of the total amount of chlorinated compounds discharged. Sweden subsequently regulated AOX discharges, an initiative adopted by several other countries.

An expert panel reviewed these results and concluded that they might not apply in North America because of the unique ecological conditions of the Baltic Sea, process upsets at the mill using Cl_2 bleaching, and the inappropriate choice of a reference unbleached pulp mill, based on different production rates and dilution kinetics in the receiving water. However, recent Canadian studies of white sucker (*Catostomus commersoni*) from both lakes and rivers receiving bleached kraft mill effluent (BKME) found results that were analogous to those observed in Sweden. These studies showed consistent induction of MFO enzymes, consistent changes in concentrations of serum sex hormones, altered stress responses, changes in energy storage and use, enhanced growth of adults, and reproductive impairment. Therefore, it appeared that the Swedish results were generally applicable in Canada. However, the Canadian studies did not demonstrate a link between levels of AOX and the responses measured, although the role of specific chlorinated compounds could not be ruled out. MFO induction and changes in concentrations of sex hormones were also found in fish exposed to effluent from unbleached pulp mills, indicating that nonchlorinated compounds could cause these responses. These data do not support an AOX regulation, and Canada has decided instead to regulate specific toxic chemicals to limit effects in receiving environments.

Pulp and paper mills are reducing their discharges of chemical wastes under pressure to protect the aquatic environment. Regulations governing the industry have been tightened in most pulp-producing countries to limit discharge of substances causing biological and chemical oxygen demand (BOD and COD), total suspended solids (TSSs), and acute lethality to aquatic biota. Concern about the potential for toxicity of chlorinated organic compounds has also prompted limitations on chlorinated dioxins and furans, chlorinated phenols, and entire classes of substances, most notably AOX.

Chlorinated compounds are formed during bleaching and alkaline extraction of wood fibre to remove residual lignin after pulping. Molecular chlorine[1] (Cl_2), chlorine dioxide (ClO_2), peroxide, oxygen, and ozone can be used alone or in combination to oxidize lignin, thereby removing color. The intensity of fiber color determines the amount of bleach used and the amount and type of byproducts. For Cl_2 bleaching, there is a vast array of chlorinated byproducts, with a wide range of molecular weights and degrees of chlorination. About 30% of the discharged waste has a molecular mass less than 1000 daltons and includes such compounds as chlorophenols, chlorinated resin and fatty acids, and chlorinated dioxins and furans (Folke et al. 1991).

Adsorbable organic halogen (AOX) is a measure of the total amount of chlorine associated with organic compounds in pulp mill effluent, after extraction by adsorption to activated charcoal. The charcoal is washed to remove free chloride ion and the total amount of chlorine retained is determined by combusting the charcoal and trapping the released hydrogen chloride in an alkaline solution. The amount of chloride is measured by microcoulometric titration and is used to estimate the amount of chlorine discharged by a pulp mill (Martinsen et al. 1988).

Regulations for AOX usually specify a maximum allowable discharge per air-dried metric ton (ADMT) of pulp manufactured. Older mills using Cl_2 for bleaching and having little waste treatment released as much as 8 to 10 kg of chlorinated compounds per ADMT. For large mills producing 1500 ADMT/d, the discharge equaled about 10 T/d of chlorinated compounds. Bleaching with Cl_2 also favored the production of tetra-chlorinated dioxins and furans in high concentrations, often with compounds in defoaming agents acting as unchlorinated substrates (Berry et al. 1991). In a modern mill, with 100% substitution of ClO_2 for Cl_2 and with primary and secondary treatment of effluent, discharge rates of AOX less than 2 kg/ADMT can be achieved, and some mills routinely discharge 1/10th this amount. Chlorinated dioxins and furans have been essentially eliminated from effluents by implementing these process modifications (Berry et al. 1991).

Table 1 Limits to AOX discharge established by several countries[1]

Country	Condition	AOX limit (kg/ADMT[2])	Year
Australia	(yearly moving average)	1.0	1990
	(weekly maximum)	2.5	1990
Austria		0.5	
Canada		No limit	1991
Finland		1.4	1995
Germany		1.0	1989
Sweden		1 to 2	1993
		1 to 0.5	1995
		0.3 to 0.5	2000

[1] adapted from Folke 1991
[2] ADMT = air-dried metric ton of pulp

[1] "Chlorine" will be used throughout as a generic term for atomic chlorine. "Cl_2" will refer to molecular chlorine gas used in bleaching.

Government regulations and the public's demand for paper made without chlorine are driving industry to lower discharges of AOX. Many countries, led by Sweden, have established limits to AOX discharge (Table 1). Some are single limits, while others specify progressive reductions over time. However, not all countries have followed suit, and Canada's federal government announced that it will not regulate AOX because there was little evidence linking AOX to effects (Environment Canada 1991), although some Canadian provinces have since established limits (Table 2).

Table 2 Limits to AOX discharge established by several provinces of Canada[1]

Province	AOX limit (kg/ADMT)[2]	Year
Alberta	1.0	all new mills
British Columbia	2.5	1992
Ontario	2.5	1992
Québec	1.5	1994

[1] adapted from Folke 1991
[2] ADMT = air-dried metric ton of pulp

Why is there such a dichotomy between countries that regulate AOX (most notably Sweden) and those that do not (most notably Canada)? The following is a brief review of the science and policy that contributed to these divergent regulations.

Scientific Basis for Swedish Regulations

Prior to 1985, most field studies of the effects[2] of pulp mill effluent focused on benthic communities through the use of community diversity indices and abundance of indicator species (see reviews by McLeay [1987] and Owens [1991]). Although effluent impacts were observed, they appeared to be due to fiber mats, color, turbidity, and low dissolved-oxygen conditions caused by decomposition of organic substances. In cases where the dataset included chemical and toxicological characterization of effluents, the observed effects were consistent with those of organic and nutrient enrichment (Owens 1991).

In the past, effluent impacts on fish and fish populations were assessed by measuring acute lethality. The major compounds contributing to lethality of whole mill effluents were resin acids and other extractives present in cooking liquors (Rogers and Mahood 1974; McLeay et al. 1979; Kovacs 1986). These substances are generally destroyed by secondary treatment (Servizi et al. 1986). Detectable levels of individual chlorinated compounds are present in specific waste streams such as spent bleachery effluents from the chlorination (C) and extraction (E) stages of bleaching. However, the contribution of chlorinated compounds to acute toxicity of whole mill effluents is debatable, since there appears to be no marked difference in toxicity between bleached and unbleached pulp mill effluent (Mehrle et al. 1989) or correlation between AOX and

[2] Where "effects" or "responses" are reported, statistically significant changes between effluent-exposed and reference populations are implied, unless otherwise stated. Not all "responses" are proven to be adverse ("effects").

final effluent toxicity (Craig et al. 1990; O'Connor et al. 1993). According to Folke (1991), spent bleach liquors are seldom acutely toxic to organisms.

This past work did not provide a strong basis for regulating the discharge of chlorinated organic compounds. Rather, the Swedish regulations evolved from field and laboratory studies of chronic toxicity of effluents to fish and from the precautionary principle for regulating mixed wastes.

Until recently, there was little published information about the sublethal effects of pulp mill effluent on wild fish populations. The Swedish "Environment/Cellulose I Project" was an innovative study of the physiological, pathological, and ecological impacts of primary-treated effluent from a bleached kraft mill on the fish communities of the Gulf of Bothnia (Södergren et al. 1988; Södergren 1989). Fish populations were sampled on 4 occasions from May 1984 to September 1985, at sites 2, 4.5, 8, and 10 km from the discharge of the mill at Norrsundet. An unexposed reference site was sampled about 120 km north of the discharge point. Fish density and diversity were low in areas less than 2 km from the discharge (Neuman and Karås 1988). At sites 2 to 4.5 km from the discharge, densities of both adults and fry of roach (*Rutilus rutilus*) and ruffe (*Gymnocephalus cernua*) increased dramatically, an effect that the authors attributed to nutrient enrichment by the effluent. In contrast, the abundance of adult and pelagic fry of perch (*Perca fluviatilis* L.), sand-goby (*Pomatoschistus minutus* [Pallas]), and herring (*Clupea harengus*) were low at the 2- and 4.5-km sites, with increasing abundance at sites further from the discharge.

A number of skin diseases and skeletal deformities were also noted, including skin sores, fungal infections, fin erosions, and deformed gills, skulls, fins, and spinal columns (Bengtsson 1988). The prevalence of these diseases was particularly high in 1984, and seemed to be exposure-dependent, i.e., more frequent at sites nearest the discharge where effluent concentrations were highest. Relative to the reference area, perch near the Norrsundet mill had smaller gonads, larger livers, higher liver enzyme activities, increased liver ascorbic acid content, marked changes in energy storage and use, impaired osmotic and ion regulation, increased red blood cell counts and hematocrits, and reduced lymphocyte counts (Andersson et al. 1988; Larsson et al. 1988; Sandström et al. 1988). The increased activity (induction) of the liver enzyme 7-ethoxyresorufin-O-deethylase (EROD), one of the mixed-function oxygenases (MFOs), was interpreted as a specific response to chemical exposure.

Mixed-function oxygenase induction is the measurable increase in activity of liver enzymes that oxygenate polyaromatic compounds. MFO enzymes belong to a family of proteins known as P450 cytochromes, since they characteristically absorb light with a wavelength of about 450 nm. The gene that codes for these proteins is named "*cyp*," and the specific inducible gene in fish that responds to chemical exposure is cyp1A1. MFO induction is measured by assays of activity, usually EROD or aryl hydrocarbon hydroxylase (AHH) (Hodson, Kloepper-Sams et al. 1991), or by measuring the CYP1A1 protein by radioimmunoassay.

EROD and AHH increase in activity when more protein is synthesized by cells after exposure to compounds such as nonchlorinated polynuclear aromatic hydrocarbons (PAHs), plant flavones, chlorinated dioxins and furans, polychlorinated biphenyls (PCBs), and chlorodiphenylethers. Only some members of these families of compounds can cause an enzyme induction, since they must have a very specific molecular size, shape, and conformation to interact with the cell receptor that turns on gene transcription, gene translation, and enzyme synthesis. Because some of these compounds are also highly toxic, an increase in MFO activity is interpreted as both chemical exposure of the organism and possible toxicity. The consistent occurrence of MFO activity in fish populations exposed to pulp mill effluent, often in concert with other physiological, pathological, and population changes, has stimulated its use as an indicator of the presence and possible effects of pulp mill effluents. Since MFO activity increases within 24 to 48 h of exposure to pulp mill effluent, it provides a rapid indicator of possible effects long before other symptoms of toxicity appear. Consequently, it is a useful tool in laboratory and field experiments for evaluating the relative potency of pulp mill effluents and their components (Hodson, Kloepper-Sams et al. 1991).

In the Swedish studies, many of the physiological changes suggested an exposure–response relationship, with responses most pronounced at the 2 sites nearest the discharge (2 and 4.5 km, representing effluent dilution factors of 160- and 330-fold, respectively). However, some responses (reduced gonad size, EROD induction, increased red blood cell count, decreased lymphocyte count, and poor chloride regulation) still occurred 10 km from the mill, at a dilution factor of more than 1100-fold.

To compare the effects of unbleached versus bleached effluent, an additional site was selected in the receiving waters of an unbleached kraft mill at Sandarne. In contrast to the perch sampled near Norrsundet, Sandarne perch showed no effects or only small effects for most parameters measured. Södergren (1989) regarded this as evidence of a relationship between the bleaching process (i.e., levels of chlorinated organic compounds) and subsequent effects in the receiving water. These conclusions were cited as the basis for using AOX to regulate pulp mills in Denmark, Sweden, and Finland (Folke 1991).

This view was supported by follow-up studies in 1989 and 1990 that showed reduced effects levels following improvements to effluent treatment, $C1O_2$ substitution, and reduced discharge of AOX (Forlin et al. 1992). However, MFO induction was still evident and more widespread than previously believed (Forlin et al. 1992), corresponding to sediment analyses of AOX and extractable organic chlorine (EOCl), which demonstrated widespread contamination by pulp mill effluents (Södergren et al. 1992).

Laboratory studies in 1982 to 1984 employed model ecosystems to evaluate the effects of different effluents on stickleback (*Gasterosteus aculeatus*) and rainbow trout (*Oncorhynchus mykiss*). In early work, impacts of 6 types of pulp mill effluent on stickleback growth, reproduction, and pathology were evaluated. The greatest effects were observed with effluent from a mill using a traditional bleaching sequence involving a high proportion of Cl_2. Much less impact was observed with bleaching sequences involving oxygen pre-bleaching and different degrees of ClO_2 substitution for Cl_2 (Lehtinen et al. 1991). The pattern of changes in hematology, ionoregulation, and MFO induction in rainbow trout was also very similar to that observed in the field studies with perch, although the responses were less intense and less consistent (Lehtinen et al. 1990). These laboratory results seemed to provide clear support for a cause–effect relationship between the discharge of chlorinated organic compounds and effects on fish.

These studies demonstrated that effluent from bleached kraft mills was likely responsible for effects on fish observed in the Baltic. However, the compounds causing the effects were unknown, although there was a strong association with chlorine bleaching; mills using Cl_2 for bleaching released higher levels of chlorinated compounds to the environment and appeared to cause a greater intensity of effects, suggesting that the majority of problems were generated by chlorinated compounds. Rather than trying to identify all toxic compounds in effluent and trying to estimate individual, synergistic, and antagonistic interactions, the Swedish Environmental Protection Agency (EPA) decided that a "precautionary approach" that regulated all chlorinated compounds would be most efficient. Accordingly, discharges of total organochlorine compounds (ToC1) were regulated, starting in 1987. AOX has since been substituted for ToCl because of advantages of the AOX analytical technique (AOX = 1.4 × ToCl) (Folke 1991). The discovery in the 1980s of chlorinated dioxins and furans in receiving waters and effluents of pulp mills using Cl_2 (Berry et al. 1991; Whittle et al. 1993) reinforced this approach. Further weight was added by the discovery that both MFO induction and contamination by chlorophenols typical of effluents from bleached kraft mills could be detected far from pulp mills in the Gulf of Bothnia (Balk et al. 1993).

Criticisms of the Swedish Approach

When these results were originally released, their applicability to the North American situation was questioned by some scientists, particularly since North American studies before 1988 had not detected the responses seen in the Baltic. A "Scientific Panel on Pulping Effluents in the Aquatic Environment" was convened to address the discrepancies between the Swedish and North American studies (Eysenbach et al. 1990). Among its conclusions (Mehrle et al. 1989) was the assertion that Swedish studies should not be extrapolated directly to North American mills because

- many physical, chemical, and biological differences exist between Swedish and North American study sites;
- the Gulf of Bothnia is a unique hydrological system, unrelated to the receiving waters of North American mills; and
- secondary biological treatment of effluents at many North American mills effectively reduces or eliminates sublethal effects in the receiving environment.

In addition, the comparison between Swedish mills with and without Cl_2 bleaching was considered inappropriate. Compared to the bleached kraft mill at Norssundet, the unbleached mill at Sandarne had a lower pulp production and a lower BOD loading, combined with a greater flushing and effluent dilution in the receiving water. Hence, fish near the unbleached mill were exposed to lower concentrations of effluent than those at the bleached mill. The panel concluded that it was not possible to draw cause–effect inferences about chlorinated organic compounds from comparisons between the 2 cases and that the toxicological database then available on the fate and effects of compounds in pulp mill effluents needed to be expanded. In particular, assessments of impacts of pulp mill effluents required the integrated use of chemical analyses, ecological field assessments, and toxicity testing to document the source of the adverse effects.

Studies in the United States by the National Council of the Paper Industry for Air and Stream Improvement (NCASI) supported the contention that secondary-treated BKME had little or no detrimental effect on receiving water ecosystems. The first of these studies was centered in New Bern, North Carolina, and represented a southern U.S. "warm-water" stream ecosystem (NCASI 1984). Concentrations of up to 18% v/v of secondary-treated effluent entered 2 artificial streams, while 2 control streams received only river water. Production, growth, survival, and reproduction of large-mouth bass (*Micropterous salmoides*), bluegill sunfish (*Lepomis macrochirus*), and golden shiner (*Notemigonus crysoleucus*) were monitored, along with a series of other ecological and physicochemical parameters. Observed effects were less pronounced and occurred at higher effluent concentrations than effects noted in the Swedish studies. However, the most sensitive response in the Swedish studies, increased activity of the MFO system, was not measured. At an effluent concentration of 18% v/v, largemouth bass production (total weight gain of all fish) was decreased relative to the control stream, but there were no other significant effects in bass, nor were there effects noted on the other 2 fish species studied. At an effluent concentration of 10% v/v, there were no effects reported for any of the 3 fish species. Although the early studies in artificial channels did not show any significant responses in fish, they did not examine many of the sublethal responses reported by the Swedish studies. Recent studies using these artificial streams have incorporated biomarkers of physiological and histopathological effects, including hematology and the determination of MFO activity in fish. However, only EROD activity and CYP1A1 concentrations were consistently elevated in both bass and catfish exposed to 8% effluent in artificial streams (Borton 1992).

Parallel "cold-water" studies of rainbow trout were conducted at Lewiston, Idaho from 1980 to 1988 (NCASI 1989; Hall et al. 1991). The first study increased the effluent concentration in treated streams at about 10-month intervals from 1.3% v/v to 5.1% v/v. In the second, effluent concentration was held at about 1.5% v/v for 3.5 y, and rainbow trout were monitored throughout their entire life cycle. At all effluent concentrations, trout from the 2 treated streams generally showed higher growth rates, average weight, and total production than fish in control streams, while survival seemed somewhat reduced by effluent exposure. There was no significant relationship among effluent exposure and liver size, hematocrit, leucocrit, or prevalence of histopathological lesions. Limited spawning occurred in both control and treated streams, but it was impossible to determine whether treatment differences in hatchability were significant because of small sample sizes. Total weight, final mean weight, mean weight gain, and total weight gain were significantly greater for fry from the treated streams than for fry from the control streams, as observed with ruffe in Sweden (Sandström et al. 1988).

In a comprehensive review of the effects of pulp mill effluents in the aquatic environment, Owens (1991) found no complete mechanistic link between exposure to contaminants in pulp mill effluents and within-organism responses, whole-organism effects, or population- or community-level effects. He called for a multidisciplinary effort to combine chemical, toxicological, and biological studies of adverse effects and to identify the responsible chemicals.

Scientific Basis for Canadian Position

The development of new Canadian regulations for pulp mills was strongly influenced by the Swedish studies and by Sweden's regulations for AOX. Questions were raised as to whether effects observed in Sweden would occur in Canada, whether these effects were typical of modern mills with secondary treatment or only of older mills, whether effects were associated with Cl_2 bleaching, and whether regulating AOX would limit effects on fish.

Are sublethal effects on fish observed in Sweden also evident in Canada?

Two Canadian field studies have examined fish in waters receiving effluents from pulp mills employing technologies similar to those of the Norrsundet mill that was the major focus of the Swedish Environment/Cellulose I Project. These Canadian studies were located in Jackfish Bay, Ontario, on the north shore of Lake Superior, and in the

St. Maurice River, Quebec. Both mills utilized only primary treatment of effluent at the time of these studies (1988 to 1989).

The Lake Superior project was prompted by a 1987/88 study by the Ontario Ministry of the Environment in which the activity of liver MFO enzymes increased in fish near a 1200- ADMT/d bleached kraft pulp mill discharging primary-treated effluent to Jackfish Bay (Smith et al. 1991). Beginning in 1988, a variety of responses were investigated in 3 species of fish. Compared to the reference site, white sucker (*Catostomus commersoni*) from Jackfish Bay exhibited higher condition factors, slower growth rates, increased age to maturity, smaller gonads, smaller egg size (females), fewer secondary sexual characteristics (males), increased liver MFO activity, and lower levels of serum sex hormones (Munkittrick et al. 1991; McMaster et al. 1991, 1992). Increased liver MFO activity and decreased serum sex hormones were also observed in longnose sucker (*Catostomus catostomus*) and lake whitefish (*Coregonus clupeaformis*) (Munkittrick, Van Der Kraak et al. 1992; Munkittrick, McMaster et al. 1992). Decreases in circulating levels of gonadal steroid hormones were associated with reductions in secondary sexual characteristics and gonadal size and with increases in age to maturity (Munkittrick et al. 1991). Similar changes in sex steroid metabolism are known to occur in mammals exposed to compounds that also induce the MFO system, although there does not seem to be a direct link between MFO induction and the reproductive effect. Instead, exposure of white sucker to effluent from bleached kraft mills appears to affect hormone release by selectively acting at multiple sites in the pituitary–gonadal axis (Van Der Kraak et al. 1992b).

The St. Maurice River study also focused on white sucker and to a lesser extent northern pike (*Esox lucius*). Fish were collected at sites upstream and downstream from a hydroelectric dam at La Tuque, Quebec. A 1400-ADMT/d bleached softwood kraft mill discharged primary-treated effluent into the river just above the dam. Downstream, white sucker showed increased liver MFO activity (Figure 1a), larger livers (Figure 1b), physiological signs of nonspecific stress, depressed serum concentrations of reproductive steroids, and subtle changes in reproductive development (Hodson et al. 1992; Gagnon et al. 1994a, 1994b). The reproductive responses were confirmed relative to a reference river and were characterized by a failure to invest more energy into reproduction (Gagnon et al. 1995). In the reference river, growth rates increased with distance downstream, presumably reflecting nutrient enrichment. In response to increased growth rate, fish matured at an earlier age and a smaller size. They showed increased gonad size and an increased slope of the fecundity–size relationship. Effluent-exposed fish in the St. Maurice River also grew faster at downstream sites, but in contrast to fish from the reference river, they showed no change in age to maturity and were larger at first maturity. They also showed reduced gonad size and a more variable fecundity; the relationship between fecundity and size was disrupted. The failure of effluent-exposed fish in the St. Maurice to benefit from increased growth rate by investing more energy in reproduction was analogous to the impairment of reproduction in fish from Jackfish Bay.

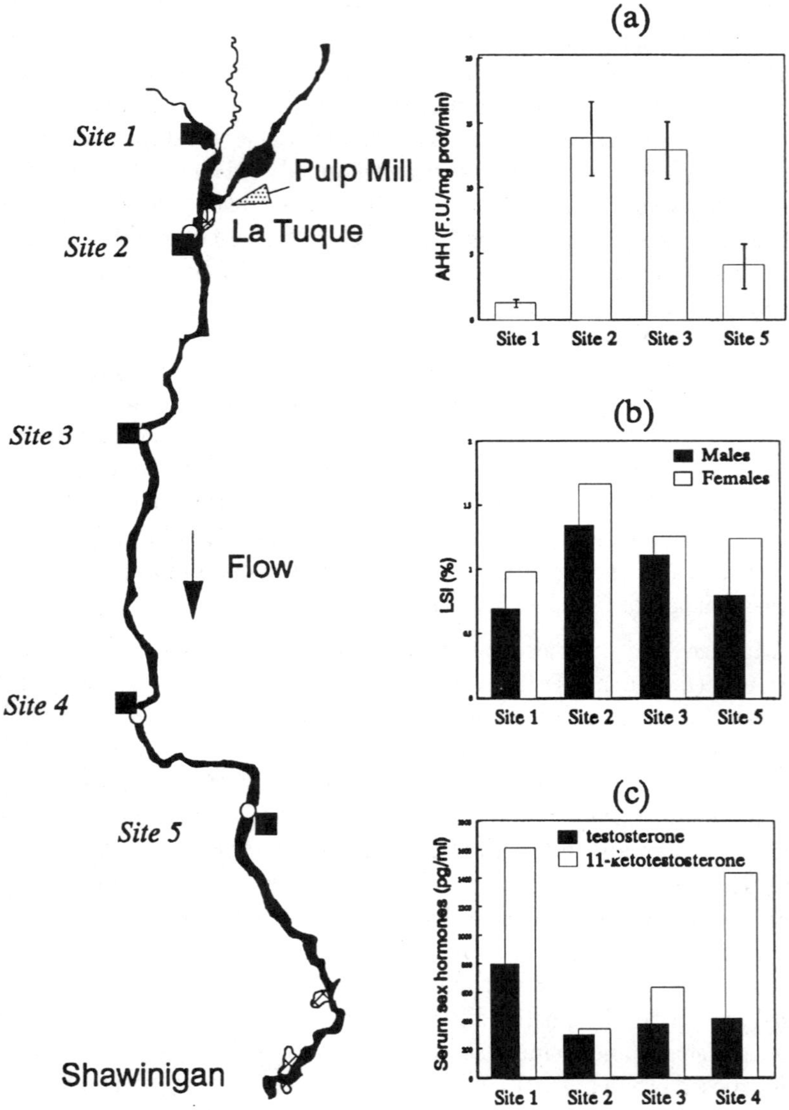

Figure 1 Sampling locations and biochemical responses in white sucker from the St. Maurice River, 1989: a) liver MFO activity; b) liver somatic index; c) male sex hormones. Relative to the mill discharge, site 1 is 10 km upstream (Hodson et al. 1992), while sites 2, 3, and 5 are 2, 32, and 95 km downstream.

These studies indicated that responses of fish to effluent from bleached kraft mills in Canada were similar to those observed in Sweden, and they supported the development of similar regulations.

Are effects on fish evident in receiving waters downstream of modernized mills?

"Modern" mills are those that substitute ClO_2 for Cl_2 as a bleaching agent and that may use oxygen, ozone, or peroxide as additional bleaching agents. Compared to older mills using chlorine bleaching, modern mills usually have better in-plant waste recovery, more efficient pulping systems, as well as full secondary treatment of effluent (Berry et al. 1991).

While sublethal effects of pulp mill effluents were first associated with mills using chlorine bleaching and having little or no waste treatment, recent Canadian studies indicate that some effects can also be found in wild fish exposed to treated effluent from modern or modernized bleached kraft mills. The sites of these studies include Jackfish Bay after the initiation of secondary treatment (Munkittrick, Van Der Kraak et al. 1992; Munkittrick, McMaster et al. 1992; Munkittrick 1992; Van Der Kraak et al. 1992a); the Spanish River, Ontario (Servos et al. 1992); the Wapiti River, Alberta (Kloepper-Sams et al. 1992; Kloepper-Sams and Swanson 1992; Swanson et al. 1992); and the Fraser River, British Columbia (Rogers et al. 1989). The most consistently measured and observed response in all studies was increased liver MFO activity associated with exposure of white sucker, longnose sucker, lake whitefish, chinook salmon (*Oncorhynchus tshawytscha*), and mountain whitefish (*Prosopium williamsoni*) to secondary-treated effluents from bleached kraft mills.

Munkittrick et al. (Munkittrick, Van Der Kraak et al. 1992; Munkittrick, McMaster et al. 1992) monitored 3 fish populations in Jackfish Bay and failed to detect improvements in MFO induction and changes in concentrations of plasma sex hormones for 3 years following the installation of secondary treatment at the bleached kraft mill in 1989. However, the extent of these physiological changes declined during mill operational shutdowns (Munkittrick, Van Der Kraak et al. 1992), and MFO induction was evident in fish exposed to this same treated effluent in the laboratory (Williams et al. 1992). The continued responses of exposed fish suggested that the causative agents were not removed by secondary treatment.

In contrast, Kloepper-Sams et al. (1992) did not find many responses of fish downstream of a modernized bleached kraft mill in Alberta on the Wapiti and Smokey Rivers. The primary responses were increased levels of tissue dioxins and increased MFO activity, relative to sites on the Saskatchewan River, a reference river unconnected to the Wapiti. The mill was located on the Wapiti River, which has no dams to impede fish movement. Exposure of fish to the mill effluent in the Wapiti was not constant but varied seasonally because of extensive fish migrations upstream and downstream of the effluent discharge and because of seasonal changes in river flow rate and effluent dilution. Fish tagging, tissue dioxins, bile metabolites of EOCl, and liver MFO induction all demonstrated exposure downstream and seasonal migration upstream of a portion of the exposed fish. As a possible consequence, correlations among these variables were not strong.

Servos et al. (1992) studied the impact on white sucker populations of the discharge of a 1200-ADMT/d bleached kraft mill employing oxygen delignification, partial ClO_2 substitution, and secondary treatment. The mill was located on the Spanish River, Ontario, just downstream from a dam that separated exposed and unexposed populations. Liver MFO activities were elevated up to 51 km downstream from the mill, and liver sizes were increased immediately below the outfall (Figure 2). Gonad sizes and plasma sex hormone concentrations were low at all sites, including reference sites, because the fish were still recovering from spawning at the time of sampling. Male white sucker sampled later in the season during the following year showed steroid hormone changes immediately downstream from the mill, consistent with those observed in Jackfish Bay and the St. Maurice River (Munkittrick et al. 1994). These studies demonstrated that effects on fish were still observable after mill modernization, with the exception of the Wapiti River, where no reproductive effects were observed, relative to a reference tributary; comparisons to upstream sites were confounded by fish migration (Kloepper-Sams et al. 1992). More population studies are needed downstream of modernized mills.

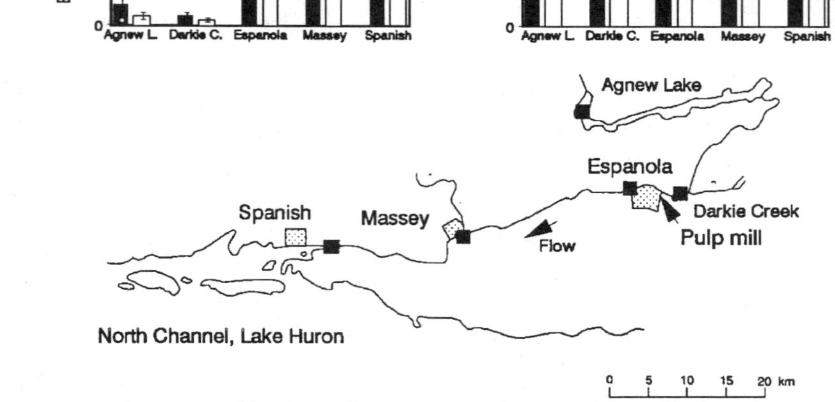

Figure 2 Sampling sites and biochemical responses in indigenous white sucker from the Spanish River, 1990: a) liver MFO activity; b) liver somatic index (Servos et al. 1992)

Are sublethal effects associated with chlorine (Cl_2) bleaching?

In the summer of 1991, samples of effluent and wild fish were collected at 12 sites in Ontario for laboratory toxicological testing, assessment of receiving water quality, and responses of fish to pulp mill effluent. Sampling was generally restricted to one species (the white sucker) and small sample sizes (n = 12 for each sex) to provide a preliminary indication of effects and to focus subsequent follow-up studies. This study included bleached kraft mills with and without secondary treatment and 3 sulfite mills with primary treatment; all mills were located in Ontario (Robinson et al. 1994).

In male white sucker collected in rivers within 0.5 km of the discharge, or at lake sites where the effluent was still visible, increased MFO activities were observed regardless of pulping process or effluent treatment (Table 3) (Munkittrick et al. 1994). Female reproductive organs were significantly smaller at all sites where sufficient numbers of fish were caught, with the exception of one mill, and plasma steroid concentrations were reduced in at least one sex at all sites except one mill (Table 3). Changes in liver size and condition factor were inconsistent but were present at sites with and without Cl_2 bleaching. There were no differences in fish responses among kraft mills with secondary treatment, a kraft mill producing primarily unbleached pulp, and mills using the sulfite process. However, fish downstream of the sulfite mills and the unbleached kraft mill had lower levels of contamination with dioxins and furans than fish near bleached kraft mills (Servos et al. 1994).

Overall, the extent of physiological disruption could not be linked to levels of chlorinated dioxins and furans near the pulp mills. Although dioxin levels in fish were correlated with MFO induction among sites, high MFO induction was observed in fish with low levels of dioxin downstream of the mill producing mostly unbleached kraft pulp, implying the presence of nonchlorinated inducers. Previous experiments with fish exposed naturally to pulpmill effluent, while held in cages or in laboratory bioassays, have shown that MFO induction disappears within 8 d of transfer to clean water (Munkittrick, Van Der Kraak et al. 1992; Munkittrick and Servos, unpublished data). Because exposure of fish to dioxins and furans in the laboratory causes a prolonged induction of MFO activity (Muir et al. 1990; Parrott et al. 1994), these results again imply the presence of inducers other than chlorinated dioxins or furans.

The lack of association of Cl_2 bleaching with physiological changes was reinforced by the follow-up study of MFO induction and other responses of fish in the Baltic. As indicated above, Balk et al. (1993) found MFO induction and elevated levels of chlorophenolics in fish bile at great distances from pulp mills, leading them to conclude that responses of fish were due to persistent compounds associated with bleaching. However, their report also included measurements of elevated MFO activity in fish with no discernable contamination by chlorophenolics, implying that induction was not due to effluent from bleached kraft mills or that nonchlorinated substances were the cause of induction.

Table 3 Summary of effects observed in a survey of 12 Ontario pulp mill effluents (Munkittrick and Van Der Kraak 1993; Munkittrick et al. 1994; Robinson et al. 1994; Servos et al. 1994; van den Heuvel et al. 1994). Data are summarized for 7-d receiving water bioassays using fathead minnow and *Ceriodaphnia* and for physiological responses recorded in free-living, adult, wild white sucker collected near the discharge: "+" = increased response; "-" = decreased response; "0" = no change; "*" = results different from only one reference. Blank indicates no measurement, due to insufficient sample size.

				Bioassays				Free-Living Adult White Suckers											
				Fathead minnow		*Ceriodaphnia*		Males					Females					AOX[2]	
Mill location	Process	Chlorine bleaching[1]	Effluent treatment	Survival	Growth	Survival	Reproduction	MFO[3]	Steriod[4]	GSI[5]	LSI[6]	TCDD[7] TEQ[s]	MFO	Steroid	GSI	LSI	TCDD TEQs	kg/ ADMT	µg/L
Fort Frances	Kraft	D	Secondary	0	0	0	0	++++[8]	--	0	+	+++	+	--	-	0	++	3.57	145
Espanola	Kraft	O,C,D	Secondary	0	0	0	+	+	-	+	0	++++	+	-	0	0	++	0.89	265
Terrace Bay	Kraft	D,C	Secondary	0	0	0	+++	+++	----	-	+	++++	+++	---	--	+	++++	1.74	684
Dryden	Kraft	C,D	Secondary	0	0	0	+											3.12	3120
RedRock	Kraft	C	Primary	-	0	0	+	++	---	*	*	--	+	*	--	+	0	2.96	15.2
Marathon	Kraft	D	Primary	0	0	0	0											7.06	36.3
Smooth Rock Falls	Kraft	C,D	Pr4imary	0	0	0	--	++++	--	0	0	++++	++	--	-	0	+++	4.34	281
Thunder Bay	Kraft	D,C	Primary	-	0	0	+	++++				++++	+++					3.23	1674
Kapuskasing	Sulfite	-	Primary	--	0	0	-	+	0	0	+	+	0	0	-	+	++		
Kenora	Sulfite	-	Primary	--	0	0	0	++	0	-	+	+	0	--	--	+	+		
Iroquois Falls	Sulfite	-	Primary	0	0	0	+												

[1] Chlorine refers to both Cl_2 (C) and ClO_2 (D); (0) = oxygen pre-bleaching
[2] Adsorbable organic halides, reported as production (kg per air-dried metric ton of pulp) and receiving water concentration. AOX was not measured at the sulfite mills because no chlorine was used for bleaching
[3] Mixed function oxygenase induction
[4] Steroid levels in plasma: 11-ketotestosterone in males; 17-estradiol in females
[5] Gonad somatic index: ratio of gonad to body weight
[6] Liver somatic index: ratio of liver to body weight
[7] Toxic equivalent (TEQ) quantities (pg/g) of tetrachlorodibenzo(p)dioxin in liver, as measured by rat H4IIE assay
[8] Size of response compared to reference: Increase (EROD and dioxin TEQs): "+" = 0- to 5-fold; "++" = 5- to 10-fold; "+++" = 10- to 20-fold; "++++" = >20-fold. Increase (all other responses): "+"= 0- to 2-fold; "++" = 2- to 4-fold; "+++" = 4- to 8-fold; "++++" = >8-fold. Decrease (all responses): "-"= 0.75 to 1.0; "--" = 0.50 to 0.74; "---" = 0.25 to 0.49; "----" = 0 to 0.24

Laboratory bioassays of MFO induction in fish have demonstrated that the most potent effluent streams within a modernized bleached kraft mill were black liquor from pulping (spent cooking liquor containing sodium hydroxide, sodium sulfide, wood extractives, and digested lignin) and effluents from the C and E stages (Cl_2 or ClO_2 used as a bleaching agent; contains wood extractives, lignin derivatives, and chlorinated compounds) (Schnell et al. 1993; Martel et al. 1994). High MFO induction in black liquor indicates that some inducing compounds are non-chlorinated extractives or lignin derivatives. Potency of black liquor appears to vary with wood furnish; black liquor from softwood mills was more potent than that from mills pulping hardwood, although some of this difference was due to relative liquor strength (water content) (Hodson et al. 1994). Because softwoods contain more resinous extractives than hardwoods, the difference in relative potency suggests that at least a portion of the inducers may be natural components of wood.

A further release of inducers with C- and E-stage effluents could represent either increased potency of weak inducers because of chlorination or further release of extractives or lignin derivatives following oxidation of residual lignin by Cl_2. Burnison et al. (1994) have demonstrated that MFO-inducing compounds can be extracted from whole effluents. While properties vary somewhat according to source, they behave more as PAH-like compounds than as dioxin-like compounds, based on solubility and chromatographic properties. These compounds may also be amenable to treatment, since enhanced biological treatment of individual waste streams or whole effluents from a modernized bleached kraft mill reduced MFO induction in rainbow trout bioassays to nondetectable levels (Schnell et al. 1993).

The presence of responses at mills without chlorine bleaching is also supported by recent field studies from Finland. Increased MFO activity was measured in rainbow trout liver cells exposed to extracts from an unbleached sulfite mill, as well as in wild perch caught in receiving waters near the mill, and in rainbow trout and whitefish caged in receiving waters near the mill discharge (Hanninen et al. 1991; Lindström-Seppä et al. 1992). The extent of induction was comparable to that observed in previous studies of effluent from bleached pulp mills.

Pesonen and Andersson (1992) compared effects on rainbow trout liver cells of dimethyl ether extracts of final effluents from 5 mills (3 bleached kraft mills, 1 unbleached kraft mill, and 1 unbleached sulfite mill). When cellular glutathione was measured as an indicator of oxidative stress, extracts from the unbleached mills were more potent than those from bleached mills at causing depletion of reduced glutathione. In addition, extracts from all 5 mills markedly increased MFO activity at concentrations well below those affecting the glutathione system. The authors concluded that bioactive substances were present in effluents from all mills, not just from those employing chlorine bleaching. A similar conclusion was reached by Gagné and Blaise (1993), who found EROD induction in fish surviving routine lethality tests with most pulp mill effluents, regardless of mill process. This study included effluents from bleached kraft, thermomechanical, thermochemical, groundwood, and sulfite chemical pulping mills.

Considering the 12-mill study and laboratory studies together, changes in steroid hormone concentrations and liver MFO responses are not associated exclusively with effluents from mills using chlorine bleaching. There is abundant evidence that some MFO-inducing compounds in pulp mill waste are nonpersistent, nonchlorinated compounds.

Would regulating AOX prevent effluent effects on fish?

In the studies of modernized mills at Jackfish Bay and on the Spanish River, responses of fish were similar to those observed at older mills. The mills at Jackfish Bay and on the Spanish River discharged less than 1.6 kg AOX/ADMT of pulp. Recent studies on the Spanish River show that MFO induction is still present at AOX discharges below 0.7 kg/ADMT (Servos and Munkittrick, unpublished data). These studies confirm that modernized mills with secondary treatment, which discharge low levels of AOX, are capable of producing the same level of induction as is evident at mills with older technology. Indeed, the relationship between effects and the discharge of AOX, which seemed clear when AOX discharges were high, disappeared when discharges were less than 2 kg/ADMT (Folke 1991).

Physiological and population responses of white sucker in the St. Maurice were greatest immediately downstream from an old technology mill and recovered progressively toward control values at sites further downstream (Hodson, Bussieres et al. 1991; Hodson et al. 1992; Gagnon et al. 1994b,1995). The effects paralleled exposure to pulp mill effluent contaminants as indicated by concentrations of chlorinated dioxins and furans and chlorinated phenolics in fish tissues and by chlorinated phenolics in river water. Both chemical contamination and MFO activity declined with distance from the mill but remained significantly elevated 95 km downstream from the mill (Hodson et al. 1992; Gagnon et al. 1994b). However, there was no correlation between the gradient of AOX levels in river water and the gradient of responses in fish: AOX showed almost no decline over the 95 km (Hodson et al. 1992).

Although receiving water acute bioassays using both fathead minnow and *Ceriodaphnia* detected toxicity near several mills in the multi-mill study cited above, there was no correlation of acute toxicity test results with the presence or absence of effects in wild fish or with concentrations of AOX in the receiving water, or with amounts of AOX discharged per ADMT (Table 3). A lack of correlation between AOX and toxicity has been reported by other authors, including Borton et al. (1990), Hall et al. (1989), Gergov et al. (1988), and O'Connor et al. (1993). There was no correlation between reproductive effects in wild fish and pulping process, dioxin concentrations in fish tissues, AOX production, AOX levels in receiving water, or the use of Cl_2 (Servos et al. 1994; van den Heuvel et al. 1994). MFO activity in male fish showed a very high correlation with effluent concentration, and consequently with AOX and other effluent-related chemicals in receiving water, but not with AOX on a kg per ADMT basis (Munkittrick et al. 1994). Similarly, one field study comparing concentrations of chlorinated organic compounds to impacts on benthic species concluded that there was no correlation between AOX exposure and effects (Craig et al. 1990).

Other recent studies involved laboratory bioassays of pulp mill effluents to test their ability to increase liver MFO activities in exposed fish. In fish exposed to effluents from a variety of mill types, Martel et al. (1994) demonstrated that ClO_2 substitution for Cl_2 as a bleaching agent was not associated with any change in MFO induction in fish. They proposed that chemicals causing physiological alterations may originate in the kraft pulping process prior to the bleaching stage. Williams (1993) has also employed lab assays to test final effluent from kraft mills employing a variety of bleaching processes, including an unbleached kraft mill. All kraft mill effluents tested caused increases in liver MFO activities, including effluent from the unbleached kraft mill. Potency for induction could be correlated to a variety of constituents in the effluent, including AOX. However, because the strongest correlation was with dissolved organic carbon (DOC), these correlates are likely surrogates for dilution, indicating the role that water use in pulp production plays in diluting all constituents, including the unknown inducing compound. Recent studies of MFO activity of carp in a lake receiving effluent from a bleached kraft mill in Australia (Ahokas et al. 1994) support this view. Correlations between MFO activity and levels of AOX ($n = 5$) in water were significant because of a strong dilution gradient over a very short distance, but little correlation was observed with extractable organic halogens (EOX) in sediments. Assuming that sample size had not caused spurious results, these correlations indicate that the inducer was waterborne and taken up across gills and that it was not a compound accumulated via the food chain from sediments. Both MFO activity and AOX in water followed the same gradient, the gradient of effluent dilution in the lake. The association between the 2 parameters cannot be used to imply cause and effect; more likely, it reflects the co-dilution of an unmeasured inducing compound.

This conclusion also has experimental support. The mesocosm studies conducted in the early 1980s by Lehtinen and co-workers were continued to compare effluents from softwood and hardwood mills, effluents with different levels of treatment, and effluents from mills using different bleaching sequences. These studies demonstrated that effects on fish were still observable in effluents from mills discharging low levels of organochlorines (OCs), measured as ToCl between 0.5 to 1.0 kg/ADMT. Furthermore, the magnitude of effects was similar to that of effluents from mills discharging ToCl of 2.0 kg/ADMT or higher (Lehtinen et al. 1991). In other words, the relationship between bulk measures of OC discharges and effects on fish was poor or nonexistent. Furthermore, liver anomalies and parasitic infections of stickleback were most severe in fish exposed to effluents from unbleached softwood or bleached hardwood without oxygen pre-bleaching (Axelsson and Norrgren 1991; Lehtinen, Tana, Härdig et al. 1992; Lehtinen, Tana, Karlsson et al. 1992).

Recent mesocosm experiments exposed stickleback and trout to treated and untreated effluents from bleached hardwood mills using 2 different sequences, one with a high degree of ClO_2 substitution and the other with oxygen delignification before bleaching. Effects on growth, liver histology, and parasitic infections were evident in both treated and untreated effluents and in a sitosterol control group (Lehtinen, Tana, Karlsson et al. 1992). No effects were observed on hematological responses or on liver MFO induction. Sitosterol, a plant sterol found at high levels in effluents, was included because it causes reproductive effects in tropical species (*Heterandria formosa; Gambusia affinis*), particularly if there are bacteria present that can transform sitosterol into androgenic compounds (Davis and Bartone 1992). This effect is expressed as a change in external sexual characteristics, in which female fish develop the secondary sexual characteristics of males. It was first observed downstream of a pulp mill discharging wastes to a small Florida river.

It is clear that reproductive effects on fish can be caused by natural extractives of wood, without the presence of chlorine. It does not rule out, however, that some chlorinated compounds may play the same role. Because AOX is correlated with responses of fish only so far as it reflects effluent dilution, reducing levels of AOX through specific effluent treatments may not necessarily remove compounds that cause physiological responses and reproductive effects in fish. As a consequence, the Canadian government chose not to regulate AOX but rather to regulate known compounds of high toxicity (i.e., chlorinated dioxins and furans).

Summary

1) Field studies of natural fish populations have demonstrated that fish downstream of pulp mills can exhibit some of the following: chemical contamination, induction of enzymes that metabolize xenobiotics, physiological changes indicative of impaired stress responses, impaired sexual maturation, reduced reproductive capability, changes in energy metabolism, pathology, changes in population demographics and recruitment, and shifts in community structure.
2) Pioneering work in Sweden associated these effects with discharges of effluent from bleached kraft mills, and subsequent field and lab studies showed that effects diminished in intensity with process changes, effluent treatment, and decreased discharges of chlorinated organic compounds. As a consequence of this association, and because individual toxic elements of complex mixtures were difficult to isolate and identify, Sweden adopted a precautionary approach and required that total chlorine discharge be limited, as measured by the bulk parameter AOX.

3) Subsequent studies have shown MFO induction and changes in sex hormone concentrations in fish when they are exposed to pulp mill effluents, whether from mills using Cl_2 bleaching or not. These responses can be present at very low effluent concentrations (<1%), can be caused by effluents from pulp mill processes not using Cl_2, are not completely removed during normal secondary treatment of effluent, and cannot be predicted from laboratory acute lethality bioassays. In receiving waters, the responses are transient and reversible and can be reduced by changes in pulp mill processes and enhanced effluent treatment.
4) MFO induction and changes in sex hormone concentrations cannot be associated with levels of AOX or of dioxins and furans, and they represent a previously unidentified category of response, distinct from the issue of chlorine use. Their previous association with the discharge of chlorinated compounds may be an anomaly caused by process changes and improved effluent treatment that simultaneously changed both total chlorine discharge and the discharge of the compounds causing effects.
5) Since the intent of effluent regulations in Canada was to eliminate environmental effects, regulations in Canada control only those chemicals known to be toxic. Canada did not regulate AOX because there was no evidence that effort expended to reduce AOX would reduce effects on fish in receiving environments. Canada has initiated research to identify and characterize the toxic chemicals causing effects, their sources within pulp mills, and methods for their elimination through effluent treatment or process change.

References

Ahokas JT, Holdway DA, Brennan SE, Goudey RW, Bibrowska HB. 1994. MFO activity in carp (*Cyprinus carpio*) exposed to treated pulp and paper mill effluent in Lake Coleman, Victoria, Australia, in relation to AOX, EOX and muscle PCDD/PCDF. *Environ Toxicol Chem* 13:41–50.

Andersson T, Förlin L, Härdig J, Larsson Å. 1988. Physiological disturbances in fish living in the receiving body of water of a kraft mill industry. *Can J Fish Aquat Sci* 45:1525–1536.

Axelsson B, Norrgren L. 1991. Parasite frequency and liver anomalies in three-spined stickleback, *Gasterosteus aculeatus* (L.) after long-term exposure to pulp mill effluents in marine mesocosms. *Arch Environ Contam Toxicol* 21:505–513.

Balk L, Förlin L, Soderstrom M, Larsson A. 1993. Indications of regional and large-scale biological effects caused by bleached pulp mill effluents. *Chemosphere* 27:631–650.

Bengtsson B-E. 1988. Effects of pulp mill effluents on skeletal parameters in fish: a progress report. *Wat Sci Technol* 20:87–94.

Berry RM, Luthe CE, Voss RH, Wrist PE, Axegard P, Gellerstedt G, Lindblad P-O, Popke I. 1991. The effects of recent changes in bleached softwood kraft mill technology on organochlorine emissions: an international perspective. *Pulp & Paper Can* 92(6):43–53.

Borton D, Miner R, Gillespie W, LaFleur W, Hall T. 1990. An examination of the relationship of AOX content in paper industry wastewaters to potential biological effects. In: Hutzinger O, Fiedler H, editors. Dioxin, Volume 3. Bayreuth, Germany: Eco-Informa Pr. p 215–218.

Borton D. 1992. An NCASI overview of biomarker research with pulp and paper industry effluents. 7th Colloquium on Pulp & Paper Mill Effluents. Toronto, ON, Canada. 16 December 1992.

Burnison BK, Hodson PV, Nuttley DJ, Efler S. 1994. Isolation and characterization of a MFO-inducing fraction from BKME. Poster presentation, 2nd International Conference on the Fate and Effect of Bleached Kraft Mill Effluents, Vancouver, BC. Canada, November 1994.

Craig GR, Orr PL, Robertson JL, Vrooman WM. 1990. Toxicity and bioaccumulation of AOX and EOX. *Pulp & Paper Can* 91(9):39–45.

Davis WP, Bartone SA. 1992. Effects of kraft mill effluent on the sexuality of fishes: an environmental early warning? In: Colborn T, Clement C, editors. Chemically-induced alterations in sexual and functional development: the wildlife/human connection. Volume XXI, Advances in modern environmental toxicology. Princeton NJ: Princeton Scientific. p 113–127.

Environment Canada. 1991. Effluents from pulp mills using bleaching. Canadian Environmental Protection Act. Priority Substances List Assessment Report No. 2. Ottawa: Environment Canada and Health and Welfare Canada.

Eysenbach EJ, Neal LW, Owens JW. 1990. Pulping effluents in the aquatic environment. Data compilation and Scientific Panel Report. *Tappi J* 104–106.

Folke J. 1991. Regulatory requirements for pulp and paper mill effluent control: scientific basis and consequences. *Wat Sci Technol* 24:19–31.

Folke J, Edde H, Lehtinen K-J. 1991. The scientific foundation of adsorbable organohalogens (AOX) as a regulatory parameter for control of organochlorine compounds. Presented at the 1991 TAPPI Environmental Conference, 7–10 April 1991, San Antonio, Texas.

Förlin L, Andersson T, Balk L, Larsson A. 1992. Biochemical and physiological effects of pulp mill effluents in fish. In: Södergren A, editor. Proceedings of an international conference on the fate and effects of pulp mill effluents. Stockholm: Swedish EPA.

Gagne F, Blaise C. 1993. Hepatic metallothionein level and mixed function oxidase activity in fingerling rainbow trout (*Oncorhynchus mykiss*) after acute exposure to pulp and paper mill effluents. *Water Res* 27:1669–1682.

Gagnon MM, Bussières D, Dodson JJ, Hodson PV. 1995. White sucker (*Catostomus commersoni*) growth, maturation and reproduction in pulp mill contaminated and reference rivers. *Environ Toxicol Chem* 14:317–327.

Gagnon MM, Dodson JJ, Hodson PV. 1994a. Ability of BKME (bleached kraft mill effluent) exposed white suckers (*Catostomus commersoni*) to synthesize steroid hormones. *Compar Biochem Physiol* 107C(2):265–273.

Gagnon MM, Dodson JJ, Hodson PV. 1994b. Seasonal effects of bleached kraft mill effluent on reproductive parameters of white sucker (*Catostomus commersoni*) populations of the St. Maurice River, Quebec, Canada. *Can J Fish Aquat Sci* 51:337–347.

Gergov M, Priha M, Talka E, Valttila O, Kangas A, Kukkonen K. 1988. Chlorinated organic compounds in effluent treatment at kraft mills. *Tappi J* 71(12):175–184.

Hall TJ, Haley RK, LaFleur LE. 1991. Effects of biologically treated bleached kraft mill effluent on cold water stream productivity in experimental stream channels. *Environ Toxicol Chem* 10:1051–1060.

Hall ER, Fraser J, Garden S, Conacchio LA. 1989. Organo-chlorine discharges in wastewaters from kraft mill bleach plants. *Pulp Pap Can* 90(11):68–72.

Hanninen O, Lindström-Seppä P, Pesonen M, Huuskonen S, Muona P. 1991. Use of biotransformation activity in fish and fish hepatocytes in the monitoring of aquatic pollution caused by pulp industry.

In: Yasuno M, Fukushima S, Sugaya T, editors. Bioindicators and environmental management. New York: Academic Pr. p 11–20.

Hodson PV, Bussières D, Gagnon MM, Dodson JJ, Couillard CM, Carey JH. 1991. Review of biochemical, physiological, pathological and population responses of white sucker (*Catostomus commersoni*) to BKME in the St. Maurice River, Quebec. In: A. Södergren, editor. Environmental fate and effects of bleached pulpmill effluent. Proceedings of a Swedish EPA conference; 1991 Nov 19–21; Stockholm, Sweden. Swedish EPA Report 4031. p 261–269.

Hodson PV, Kloepper-Sams PJ, Munkittrick KR, Lockhart WL, Metner DA, Luxon L, Smith IR, Gagnon MM, Servos M, Payne JF. 1991. Protocols for measuring mixed function oxygenases of fish liver. *Can Tech Rep Fish Aquat Sci* #1829: 49 p.

Hodson PV, McWhirter M, Ralph K, Gray B, Thivierge D, Carey J, Van Der Kraak G, Whittle DM, Levesque M-C. 1992. Effects of bleached kraft mill effluent on fish in the St. Maurice River, Quebec. *Environ Toxicol Chem* 11:1635–1651.

Hodson PV, Maj MM, Efler S, Schnell A, Carey J. 1994. Kraft black liquor as a source of MFO inducers [poster presentation]. 2nd International Conference on the Fate and Effect of Bleached Kraft Mill Effluents; 1994 Nov; Vancouver, BC, Canada.

Kloepper-Sams P, Swanson S. 1992. Bioindicator field monitoring: use of fish biochemical parameters at a modern bleached kraft pulp mill site. *Mar Env Res* 34:163–168.

Kloepper-Sams P, Marchant T, Bernstein J, Swanson S. 1992. Use of fish biomarkers and exposure measures to assess fish health at a Canadian bleached kraft mill site. In: Södergren A, editor. Environmental fate and effects of bleached pulpmill effluent. Proceedings of a Swedish EPA conference; 1991 Nov 19–21; Stockholm, Sweden. Swedish EPA Report 4031. p 283–292.

Kovacs T. 1986. Effects of bleached kraft mill effluent on freshwater fish: a Canadian perspective. *Water Poll Res J Can* 21:91–118.

Larsson L, Andersson T, Förlin L, Härdig J. 1988. Physiological disturbances in fish exposed to bleached kraft mill effluents. *Water Sci Technol* 20:67–76.

Lehtinen K-J, Tana J, Karlsson P, Engstrom C, Mattsson K, Hemming S, Hemming J, Fugleberg A-L. 1992. Effects in mesocosms exposed to effluents from bleached hardwood kraft pulp mill. Part 2. Chemical characterization and effects in mesocosms from bleached hardwood kraft pulp production. In Study No 10. Finland Board of Waters and the Environment. p 57–108.

Lehtinen K-J, Kierkegaard A, Jakobsson E, Wandell A. 1990. Physiological effects in fish exposed to effluents from mills with six different bleaching processes. *Ecotox Environ Safety* 19:33–46.

Lehtinen K-J, Tana J, Härdig J, Mattson K, Hemming J, Lindström-Seppä P. 1992. Effects in mesocosms exposed to effluents from bleached hardwood kraft pulp mill. Part 1. Effects on survival, growth, parasites and physiological status in fish exposed in mesocosms to effluents from bleached hardwood pulp production. Study No 10. Helsinki: Finland Board of Waters and the Environment. p 3–54.

Lehtinen K-J, Axelsson B, Kringstad K, Strömberg L. 1991. Characterization of pulp mill effluents by the model ecosystem approach. SSVL-investigations in the period 1982 to 1990. *Nordic Pulp & Paper Res J* 2:81–88.

Lindström-Seppä P, Huuskonen S, Pesonen M, Muona P, Hanninen O. 1992. Unbleached pulp mill effluents affect cytochrome P-450 mono-oxygenase enzyme activities. *Mar Env Res* 34:157–161.

Martel PH, Kovacs TG, O'Connor BI, Voss RH. 1994. A survey of pulp and paper mill effluents for their potential to induce mixed function oxidase enzyme activity in fish. *Water Res* 28:1835–1844.

Martinsen K, Kringstad A, Carlberg GE. 1988. Methods for the determination of sum parameters and characterization of organochlorine compounds in spent bleach liquors from pulp mills and water, sediment and biological samples from receiving waters. *Water Sci Technol* 20:13–24.

McLeay DJ, Walden CC, Munro JR. 1979. Effect of pH on toxicity of kraft pulp and paper mill effluent to salmonid fish in fresh and seawater. *Water Res* 13:249–254.

McLeay DJ. 1987. Aquatic toxicology of pulp and paper mill effluent: a review. Ottawa ON: Environment Canada. EPS 4/PF/1.

McMaster ME, Portt CB, Munkittrick KR, Dixon DG. 1992. Milt characteristics, reproductive performance and larval survival and development of white sucker exposed to bleached kraft mill effluent. *Ecotoxicol Environ Safety* 23:103–117.

McMaster ME, Van Der Kraak GJ, Portt CB, Munkittrick KR, Sibley PK, Smith IR, Dixon DG. 1991. Changes in hepatic mixed-function oxygenase (MFO), plasma steroid levels and age at maturity of a white sucker (*Catostomus commersoni*) population exposed to bleached kraft mill effluent. *Aquat Tox* 21:199–218.

Mehrle PM, Dickson K, Hartung R, Huggett R, McLeay D, Oikari A, Sprague J. 1989. Pulping effluents in the aquatic environment. Cincinnati OH: Report prepared for Procter & Gamble Co.

Muir DCG, Yarechewski AL, Metner DA, Lockhart WL, Webster GBR, Friesen KJ. 1990. Dietary accumulation and sustained hepatic mixed function oxidase enzyme induction by 2,3,4,7,8-pentachlorodibenzofuran in rainbow trout. *Environ Toxicol Chem* 9:1463–1472.

Munkittrick KR. 1992. Recent North American studies of bleached kraft mill impacts on the reproduction and growth of fish populations. In: Södergren A, editor. Environmental fate and effects of bleached pulpmill effluent. Proceedings of a Swedish EPA conference; 1991 Nov 19–21; Stockholm, Sweden. Swedish EPA Report 4031. p 347–356.

Munkittrick KR, Van Der Kraak GJ, McMaster ME, Portt CB. 1992. Relative benefit of secondary treatment and mill shutdown on mitigating impacts of bleached kraft mill effluent (BKME) on MFO activity and serum steroids in fish. *Environ Toxicol Chem* 11:1427–1439.

Munkittrick KR, Portt C, Van Der Kraak GJ, Smith IR, Rokosh DA. 1991. Impact of bleached kraft mill effluent on liver MFO activity, serum steroid levels and population characteristics of a Lake Superior white sucker population. *Can J Fish Aquat Sci* 48: 1371–1380.

Munkittrick KR, McMaster ME, Portt CB, Van Der Kraak GJ, Smith IR, Dixon DG. 1992. Changes in maturity, plasma sex steroid levels, hepatic MFO activity and the presence of external lesions in lake whitefish exposed to bleached kraft mill effluent. *Can J Fish Aquat Sci* 49:1560–1569.

Munkittrick KR, Van Der Kraak GJ. 1993. Receiving water environmental effects associated with discharges from Ontario pulp mills. Proc 79th Annual Meeting, Technical Section, Canadian Pulp and Paper Association; 1991; Montreal, Quebec. p. A341–A345.

Munkittrick KR, Van Der Kraak GJ, McMaster ME, Portt C, van den Heuvel MR, Servos MR. 1994. Survey of receiving water environmental impacts associated with discharges from pulp mills. II. Gonad size, liver size, hepatic EROD activity and plasma sex steroid levels in white sucker. *Environ Toxicol Chem* 13:1089–1101.

[NCASI] National Council of the Paper Industry for Air and Stream Improvement. 1989. Effects of biologically treated bleached kraft mill effluent on cold water stream productivity in experimental streams - fifth progress report. New Bern NC: NCASI Tech Bull 566.

[NCASI] National Council of the Paper Industry for Air and Stream Improvement. 1984. Effects of biologically treated bleached kraft mill effluent on warm water stream productivity in experimental streams - third progress report. New Bern NC: NCASI Tech Bull 414.

Neuman E, Karås P. 1988. Effects of pulp mill effluent on a Baltic coastal fish community. *Water Sci Technol* 20:95–106.

Owens JW. 1991. The hazard assessment of pulp and paper effluents in the aquatic environment: a review. *Environ Toxicol Chem* 10:1511–1540.

O'Connor BI, Kovacs TG, Voss RH, Martel PH. 1993. A study of the relationship between laboratory response and AOX content for pulp mill effluent. *J Pulp Pap Sci* 19: J33–39.

Parrott JL, Hodson PV, Servos MR, Huestis SL, Dixon DG. 1994. Relative potencies of polychlorinated dibenzo-p-dioxins and dibenzofurans for inducing mixed function oxygenase activity in rainbow trout. *Environ Toxicol Chem* 14:1041–1050.

Pesonen M, Andersson T. 1992. Toxic effects of bleached and unbleached paper mill effluents in primary cultures of rainbow trout hepatocytes. *Ecotoxicol Environ Saf* 24:63–71.

Robinson RD, Carey JH, Solomon KR, Smith IR, Servos MR, Munkittrick KR. 1994. Survey of receiving water environmental impacts associated with discharges from pulp mills. I. Mill characteristics, receiving water chemical profiles and laboratory chronic toxicity tests. *Environ Toxicol Chem* 13:1075–1088.

Rogers IH, Mahood HW. 1974. Removal of fish-toxic organic solutes from whole kraft effluent by biological oxidation and the role of wood extractives. Fish Res Board Can Tech Rep 434.

Rogers IH, Levings CD, Lockhart WL, Norstrom RJ. 1989. Observations on over-wintering juvenile chinook salmon (*Oncorhynchus tshawytscha*) exposed to bleached kraft mill effluent in the upper Fraser River, British Columbia. *Chemosphere* 19:1853–1868.

Sandström O, Neuman E, Karås P. 1988. Effects of a bleached pulp mill effluent on growth and gonad function in Baltic coastal fish. *Wat Sci Technol* 20:107–118.

Schnell A, Hodson PV, Steel P, Melcer H, Carey JH. 1993. Optimized biological treatment of bleached kraft mill effluents for the enhanced removal of toxic compounds and MFO induction response in fish. Proceedings of the 1993 Environment Conference of the Canadian Pulp and Paper Association, Thunder Bay, Oct. 26–28, 1993. 15 p.

Servizi JA, Martens DW, Gordon RW, Kutney JP, Singh M, Dimitriadis E, Hewitt GM, Salisbury PJ, Choi LSL. 1986. Microbiological detoxification of resin acids. *Wat Pollut Res J Can* 21:119–129.

Servos MR, Huestis SY, Whittle DM, Van Der Kraak GJ, Munkittrick KR. 1994. Survey of receiving water environmental impacts associated with discharges from pulp mills. III. Polychlorinated dioxins and furans in muscle and liver of white sucker, *Catostomus commersoni*. *Environ Toxicol Chem* 13:1103–1115.

Servos M, Carey J, Ferguson M, Van Der Kraak G, Ferguson H, Parrott J, Gorman K, Cowling R. 1992. Impact of a modern bleached kraft mill on white sucker populations in the Spanish River, Ontario. *Wat Pollut Res J Canada* 27:427–438.

Smith IR, Portt C, Rokosh DA. 1991. Hepatic mixed function oxidases are induced in populations of white sucker, *Catostomus commersoni*, from areas of Lake Superior and the St. Mary's River. *J Great Lakes Res* 17:382–393.

Södergren A, Adolfsson-Erici M, Bengtsson B-E, Jonsson P, Lagergren S, Rahm L, Wulff F. 1992. Environmental effect of bleached pulp mill effluents discharged into the Baltic Sea. In: Södergren A, editor. Proceedings of an International Conference on the Fate and Effects of Pulp Mill Effluents; 1991 Nov; Stockholm, Sweden. Stockholm: Swedish EPA.

Södergren A. 1989. Biological effects of bleached pulp mill effluents. Stockholm: National Swedish Environmental Protection Board Report 3558.

Södergren A, Bengtsson B-E, Jonsson P, Lagergren S, Larsson Å, Olsson M, Renberg L. 1988. Summary of results from the Swedish project "Environment/Cellulose." *Wat Sci and Technol* 29:49–60.

Swanson S, Shelast R, Schryer R, Kloepper-Sams P, Marchant T, Kroeker K, Bernstein J, Owens JW. 1992. Fish populations and biomarker responses at a Canadian bleached kraft mill site. *Tappi J* (1992): 139–149.

van den Heuvel MR, Servos MR, Munkittrick KR, Portt CB, Van Der Kraak GJ, McMaster ME, Dixon DG. 1994. Survey of receiving water environmental impacts associated with discharges from pulp mills. IV. Bioassay derived 2,3,7,8-tetrachlorodibenzo-p-dioxin toxic equivalent concentration in white sucker, *Catostomus commersoni*, in relation to biochemical indicators of impact. *Environ Toxicol Chem* 13:1117–1126.

Van Der Kraak GJ, Munkittrick KR, McMaster ME, Portt CB, Chang JP. 1992a. Exposure to bleached kraft pulp mill effluent disrupts the pituitary-gonadal axis of white sucker at multiple sites. *Toxicol Appl Pharmacol* 115:224–233.

Van Der Kraak GJ, Munkittrick KR, McMaster ME, Portt CB, Chang JP. 1992b. Assessment of reproductive fitness in white sucker exposed to bleached kraft mill effluent. *Toxicol Appl Pharmacol* 115:224–233.

Whittle DM, Mageau C, Duncan RK, Sergeant DB, Nassichuk MD, Morrisson J, Piuze J. 1993. Canadian national dioxin sampling program: dioxins and furans in biota near 46 pulp and paper mills using the chlorine bleaching process. *Chemosphere* 27:279–286.

Williams TG, Servos MR, Carey JH, Dixon DG. 1992. Biological response monitoring of pulpmill effluent. In: Södergren A, editor. Environmental fate and effects of bleached pulpmill effluent. Swedish EPA Report 4031. Proceedings of a Swedish EPA conference; 1991 Nov 19–21; Stockholm, Sweden. p 391–394.

Williams TG. 1993. A comparative laboratory based assessment of EROD enzyme induction in fish exposed to pulp mill effluents [masters thesis]. Waterloo, Ontario: Univ of Waterloo, Dept of Biology, 85 p.

Appendix 3

Organochlorine Concentrations and Trends in North American Fish and Wildlife

Donald E. Tillitt and Christopher J. Schmitt,
U.S. Geological Survey, Columbia MO

D. Vaughn Weseloh, Canadian Wildlife Service,
Burlington, Ontario, Canada·

Christine M. Bunck,
U.S. Geological Survey, Madison WI

Synthetic organochlorine (OC) chemicals have been used advantageously in agriculture and industry. Often the advantage has come from the reactivity and resistance of OC compounds to degradation. However, these same properties have resulted in the environmental toxicity and persistence of some OC chemicals. Some OC compounds bioconcentrate in fish and biomagnify through food chains. The results are sometimes devastating, with complete elimination of top predator species in heavily affected areas. Concerns about the effects of some persistent, toxic OCs on fish and wildlife led to the development of biomonitoring programs in the United States, Canada, and around the world.

The objectives of this presentation were to discuss trends of OC contaminants in fish and wildlife over the last 20 to 30 years, as exemplified by 2 federal monitoring programs (1 in the U.S. and 1 in Canada), highlight the major findings of these programs, and identify potentially affected fish and wildlife species and pathways of OC exposure.

Environmental controls imposed over the last 20 years appear to have reduced the amount of OC compounds entering the environment, which has resulted in lower concentrations in fish and wildlife. Concentrations of OC chemicals are significantly lower in most areas, with the exception of remote (i.e., Arctic) environments where monitoring results are controvertible. The general decreases in continental areas nearer to pollution sources have followed first-order reductions to the current concentrations; however, based on these trends, further reductions are expected to be small. Therefore, we must now assess whether existing concentrations are above or below thresholds for effects in fish and wildlife populations.

Chlorine is electronegative, which makes its chemistries with carbon particularly reactive and results in stable bonds and compounds. The reactivity of organochlorine (OC) compounds and their resistance to degradation were viewed as advantageous properties for pesticides. The ability of manufacturers to synthesize a wide variety of similar chemical structures with only slightly different configurations allowed the screening and production of potent and efficacious pesticides. The same properties of thermal stability and resistance to degradation of certain OC compounds also led to their industrial application, e.g., the use of polychlorinated biphenyls (PCBs) as heat transfer fluids in electrical components. Unfortunately, these same properties cause OC compounds to persist in the environment and to bioaccumulate through food chains (Mackay et al. 1983; Hayes and Laws 1991). Consequently, concentrations of some OC pesticides and industrial compounds reached toxic amounts in organisms at the top of the food chain (Gilbertson and Fox 1977; Gilbertson 1983, 1989; Peakall 1988; Kubiak et al. 1989; Giesy et al. 1994). The classic example is the insecticide 2,2-

bis(*p*-chlorphenyl)-1,1,1-trichloroethane (DDT) and its primary degradation product, 2,2-bis(*p*-chlorophenyl)-1,1-dichloroethane (DDE). The efficacy of DDT to control mosquitoes, and thus the spread of malaria, was unrivaled (Hayes and Laws 1991). The effects on nontarget organisms, however, were not anticipated. Bird populations were devastated by DDT and DDE because of the abilities of these compounds to disrupt the deposition of calcium carbonate in eggshells. This consequently caused eggshell thinning and, ultimately, reproductive failure in affected populations (Pearce et al. 1979; Wiemeyer et al. 1993). The nontarget effects were not predicted due to either lack of testing or testing of species that were inappropriate with regard to the potential exposures.

Many OC compounds that have been used extensively in the 1950s and 1960s in North America and elsewhere were banned from use or severely limited in use in the U.S. and Canada during the late 1970s and early 1980s. Production of some, but not all, of these OC compounds was also banned in the U.S.; heptachlor and methoxychlor, e.g., are still manufactured for export. Developing nations, including many in South and Central America, perceive that the benefits of OC insecticides as disease-vector control agents outweigh the environmental costs (Hayes and Laws 1991). However, some countries, including many in Africa, have adopted alternative control mechanisms for disease vectors. Mainland African nations control malaria mostly through chemotherapy and chemoprophylaxis programs directed toward the most vulnerable groups (World Health Organization [WHO] 1983, 1984). Decisions to this end must carefully balance the potential effects of pesticide and the benefits gained from its use. The threat to human life attributable to insect-borne diseases is less severe in developed nations than in some developing countries, and therefore, the cost–benefit analysis is not as straightforward. The real and potential effects of OC compounds on the environment are often insidious; consequently, their resultant costs are more difficult to quantify than is the loss of human life to disease or crop loss to pests.

The adverse effects that intentional release of some OC pesticides and inadvertent release of some chlorinated industrial compounds have had on fish and wildlife species around the world has been well documented. Bald eagle (*Haliaeetus leucocephalus*) populations were extirpated from some areas, such as the Great Lakes of the U.S. and Canada (Grier 1982). Similarly, populations of other fish-eating birds, including Caspian tern (*Sterna caspia*), Forster's tern (*Sterna forsterii*), double-crested cormorant (*Phalacrocorax auritis*), brown pelican (*Pelicanus occidentalis*), and osprey (*Pandion haliaetus*), declined precipitously concomitant with increasing amounts of several OC pesticides and industrial compounds (Cooke 1973; Peakall and Fox 1987; Gilbertson 1989; Kubiak et al. 1989; Tillitt et al. 1992). Terrestrial birds associated with some agricultural uses of OC pesticides also experienced reductions in their populations (Wuster et al. 1965; Blus et al. 1983), and populations of the American robin (*Turdus migratorius*) and other species were decimated by DDT spraying for Dutch elm disease control (Cooper 1991). Adverse effects have also been demonstrated in populations of fish-eating mammals, such as mink (*Mustela vison*) and otter (*Lutra canadense*), which obtain disproportionately large amounts of these chemicals through aquatic food chains (Wren 1991). Fishes can also be very sensitive to the adverse effects of many OC compounds, which are believed to be linked to the reproductive failure of lake trout (*Salvelinus namaycush*) in the Great Lakes (U.S. Department of the Interior [USDI] 1981) and in Lake George, NY (Burdick et al. 1964), and of landlocked Atlantic salmon (*Salmo salar*) in Sebago Lake, ME (Anderson and Everhart 1966). Additionally, teratogenic and reproductive problems attributable to OC pesticides have been documented in reptiles, including snapping turtles (*Chelydra serpentina)* (Bishop et al. 1991) and alligators (*Alligator mississippiensis*) (Guillette et al. 1994). Many of these environmental effects became evident in the late 1960s and early 1970s, as the phenomena of bioaccumulation and biomagnification of persistent OC compounds were discovered.

Global contamination through atmospheric transport of OC compounds has been a concern and the subject of intense research (Tanabe, Iwata et al. 1994). The open waters of the oceans, and particularly the Arctic waters, serve as a sink for the persistent OC compounds (Muir et al. 1992; Barrie et al. 1992; Tanabe, Iwata et al. 1994). As a result, the concentrations of persistent OCs in marine mammals, and in particular cetaceans, have been high because of the global distribution of these compounds (Norstrom et al. 1990; Tanabe and Tatsukawa 1991; Muir et al. 1992; O'Shea and Brownell 1994; Tanabe et al. 1994). However, the information on trends in concentrations of persistent OC compounds in these remote areas, such as the Arctic, is controvertible at this time (Muir et al. 1992; Norstrom and Muir 1994), and more information is required. A complete discussion of OC contamination in marine mammals and the global atmospheric transport of these chemicals is beyond the scope of this paper.

Environmental monitoring programs for OCs have been developed in many industrialized countries. These programs are designed to identify time trends in OC concentrations, study the relationships of chemical concentrations among various trophic levels, assess the ongoing needs of chemical and biological monitoring programs, and determine the potential for environmental effects attributable to OC chemicals, based on their presence in fish and wildlife. Two are North American monitoring programs: the Herring Gull Monitoring Program (HGMP) in the Great Lakes, established and maintained by the Canadian Wildlife Service (CWS), and the National Contaminant Biomonitoring Program (NCBP), established by the U.S. Fish and Wildlife Service (USFWS) and now maintained by the National Biological Service (NBS). The HGMP and NCBP are long-term programs that span much of the history of OC chemical use throughout North America. Excellent monitoring programs were also developed in Sweden, Japan, the United Kingdom, and other industrialized nations to address the same concerns as do the HGMP and NCBP. Other, more localized programs have been developed by states and provinces in the U.S. and Canada and are designed to meet more specific needs. Newer U.S. national programs include the National Water Quality Assessment program being implemented by the U.S. Geological Survey; the National Status and Trends program, an estuarine program administered by the National Oceanic and Atmospheric Administration (NOAA), U.S. Department of Commerce; the Environmental Monitoring and Assessment Program, which was recently initiated by the U.S. Environmental Protection Agency (USEPA); and the Biomonitoring Environmental Status and Trends program, initiated by the USFWS and now under continuing development and implementation by NBS. Canadian programs include the Environmental Effects Monitoring Program conducted by the Environmental Conservation Service, Environment Canada, for downstream pollutants at pulp and paper mills and the Shellfish Monitoring Program on the Atlantic and Pacific coasts which is administered by Environment Canada.

It is the objective of this paper to describe the trends of OC contaminants in fish and wildlife over the last 20 to 30 years as exemplified by the HGMP and NCBP, highlight the major findings of these programs, and identify potentially affected fish and wildlife species and pathways of OC exposure. We have focused this paper on the results and conclusions of the HGMP and NCBP because the major trends observed by these programs are comparable to the findings of the other excellent national programs in Sweden (Olsson 1986; Bignert et al. 1992, 1995; Järnberg et al. 1993), Japan (Tanabe et al. 1989; Tanabe, Iwata and Patsuckawa. 1994; Tanabe, Sung et al. 1994), and the United Kingdom (Jones et al. 1992).

Herring Gull Monitoring Program, Canadian Wildlife Service: OC Chemicals in Great Lakes Herring Gulls—1974 to 1993

In the late 1960s and early 1970s, fish-eating birds (e.g., herring gulls [*Larus argentatus*], terns, cormorants, herons) inhabiting the North American Great Lakes experienced extremely thin eggshells, reproductive failure, and declining populations (Keith 1966; Ludwig and Tomoff 1966; Hickey and Anderson 1968; Gilberton and Reynolds 1972; Gilbertson 1974; Gilbertson and Hale 1974a, 1974b; Gilbertson 1975). By 1974, the cause had been identified as high dietary concentrations of toxic chemicals being passed into their eggs, and an annual monitoring program, the HGMP, was established by the CWS (Gilman et al. 1979). The HGMP has been continued and now constitutes one of the world's longest-running annual monitoring programs for organic contaminants in wildlife. The objective of the program is to monitor temporal and spatial trends in contaminants and their effects on wildlife in the 5 Great Lakes.

HGMP sample collection

The herring gull was selected by the HGMP as the indicator species primarily because it is nonmigratory as an adult on the Great Lakes, it accumulates high levels of contaminants, and it breeds on all 5 Great Lakes (Gilman et al. 1979; Mineau et al.

1984). Fresh eggs are collected annually from each of up to 16 colonies situated on the Great Lakes and connecting channels (Figure 1) and, at present, are analyzed for as many as 78 chemicals including OCs, pesticides, individual PCB congeners, and selected polychlorinated dibenzo-dioxins (PCDDs) and -furans (PCDFs) (Peakall et al. 1986; Bishop et al. 1992; Pettit et al. 1994; Hebert et al. 1994). In addition to the portions collected for chemical analysis, egg tissue is stored in the CWS tissue bank for subsequent use (Elliott 1984). Biological effects, including eggshell thinning, reproductive success, reproductive behavior, and physiological biomarkers, are also measured (Mineau et al. 1984; Fox 1993).

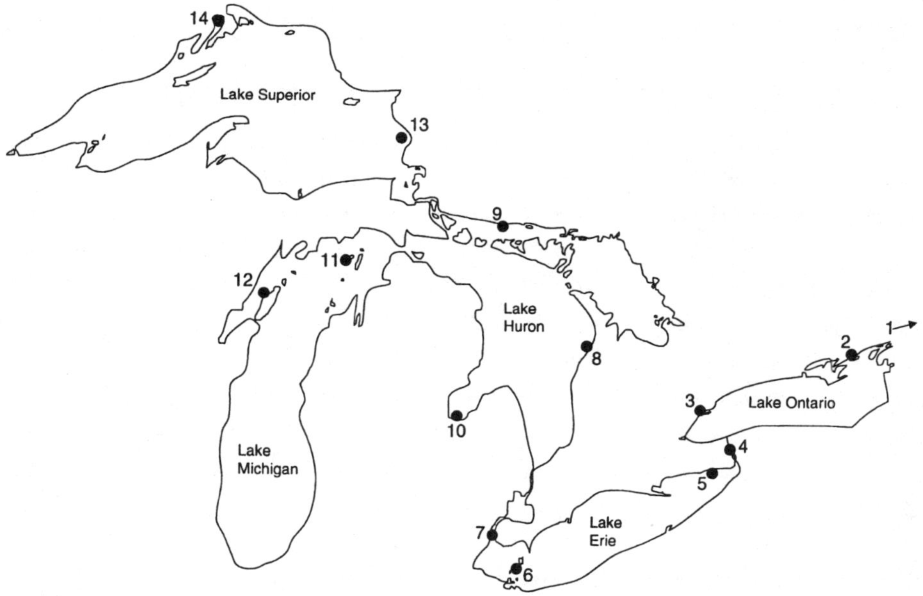

Figure 1 Location of collection sites for the Herring Gull Monitoring Program of the Canadian Wildlife Service. The sites are 1) Strachan Island, St. Lawrence River (located approximately 100 km east of site 2; 2) Snake Island; 3) Toronto Harbor; 4) Niagara River; 5) Port Colbourne; 6) Middle Island; 7) Fighting Island, Detroit River; 8) Chantry Island; 9) Double Island; 10) Channel-Shelter Island, Saginaw Bay; 11) Gull Island; 12) Big Sister Island; 13) Agawa Rocks; and 14) Granite Island.

Temporal trends of OCs in HGMP—1974 to 1993

Perhaps no question about the HGMP is asked as often as "Are the levels of OC chemicals increasing or decreasing?" To answer this question concisely, we have depicted the results of the analysis for 6 major compounds, monitored since the beginning of the program, at 13 sites for the period 1974 to 1993; the chemical results were analyzed by change-point regression (Draper and Smith 1981).

Prior to the respective change points, most of which occurred in the 1980s, 78% of the colony-contaminant trends at the various colonies were significantly declining, 21% showed no change, and 1% were increasing (Table 1). After the change points, 56% continued to show a significant decline (though many at a slower rate), 37% showed no change, and 6% showed a significant increase (Table 1) (see Pekarik and Weseloh 1995 for more details). Most of the decreases in specific compounds at individual sites occurred between 1974 and 1980/81. Since then, the rate of decrease of most compounds has been much slower but nevertheless statistically significant in many instances (Pekarik and Weseloh 1995).

Table 1 Summary of change-point regression analysis of the colony-contaminant comparisons for the 6 major contaminants in the Herring Gull Monitoring Program, 1974–1993[1]

Water body	Colony location[2]	DDE	Mirex	HCB	Dieldrin	PCB[3]	HE
Lake Ontario	Snake Island	(–/+) '92	(–/–) '82	(–)	(–/+) '92	(–/0) '92	(–/–) '82
	Mugg' s Island	(–/–) '90	(–)	(–/–) '84	(–/–) '90	(–/0) '88	(–/–) '90
Niagara River	Niagara Falls	(0/0) '88	(–)	(–/0) '91	(–/0) '91	(–/0) '88	(0/–) '85
Lake Erie	Port Colbourne	(–/–) '82	(–/–) '82	(–)	(–)	(–/–) '82	(0/0) '91
	Middle Island	(–/–) '79	(0/0) '79	(–/–) '79	(0/0) '88	(–)	(0/0) '88
Detroit River	Fighting Island	(–/0) '92	(+/0) '87	(–)	(0/0) '87	(–)	(0/0) '90
Lake Huron	Chanty Island	(–/0) '91	(–/–) '89	(–/–) '78	(–)	(–/0) '88	(0)
	Double Island	(0/0) '80	(–/–) '82	(–/0) '88	(–)	(–/–) '79	(0)
	Channel-Shelter Isl.	(–/+) '89	(0)	(–)	(–)	(–/0) '89	(–)
Lake Michigan	Gull Island	(0/0) '84	(–/+) '90	(–/–) '89	(–/–) '91	(–/–) '91	(0/–) '92
	Big Sister Island	(–/–) '91	(–)	(–)	(–)	(–/–) '91	(–/+) '92
Lake Superior	Agawa Rocks	(–/0) '85	(–/–) '78	(–)	(–)	(–/0) '85	(0/–) '89
	Granite Island	(–/0) '86	(–/0) '88	(–)	(–)	(–/0) '87	(0)

[1] Results of fitting herring gull egg contaminants data to change-point model (Draper and Smith 1981). When a single regression line model provided the best fit for the data, i.e., there was no change point, the direction of the regression slope is indicated (+ = increase, – = decrease, and 0 = no trend). When a 2-line regression model was the best fit and a significant change point was calculated, the directions of the regression slopes before and after the change point are given, along with the year of the change point (e.g., PCBs in Mugg's Island, Lake Ontario (–/0) '88; from Pedarik and Weseloh 1995).

[2] Locations given in Figure 1.

[3] PCBs quantitated against 1:1, Aroclor 1254:1260 mixture (see Bishop et al. 1992; Pettit et al. 1994).

Several chemicals that were not monitored until later in the HGMP program (e.g., oxychlordane, photomirex, HCH, and 2,3,7,8-TCDD; data not presented) have also decreased significantly (CWS unpublished data; Hebert et al. 1994). Thus, as with other contaminant monitoring studies on the Great Lakes (e.g., DeVault et al. 1986; Baumann and Whittle 1988; Stow et al. 1994), the HGMP confirms that in recent years the number of declining trends has decreased, that more contaminants are showing "no significant change," and that the trend for some compounds has reversed so that they are now increasing. The general patterns of these trends in herring gull eggs are depicted in Figures 2 through 5. Similar OC contaminant trends have been documented in Great Lakes species other than the herring gull species, such as Caspian terns (Figure 2) (Ewins et al. 1994), common terns (Weseloh et al. 1989), and double-crested cormorants (Price and Weseloh 1986; Weseloh et al., unpublished), and in ospreys (Ewins et al. 1994).

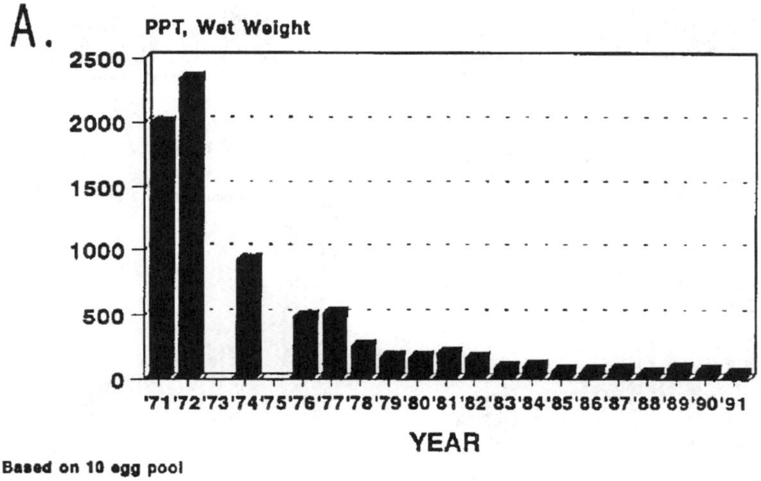

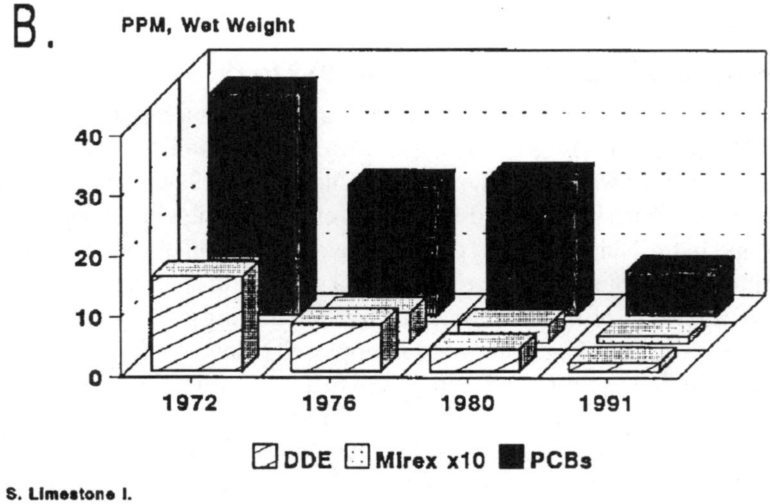

Figure 2 A) concentrations of 2,3,7,8-TCDD (pg/g, wet weight) in herring gull eggs (10 egg composites) from Canadian Wildlife Service monitoring program in Lake Ontario from 1971–1991 and B) concentrations (μg/g) of DDE, mirex, and total PCBs in Caspian tern eggs collected from Lake Ontario 1972–1991, by the Canadian Wildlife Service (data from Bishop et al. 1992 and Pettit et al. 1994).

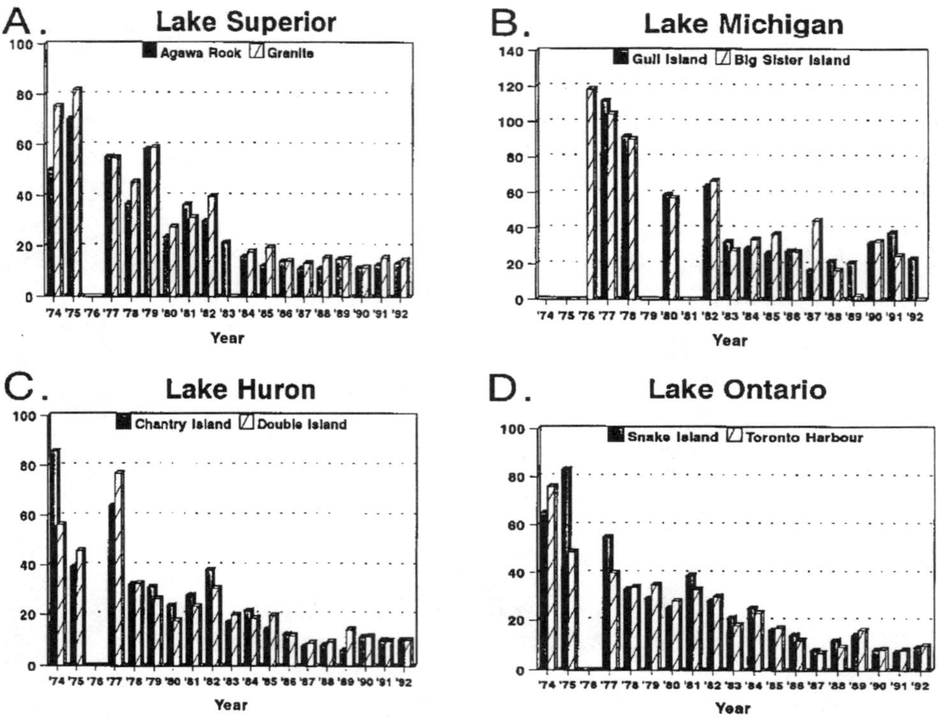

Figure 3 Mean concentrations (μg/g, wet weight) of total PCBs (based on a summation of individual congeners) in herring gull eggs collected from Lakes Michigan, Superior, Huron, and Ontario, 1974–1992, by the Canadian Wildlife Service (data from Bishop et al. 1992 and Pettit et al. 1994).

Temporal patterns of OCs in HGMP — 1989 to 1993

Concentrations of DDE, PCBs, mirex, HCB, HE, and dieldrin were compared at each site over a 5-year period (1989 to 1993). For 60% of the contaminant-colony combinations, concentrations were lower (i.e., > 10% reduction) in 1993 than in 1989, 27% had greater concentrations in 1993, and 10% had no change. At sites on Lakes Erie and Ontario, and in the Niagara and St. Lawrence rivers, nearly all of the concentrations of the monitored compounds decreased from 1989 to 1993. In contrast, more than half of the concentrations of the monitored chemicals in herring gull eggs from the Upper Lakes (i.e., Huron, Michigan, and Superior) appeared to increase from 1989 to 1993. When all of the sites in the HGMP and these 6 major OC compounds are considered, DDE increased at more than half of the sites; while the other 5 compounds (PCBs, mirex, HCB, HE, and dieldrin) decreased at more than half the sites. Comparison of the 2 most recent years for which data are available (1992 and 1993) shows an apparent decline (i.e., > 10% decrease) in concentrations of these 6 OC chemicals at 44% of the site–chemical combinations, while 32% had apparent increases (> 10%). Most of the increases in concentrations of OC compounds were related to mirex and heptachlor epoxide in eggs from the 3 Lake Huron sites and Snake Island, Lake Ontario.

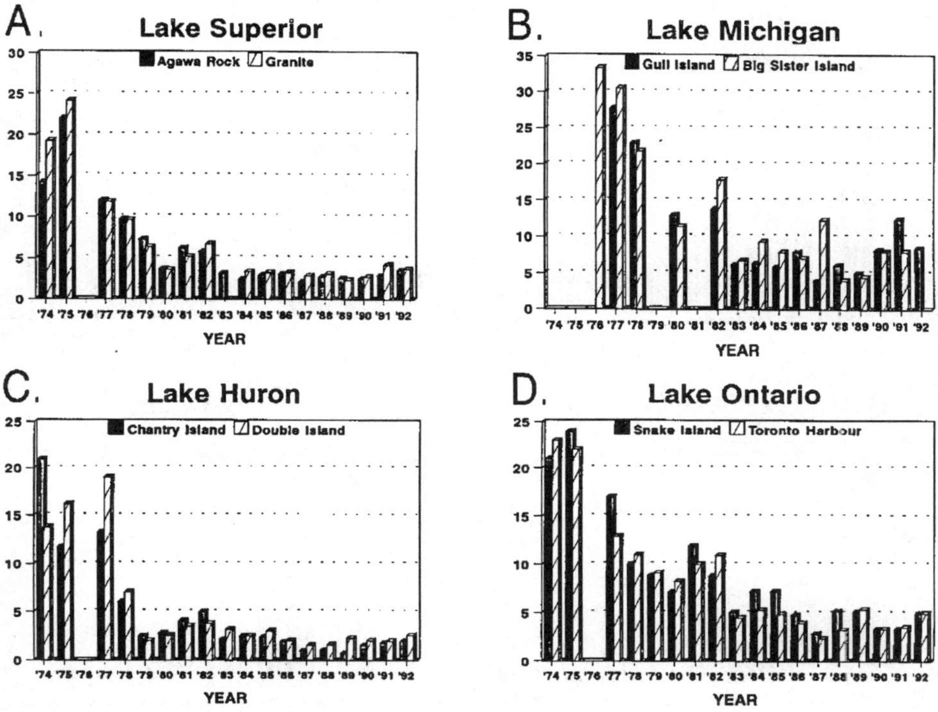

Figure 4 Mean concentrations (µg/g, wet weight) of DDE in herring gull eggs collected in Lakes Michigan, Superior, Huron, and Ontario, 1974–1992, by the Canadian Wildlife Service (data from Bishop et al. 1992 and Pettit et al. 1994).

Spatial trends of OCs in HGMP

Another important objective of the HGMP has been to determine spatial trends, i.e., ascertain where gulls are exposed to the greatest amounts of various contaminants and thus are at the greatest risk. The spatial patterns in concentrations of selected compounds (e.g., PCBs, DDE, mirex, dieldrin) at 14 of the monitoring sites are illustrated in Figure 6. In general, the amounts of PCBs in gull eggs have characteristically been elevated in the colonies from Saginaw Bay in Lake Huron, the Detroit River, western Lake Erie, and the St. Lawrence River; DDE concentrations have been elevated in Lake Michigan, Saginaw Bay in Lake Huron, Lake Ontario, and the St. Lawrence River; and elevated mirex occurs almost exclusively in Lake Ontario (Figure 6). Both oxy-chlordane and heptachlor epoxide, metabolites of *cis*-chlordane and heptachlor, respectively, were greatest in Lake Superior and Lake Michigan (Figure 6). Spatial distributions for such compounds illustrate that the greatest contaminant concentrations do not necessarily occur at the most heavily developed areas (Lake Erie and Lake Ontario). Now that most point sources of contaminants have been eliminated or regulated, sediment resuspension and atmospheric transport remain major avenues of

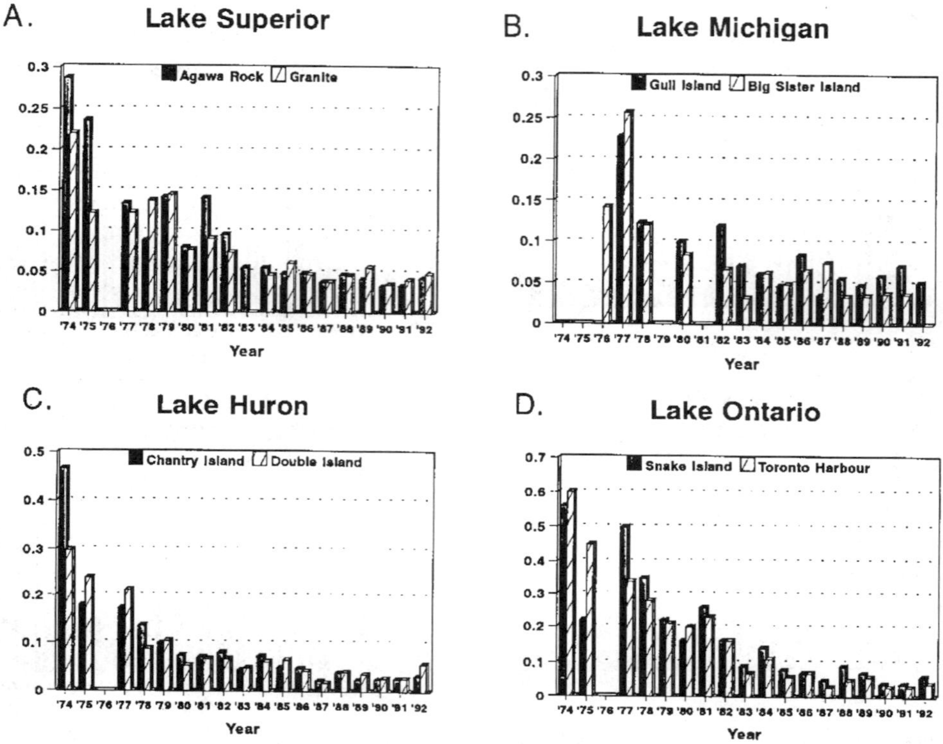

Figure 5 Mean concentrations (μg/g, wet weight) of hexachlorobenzene (HCB) in herring gull eggs collected at various locations in Lake Michigan, Lake Superior, Lake Huron, and Lake Ontario, 1974–1992, by the Canadian Wildlife Service (data from Bishop et al. 1992 and Pettit et al. 1994).

contaminant input to and transport within the Great Lakes ecosystem (e.g., Baker and Eisenreich 1990; Hoff et al. 1992; Mackay et al. 1994).

The spatial distribution of PCBs in gull eggs differed in 1991 compared to that observed in 1974 (Figure 7A). Concentrations of PCBs have declined at all of the 14 collection sites; however, the rates of decline at each of the locations were not similar (Figure 7A). A comparison of the distribution of PCB concentrations (Figure 7A) with those of 2,3,7,8-TCDD (Figure 7B) in gull eggs from the same collection sites exemplifies the unique patterns of 2 types of contaminants. Elevated amounts of PCBs were found in Lake Michigan, in Saginaw Bay (Lake Huron) and in western Lake Erie, while concentrations of 2,3,7,8-TCDD were greatest at Saginaw Bay (Lake Huron), and Lake Ontario collection sites (Figure 7). These data illustrate the important trends and distinctions that can be made from biomonitoring programs such as the HGMP of the CWS.

The CWS HGMP database has been used for numerous studies on the uptake, mass balance, trophodynamics, and bioenergetics modeling of toxic chemicals in the Great

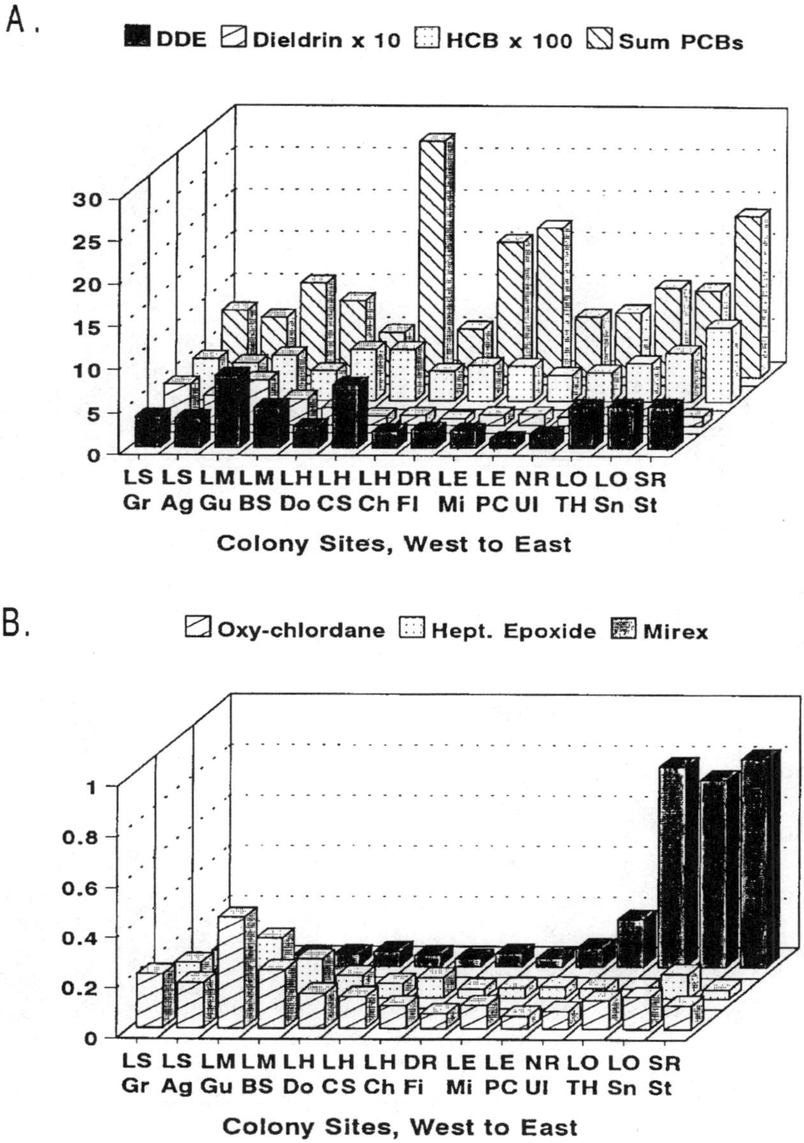

Figure 6 Spatial distribution of mean concentrations (μg/g, wet weight) of A) DDE, dieldrin (× 10), HCB (× 100), and total PCBs; and B) oxychlordane, heptachlor epoxide, and mirex in herring gull eggs collected in 1992 from 14 sites on the Great Lakes and connecting channels by the Canadian Wildlife Service (data from Pettit et al. 1994). Colony sites correspond to those presented in Figure 1.

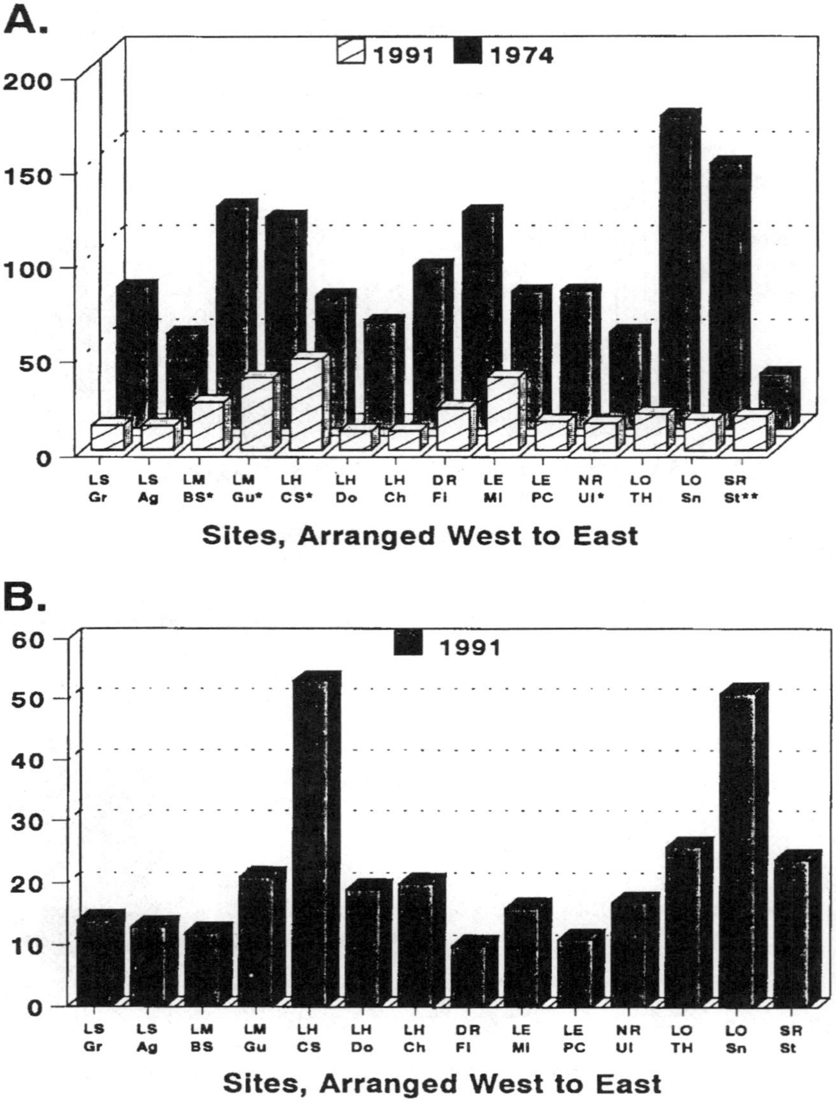

Figure 7 A) Spatial and temporal distribution of total PCBs (µg/g, wet weight) in herring gull eggs collected in 1974 and 1991; and B) spatial distribution of 2,3,7,8-TCDD (pg/g, wet weight) in herring gull eggs collected in 1991 from 14 sites on the Great Lakes and connecting channels by the Canadian Wildlife Service (data from Pettit et al. 1994). Colony sites correspond to those presented in Figure 1.
* The early year of these comparisons occurred between 1976 and 1980.
** The early year of this comparison was 1986.

Lakes over the last 2 decades. For pertinent studies, the reader is encouraged to consult Braun and Norstrom (1989), Comba et al. (1993), Hebert et al. (1994), and other references cited in this appendix.

The major findings of this program include documenting the biological effects in fish-eating birds associated with contaminant burdens, the dramatic decline in contaminant concentrations in fish-eating birds during the late 1970s and early 1980s, the discovery of PCDDs and PCDFs in birds' eggs, the maintenance of the CWS Tissue Bank, the derivation of the herring gull bioenergetics model (Norstrom et al. 1986), and the cell culture bioassay system (Kennedy et al. 1993).

National Contaminant Biomonitoring Program, U.S. Department of Interior: OC Chemical Residues in U.S. Freshwater Fish and Avian Wildlife—1967 to 1986

In the mid 1960s, the U.S. government responded to the potential for residues of DDT and other OC pesticides to accumulate in and harm fish and wildlife by establishing a multi-agency monitoring program. This effort, the National Pesticide Monitoring Program, comprised 3 networks through which concentrations of pesticides and, later, other long-lived toxic contaminants were monitored in all segments of the environment. The USFWS participated by periodically measuring residue concentrations in freshwater fish and avian wildlife. Fish were selected for monitoring aquatic ecosystems; the omnivorous, widely distributed European starling (*Sturnus vulgaris*) was chosen for terrestrial habitats; and the wings of hunter-killed ducks were selected for monitoring contaminants in duck populations of the major flyways, thereby also providing an assessment of contaminant levels in wetlands (Heath and Prouty 1967; Johnson et al. 1967). The USFWS maintained the now-defunct NCBP's 3 networks into the 1980s, with the objective of continuing the documentation of temporal and geographic trends in contaminant concentrations (Prouty and Bunck 1986; Bunck et al. 1987; Schmitt et al. 1990; Schmitt and Brumbaugh 1990).

NCBP sample collection

Freshwater fish were collected regularly from a U.S. NCBP network of fixed stations located at key points, such as at the confluences of major tributaries and below dams, in the major rivers of the 50 states, and in the open waters of the Great Lakes. Beginning in 1967, fish were collected in spring and fall from 50 stations. In 1970, an additional 50 stations were added, but sampling was restricted to the fall. Some stations were co-located with water quality monitoring stations established by the U.S. Geological Survey. Beginning in 1976, sampling was further reduced to every other year, and 17 stations of interest to the USFWS were added. Samples comprised 5 whole, adult fish of a single-species composite for analysis; 1 bottom-dwelling and 1 piscivorous species were analyzed from each site, with the bottom-dwelling species replicated.

The starling network was also based on fixed sites. These were selected by first dividing each of the conterminous (i.e., lower 48) states into blocks of 5 degrees latitude and longitude. As many as 4 locations were randomly selected and sampled from each block. Sites were relocated within the 5-degree blocks if no starlings were found during the initial collections in the summer of 1967 and in the winter and fall of 1968, but site locations remained fixed thereafter. Initially there were 139 sites; several were added in the early 1980s to provide a minimum of 2 sites in each 5-degree block. One composite sample of 10 starlings was collected from each site.

Duck wing sampling was not based on pre-selected locations. Instead, samples representing duck populations from each of the 4 major waterfowl flyways (Atlantic, Mississippi, Central, and Pacific) were obtained from hunters as part of annual USFWS harvest surveys. Hunters throughout the U.S. participated in these surveys and voluntarily submitted wings from ducks they killed. At 2- to 3-year intervals, wings from mallards (*Anas platyrhynchos*) and American black ducks (*A. rubripes*) were aggregated by state within each flyway, and groups of 25 were randomly selected from each state for chemical analysis.

All fish and bird samples were analyzed for OC chemicals—pesticides and selected toxic industrial chemicals such as PCBs—by gas chromatography (Ribick et al. 1982). Fish samples were also analyzed for metals and other inorganic contaminants.

Spatial and temporal trends of OC chemicals in NCBP

During the 2 decades spanned by USFWS contaminant monitoring, the use of persistent insecticides such as DDT was greatly curtailed, and concentrations in fish and wildlife declined. This reduction was reflected in downward trends for DDE, the most persistent metabolite of DDT, in all 3 NCBP networks (Figure 8), indicating that the total DDT burden in North America declined. In fish, DDE increased slightly from about 70% of total DDT in 1976 to about 75% in 1986 (Figure 9) (Schmitt et al. 1990). The proportional change in fish provides additional evidence of reduced inputs to North American ecosystems. A similar trend toward increasing percentages of DDE relative to DDT has been noted elsewhere (Aguillar 1984), indicating that the global DDT burden is also declining.

In the U.S., the bioaccumulation of DDT led to eggshell thinning in fish-eating birds, including the bald eagle. The resulting decline in recruitment of young to bald eagle populations caused the near extirpation this species and its subsequent listing as endangered in the conterminous states (USDI 1978, 1994). The downward trend of DDT concentrations documented in fish, starlings, and duck wings (Figures 8 and 9) was paralleled by declining DDE concentrations in bald eagle eggs, and eagle eggshell thickness increased (Wiemeyer et al. 1993). Corresponding increases in recruitment

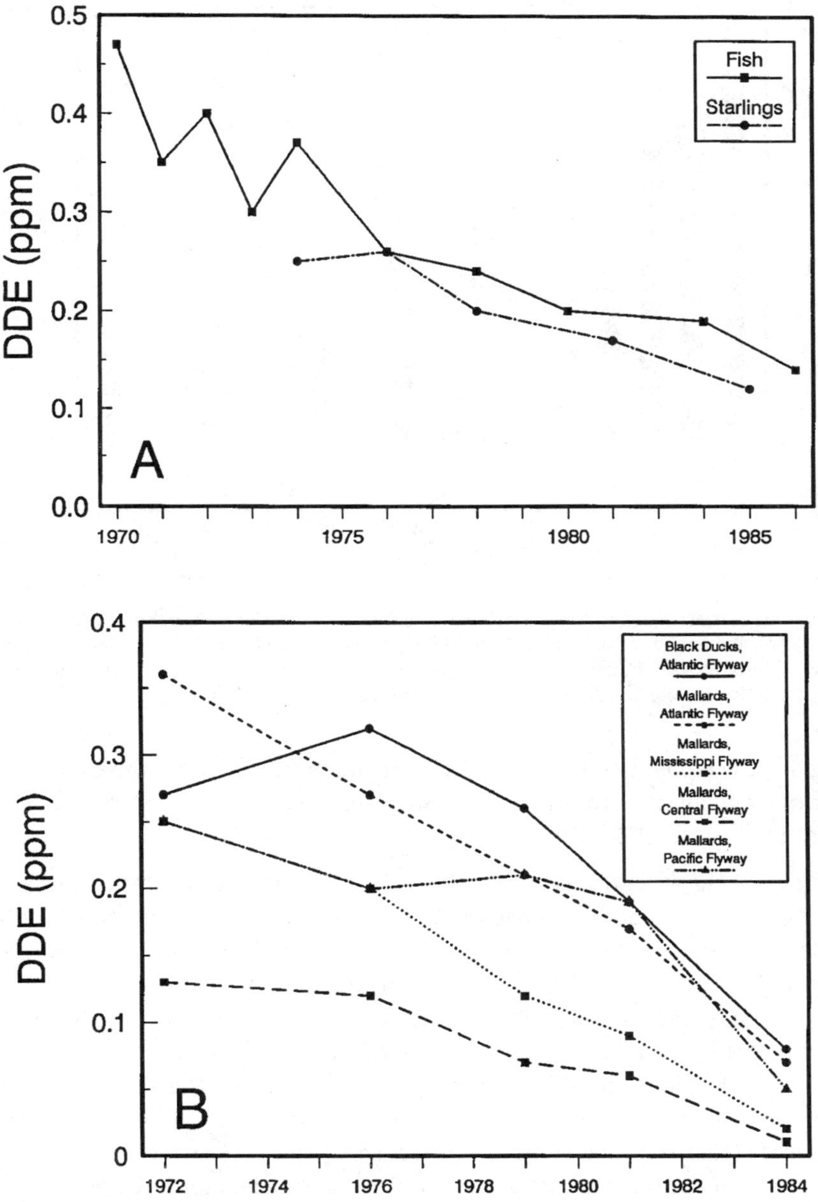

Figure 8 Mean concentrations (μg/g, wet weight) of DDE in A) fish and starlings collected from 1970 to 1986 and B) flyway populations of mallards and black ducks collected from 1972 to 1984 in the U.S. Fish and Wildlife Service monitoring networks.

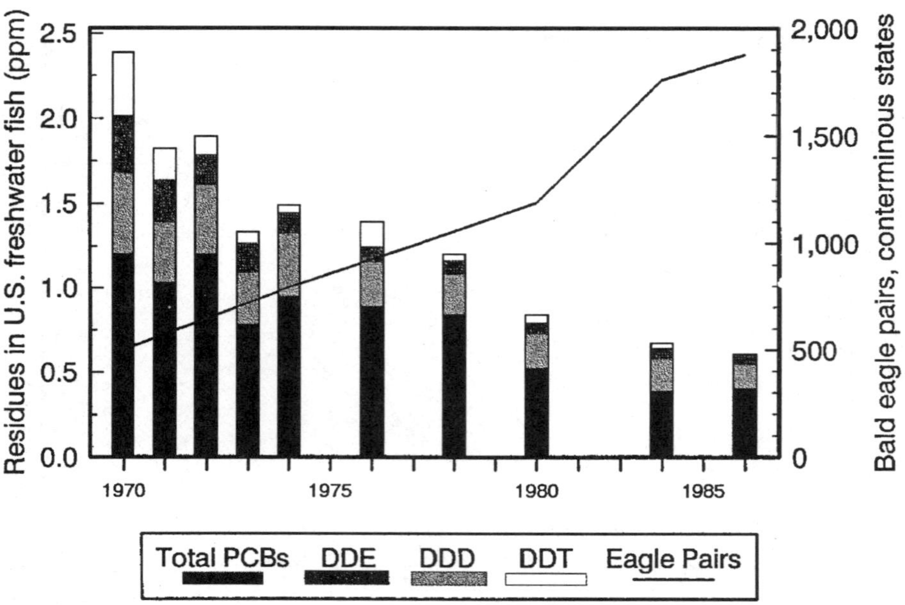

Figure 9 Mean concentrations (μg/g, wet weight) of DDT and its primary metabolites, DDE and DDD (dichlorodiphenyldichloroethane), and of total PCBs, in fish collected from 1970 to 1986 by the U.S. Fish and Wildlife Service (USFWS). Also shown are the numbers of territories occupied by bald eagle pairs in the conterminous U.S. during the same period (USFWS, unpublished data).

have led to the repopulation of many areas (Figure 9) and the bald eagle's reclassification, from "endangered" to "threatened," has been proposed for most areas of the conterminous states (USDI 1994).

In addition to the well-documented effects of DDT and its metabolites on eggshell thickness, these compounds, as well as PCBs and other contaminants, have been reported to interfere with other reproductive and maturation processes in fish and wildlife (e.g., Aulerich et al. 1971; 1973; Fry and Toone 1981; Peterson et al. 1993; Guillette et al. 1994). Although overall concentrations have declined, residues of DDT, other insecticides, and PCBs remain widespread, and problem areas are still evident. In the U.S., concentrations of DDT (mostly as DDE) remain highest in fish and wildlife from areas in the South, Southwest, and Northwest, where DDT was used to protect cotton and orchards from insects; in the Northeast, where it was used to control mosquitoes; and near former centers of DDT production and formulation. Areas affected by former production centers of DDT include northern Alabama, near the former Redstone Arsenal—now Wheeler National Wildlife Refuge (O'Shea et al. 1980)—and the Arkansas, Tombigbee, Alabama, and Tennessee rivers (Figures 10 and 11). Orchards in the northwest United States and Canada, where DDT was used extensively, constitute areas in which DDT still accumulates to elevated amounts in the eggs of resident birds (Blus et al. 1987; Elliot et al. 1994).

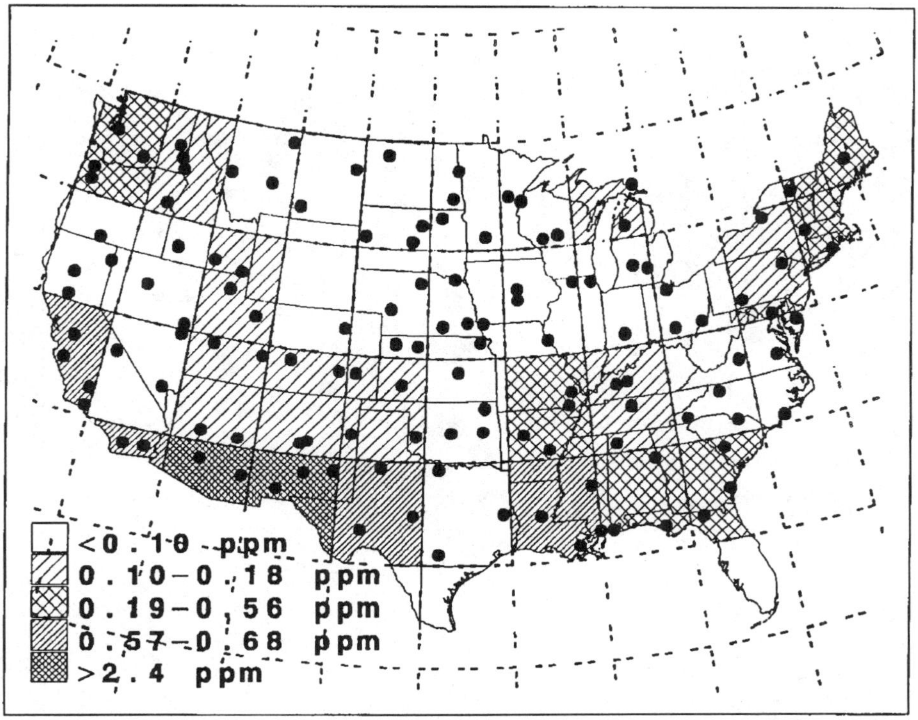

Figure 10 Geographic distribution of DDE concentrations (µg/g, wet weight) in starlings collected in 1985. Also shown are boundaries of the 5-degree (latitude and longitude) sampling blocks (dashed lines) and collection sites (solid circles) of USFWS starling monitoring network.

Concentrations of cyclodiene insecticides—heptachlor, dieldrin, endrin, and chlordane—have also declined in all 3 NCBP networks (Prouty and Bunck 1986; Bunck et al. 1987; Schmitt et al. 1990). Nevertheless, residues of chlordane remain sufficiently high in fish from some areas of the Midwest to warrant the issuance of human consumption advisories by state health agencies. Concentrations are also high in Hawaii, where chlordane and other cyclodiene compounds were used against termites and agricultural pests. Nationally, the proportion of *cis*-chlordane, the active ingredient in technical chlordane found in fish decreased, while the proportion of the more recalcitrant *trans*-nonachlor increased during the 1980s, indicating an overall decline (Schmitt et al. 1990).

Concentrations of toxaphene, an insecticide that replaced DDT in cotton farming and many other applications, have also declined in fish since 1980, when its registration was canceled (Schmitt et al. 1990). In the 1980s, elevated concentrations of toxaphene were discovered in lake trout collected from the upper Great Lakes (Ribick et al. 1982). Because little toxaphene had been used in the Great Lakes region, atmospheric

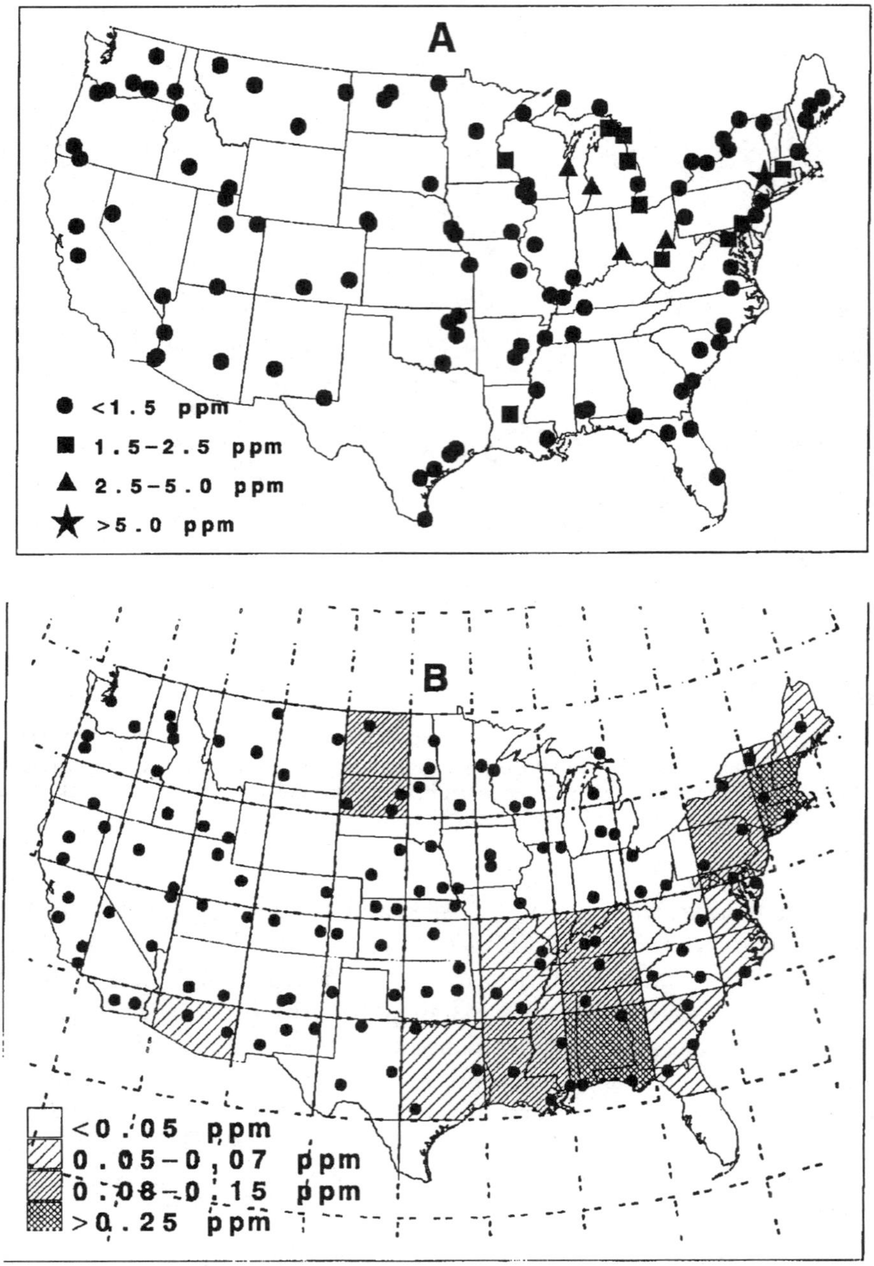

Figure 11 Geographic distribution of PCB concentrations (µg/g, wet weight) in A) fish collected in 1986 from NCBP sites. Not shown are stations in Alaska and Hawaii, at which PCB concentrations were < 1.5 ppm at all sites, and B) starlings collected in 1985 from the USFWS monitoring networks. Also shown are boundaries of the 5-degree (latitude and longitude) sampling blocks (dashed lines) and collection sites (solid circles) of the USFWS starling monitoring network.

transport from the cotton-growing regions to the south and southwest was determined to be the source. Toxaphene is a complex mixture of chlorinated terpenes; although it degrades more rapidly than DDT and many other OC compounds, the most toxic components of toxaphene resist degradation (Harder et al. 1983). Concentrations of these toxic components remain elevated in Great Lakes lake trout (Gooch and Matsamura 1987), as do total concentrations in Great Lakes lake trout compared to other areas.

Other OC pesticides monitored by the NCBP were less widely distributed than the compounds discussed above. Mirex was characteristically found at low concentrations only in the Southeast, where it was used extensively to control the red imported fire ant (*Solenopsis invicta*), and in Lake Ontario and the St. Lawrence River as a result of industrial discharges (Kaiser 1987; Comba et al. 1993). Residues of other pesticides, such as the herbicide Dacthal^{R} (dimethyl tetrachloroterephthalate [DCPA]) and pentachloroanisole (PCA) (the most stable metabolite of the wood preservative and fungicide pentachlorophenol [PCP]), are much more localized in their distributions. PCA residues were found at 30% of the NCBP sampling stations, but most concentrations were <0.02 ng/g (Schmitt et al. 1990). Pentachloroanisole residues were greatest at locations near sites of PCP manufacture and industrial use, such as the Merrimack River in Massachusetts, the Alabama River, and the Raritan River in New Jersey (Schmitt et al. 1990).

Large quantities of PCBs—complex mixtures comprising as many as 209 individual isomers and congeners—were discharged directly to major U.S. waterways; most notable of the affected inland waterways are Lakes Michigan and Ontario and the Hudson, Mississippi, Kanawha, and Ohio Rivers. In addition, PCBs are often present in landfills and urban runoff. These discharge and disposal patterns are reflected in the geographic trends for PCBs in fish and wildlife; greatest concentrations generally occur in the urban-industrial regions of the Midwest and Northeast (Figure 11). By 1980, the direct discharge of PCBs to waterways had been greatly restricted, and total PCB concentrations generally declined in U.S. fish and wildlife (Prouty and Bunck 1986; Bunck et al. 1987; Schmitt et al. 1990). Residual PCBs nevertheless remain a problem in some areas, as evidenced by human consumption advisories in effect for fish from the Great Lakes, Lake Champlain, the Hudson River, and elsewhere.

Some of the more potent PCB congeners (both Ah-receptor agonists and the carcinogenic PCB classes) are long-lived and selectively accumulated by aquatic organisms (Tanabe et al. 1987; Tanabe 1988; de Boer et al. 1993). NCBP fish samples collected in 1988 from some regions, especially the Great Lakes, contained planar PCBs at concentrations great enough to be harmful to fish-eating birds (NBS, unpublished data 1993). Indeed, PCBs and other contaminants in Great Lakes fishes are believed to limit the reproduction of bald eagles and other fish-eating birds, mink, and otters in coastal areas of the Great Lakes (Wren 1991; Tillitt et al. 1992; Giesy et al. 1994). PCBs, along with DDE and other contaminants—including chlorinated dioxins—may also be involved in the failure of lake trout to reproduce naturally in Lake Michigan (USDI 1981; Spitsbergen et al. 1991; Walker et al. 1991). In spite of discharge restrictions, the concentrations of PCBs and chemically similar compounds in the Great Lakes will likely remain elevated due to atmospheric transport and the internal cycling of contaminants already present in the lakes (e.g,. Baker and Eisenreich 1990; Hoff et al. 1992; Mackay et al. 1994).

Although problem areas remain, U.S. concentrations of OC pesticides and PCBs in fish and wildlife are generally lower now than at any time for which accurate data exist. These results imply that direct inputs of many such substances to the environment have been reduced through the regulation of industrial discharges and pesticide use. Declining concentrations of DDT and other contaminants in North America have permitted the return of predatory birds, such as bald eagles, to some areas from which they had been eliminated (Figure 9). The persistence of contaminant problems, despite curtailment of direct discharges to waterways and restrictions on the uses of persistent pesticides, has highlighted the importance of global and ecosystem processes, such as atmospheric transport and internal cycling. Atmospheric transport represents a significant source of PCBs, toxaphene, and other pesticides to the Great Lakes.

The continuing use of OC pesticides in South and Central America constitutes a source of exposure to migratory species, including the peregrines (*Falco perigrinus*), on their wintering grounds (Henny et al. 1982). Concentrations of OC pesticides in returning migratory birds sampled in South Texas generally declined over the last decade (Mora 1995), indicating that the magnitude of this threat has also declined. Additionally, there are no long-term U.S. monitoring data on chlorinated dioxins, which may affect fish and wildlife at extremely low concentrations (Elliot et al. 1989; Sanderson et al. 1994). In the Canadian Great Lakes, the CWS has monitored chlorinated dioxins and furans since 1981 (Hebert et al. 1994).

Exposure Pathways of Concern and Potentially Affected Species

The critical pathways of OC compound movements in the environment are those leading to organisms at the top of aquatic food chains, as well as those directly associated with aquatic sediments. As regulatory measures to control the release of OC compounds into the environment have been put into place and the programs have become successful, the amounts of these compounds entering the environment from point and nonpoint sources have been reduced. Contaminated sediments, which long have been a repository of persistent, bioaccumulative OC chemicals, have now become a major source of these compounds in many areas (Baudo and Muntua et al. 1990).

Therefore, critical pathways of exposure of fish and wildlife species often include benthic organisms as well as organisms that reside at the top of the food web. Species in both categories will receive the greatest dose of OC compounds and are therefore at the greatest risk for adverse effects.

In addition to species that may receive the greatest exposure, species that are particularly sensitive to the toxic effects of OC compounds must be included in a high risk group, which may be true even though in some instances their exposure may be substantially lower than that of other, less sensitive species. The relative sensitivity of avian species to dioxins and dioxin-like OC compounds, e.g., varies widely among species (Bosveld and Van den Berg 1994; Giesy et al. 1994). Chickens are the most sensitive species tested to date (Bosveld and Van den Berg 1994; Giesy et al. 1994), with an LD50 for TCDD embryo lethality approximately one tenth that of ring-necked pheasants (*Phasianus colchicus*) (reviewed by Giesy et al. 1994). Herring gulls appear to be even less sensitive than pheasants to the effects of dioxins and related compounds, whereas other fish-eating water birds are intermediate to the chicken and the herring gull (Bosveld and Van den Berg 1994; Giesy et al. 1994). A similar wide range in sensitivities to the reproductive toxicity of dioxins was observed among mammals and fish (reviewed by Peterson et al. 1993). In this regard, many issues related to individual species sensitivities, particularly those of fish and wildlife species, remain unresolved. Large differences in species sensitivities can occur, yet many species remain untested. Testing of some species has been difficult or impossible because of a lack of knowledge on animal husbandry for many wild species; however, in vitro testing protocols with cell cultures from the species of interest may allow better assessments of their relative sensitivities to chemicals (Kennedy et al. 1993; Sanderson and Bellward 1995).

Organism with the greatest exposure and consequently the greatest risk from biomagnification include top carnivores in an aquatic food chain: fish-eating raptors such as eagles and ospreys; fish-eating colonial water birds, including terns, cormorants, herons, and gulls; and peregrine falcons, which may in certain areas feed upon colonial water birds, and, which may also be exposed to pesticides on their wintering grounds. Insectivorous birds that feed on emergent aquatic insects can also ingest great amounts of these compounds in areas where the sediments are contaminated. Tree swallows (*Tachycineta bicolor*) are an example of such a species; they feed on emerging aquatic insects along rivers and bays (Quinney and Ankney 1985). Contaminant uptake studies have demonstrated that tree swallows receive and deposit into their eggs nearly the same amounts of hydrophobic OC contaminants as do piscivorous birds at the same location (Ankley et al. 1993). In these same studies, red-wing blackbirds (*Agelaius phoeniceus*), insectivores that feed primarily on upland food sources, had significantly lower contaminant uptake rates and body burdens than did tree swallows nesting in the same locations (Ankley et al. 1993; Bishop et al. 1995).

Global Applicability of NCBP and HGMP Conclusions

Concentrations of OCs and many of the traditionally monitored compounds in most trophic levels of the environment have declined. These trends are exemplified in the Great Lakes (DeVault et al. 1986; Baumann and Whittle 1988; Bishop et al. 1992; Pettit et al. 1994; Hebert et al. 1994; Stow et al. 1994), but the same general patterns occur across North America and around the world. Sweden has had longstanding efforts to monitor OCs in both marine and freshwater species. Their monitoring dates back to 1967 (Jensen et al. 1969) and has focused on the freshwater pike (*Esox lucius*), the Atlantic herring (*Clupea harengus*) as a marine fish, and eggs of guillemot (*Uria aalge*) as an avian top predator. Temporal trends for concentrations of PCBs and DDTs in all 3 species in the Swedish monitoring program (Olsson 1986; Olsson and Reutergårdh 1986; Bignert et al. 1992,1995) parallel the declines documented in North America by the HGMP and the NCBP since concentrations peaked in the late 1970s to early 1980s. Concentrations of non-ortho-substituted PCBs and polychlorinated naphthalenes (PCNs), as monitored in Sweden, declined comparably in these same species (Järnberg et al. 1993). In Japan, monitoring based on preserved samples of northern fur seals (*Callorhinus ursinus*), collected along the Pacific coast since 1973, revealed a continuous decline since PCB and DDT concentrations peaked in 1976 (Tanabe, Sung et al. 1994). Concentrations of most OC pesticides and industrial compounds such as PCBs measured by these national monitoring programs have declined according to a first-order decay function.

The reductions in OC concentrations in the Great Lakes are thought to be primarily due to reductions in input, while the current status is maintained by atmospheric input and internal recycling (Baker and Eisenreich 1990; Hoff et al. 1992; Mackay et al. 1994). The net result is that the dramatic declines in concentrations observed over the last 2 decades, especially in areas near pollution sources, are not expected to recur. Further decreases will only result from international cooperation and agreements on continued regulatory control of manufacture, distribution, use, and discharge of these compounds into the environment. These decreases, however, are only expected to be small. In fact, increases in the amounts of many of the traditional OC compounds were noted at HGMP colonies between 1991 and 1993. This was due to relatively small increases at the monitored colonies, some of which may have been linked to disturbances of in-place pollutants (i.e., sediments) (Ludwig et al. 1993). Another theory for the increases (or reduced rates of decline) in contaminant concentrations in the Great Lakes is that zebra mussels (*Dreissena polymorpha*) are recycling contaminants from the sediments. By resuspending them into the water column, while at the same time reducing the amount of suspended solids in the water column, mussels increase bioavailability of the contaminants (Brieger and Hunter 1993; Fisher et al. 1993). Zebra mussels were inadvertently introduced into the Great Lakes and other waters of the U.S. and Canada in ballast water discharge from oceanic freight ships that had visited or originated from Europe. In support of this theory, as zebra mussel numbers have increased, water clarity has increased substantially in many areas of the Great Lakes (Nalepa and Schloesser 1993).

In more remote areas, and in some longer-lived species of marine mammals, the decreases in concentrations of OC compounds reported in the national monitoring programs (HGMP, NCBP, Swedish and Japanese) have not been consistently observed (Tanabe, Iwata and Tatsukawa 1994; Muir et al. 1990; Norstrom and Muir 1994). The temporal trends of OC compounds in Arctic marine mammals have been somewhat contradictory (Norstrom and Muir 1994), largely due to a paucity of information. Increases in the concentrations of OC compounds have been observed in the most recent data for beluga whales (*Delphinapterus leucas)* (Muir et al. 1990), and polar bears (*Ursus maritimus)* (Norstrom and Muir 1994) from the Arctic. Concentrations of OC compounds in marine mammals from the open waters of the oceans also have not declined in all cases (Tanabe and Tatsukawa 1991; Tanabe, Iwata and Tatsukawa 1994). For example, concentrations of PCBs and DDTs in striped dolphins (*Stenella coeruleoalba*) collected from the western north Pacific Ocean did not change appreciably from 1978/79 to 1986, and concentrations of hexachlorocyclohexane (HCH) and HCB declined only slightly (Loganathan et al. 1990). Therefore, the temporal trends in OC concentrations in areas remote from the contaminant sources (i.e., in the Arctic and the open oceans) are controvertible at this point.

In some localized areas, such as specific bays, rivers, or impoundments where historic pollution was or is a problem, weather and human events that disturb aquatic sediments can liberate and make available the OC compounds, as noted above in the discussion of recent HGMP increases. This is because sediments have become a repository for many of these hydrophobic, persistent chemicals (Baudo et al. 1990). In many instances, sediments have become the sources of the contaminants. Meteorological events, such as flooding, can resuspend bottom sediments in areas that may normally trap or collect sediments. Once resuspended, the sediments, which can contain large amounts of OCs from 20 to 40 years ago, partition these OCs into the water column and thus make them available for uptake in biota. The results can be devastating, as was the situation in 1986 when a 100-year flood disrupted PCB and TCDD-containing sediments throughout the Saginaw River, Michigan. Caspian terns, which nest on an island in Saginaw Bay at the mouth of the Saginaw River, experienced complete reproductive failure that year (Ludwig et al. 1993).

Conclusions

Despite some setbacks, the environmental controls imposed over the last 20 years appear to have reduced the amount of OC compounds entering the environment, which has resulted in a general lowering of concentrations in fish and wildlife. The troublesome side of the situation in the Great Lakes and in other areas of North

America is that the concentrations of some of these OCs remain greater than the threshold for adverse effects in some species. In some other areas, it appears that the environmental concentrations are at or near the threshold. An example of a contaminant that is at or above a threshold in a particular species is PCB in some Great Lakes bald eagle populations (Bowerman et al. 1995). In this and other species, the question that must then be addressed is whether the current concentrations of these OCs are above or below the threshold for adverse effects on the populations that are exposed. In addition to populations that are at or above a threshold for adverse effects on the population, there are also instances, like the concentrations of TCDD in herring gull eggs from many Great Lakes locations, in which the environmental concentrations are below what is thought to be the threshold for adverse effects. The predicament that we are faced with is that the concentrations of many of the OCs are not expected to decline below present levels for an extended time. Thus, it becomes imperative to answer the question: Are we currently above or below a threshold for adverse effects of these OC compounds for species that are potentially at risk?

References

Aguillar A. 1984. Relationships of DDE/DDT in marine mammals to the chronology of DDT input into the ecosystem. *Can J Fish Aq Sci* 21:840–844.

Anderson RB, Everhart WH. 1966. Concentrations of DDT in landlocked salmon (*Salmo salar*) at Sebago Lake, Maine. *Trans Am Fish Soc* 95:160–164.

Ankley GT, Niemi GJ, Lodge KB, Hallet HJ, Beaver DL, Tillitt DE, Schwartz TR, Giesy JP, Jones PD, Hagley C. 1993. Uptake of planar polychlorinated biphenyls and 2,3,7,8-substituted polychlorinated dibenzofurans and dibenzo-*p*-dioxins by birds nesting in the lower Fox River and Green Bay, Wisconsin, USA. *Arch Environ Contam Toxicol* 24:332–344.

Aulerich RJ, Ringer RK, Seagran HL, Youatt WG. 1971. Effects of feeding Coho salmon and other Great Lakes fish on mink reproduction. *Can J Zool* 49:611–616.

Aulerich RJ, Ringer RK, Iwamoto S. 1973. Reproductive failure and mortality in mink fed Great Lakes fish. *J Reprod Fert* (Supplement) 19:365–376.

Baker JE, Eisenreich SJ. 1990. Concentrations and fluxes of poly aromatic hydrocarbons and polychlorinated biphenyls across the air water interface of Lake Superior. *Environ Sci Technol* 24:342–352.

Barrie LA, Gregor D, Hargrave B, Lake R, Muir D, Shearer R, Tracey B, Bidleman T. 1992. Arctic contaminants: sources, occurrence and pathways. *Sci Total Environ* 122:1–74.

Baudo R, Muntau H. 1990. Lesser known in-place pollutants and diffuse source problems. In: Baudo R, Giesy JP, Muntau H, editors. Sediments: chemistry and toxicity of in-place pollutants. Chelsea MI: Lewis. p 1–14.

Baumann PC, Whittle DM. 1988. The status of selected organics in the Laurentian Great Lakes: an overview of DDT, PCBs, dioxins, furans, and aromatic hydrocarbons. *Aquat Toxicol* 11:241–257.

Bignert A, Jensen S, Olsson M, Persson W, Roos A. 1992. Time series of contaminants in the marine environment along the Swedish coasts. In: Bignert A, Odsjö T, Olsson M, editors. Övervakning av miljögifter i levande organismer. Naturvårdsverket Rapport 4314, Solna, Sweden. Appendix 3.

Bignert A, Litzèn K, Odsjö T, Olsson M, Persson W, Reutergårdh L. 1995. Time-related factors influence the concentrations of DDTs, PCBs and shell parameters in eggs of Baltic guillemot (*Uria aagle*), 1861–1989. *Environ Pollut* 89:27–36.

Bishop CA, Brooks RJ, Carey JH, Ng P, Norstrom RJ, Lean DRS. 1991. The case for a cause-effect linkage between environmental contamination and development in eggs of the common snapping turtle (*Chelydra serpentina*) from Ontario, Canada. *J Toxicol Environ Health* 33:521–547.

Bishop CA, Koster MD, Chek AA, Hussell DJT, Jock K. 1995. Chlorinated hydrocarbons and mercury in sediments, red-winged blackbirds (*Agelaius phoeniceus*) and tree swallows (*Tachycineta bicolor*) from wetlands in the Great Lakes-St. Lawrence River basin. *Environ Toxicol Chem* 14:491–501.

Bishop CA, Weseloh DVC, Burgess NM, Stuger J, Norstrom RJ, Turle R, Logan KA. 1992. An atlas of contaminants in eggs of fish-eating colonial birds of the Great Lakes (1970–1988). 2 Vols. Technical Report Series Nos. 152 and 153. Burlington, Ontario: Ontario Region, Canadian Wildlife Service.

Blus LJ, Henny CJ, Kaiser TE, Grove RA. 1983. Effects on wildlife from use of endrin in Washington state orchards. *Trans North Am Wild Nat Resour Conf* 48:159–174.

Blus LJ, Henny CJ, Stafford CJ, Grove RA. 1987. Persistence of DDT and metabolites in wildlife from Washington state orchards. *Arch Environ Contam Toxicol* 16:467–476.

Bosveld ATC, Van den Berg M. 1994. Effects of polychlorinated biphenyls, dibenzo-*p*-dioxins, and dibenzofurans on fish-eating birds. *Environ Rev* 2:147–166.

Bowerman WW, Giesy JP, Best DA, Kramer VJ. 1995. A review of the factors affecting productivity of bald eagles in the Great Lakes Region: implications for recovery. *Environ Health Perspec* 103(Suppl. 4):51–59.

Braune BM, Norstrom RJ. 1989. Dynamics of OC compounds in herring gulls: III. Tissue distribution and bioaccumulation in Lake Ontario gulls. *Environ Toxicol Chem* 8:957–968.

Brieger G, Hunter RD. 1993. Uptake and depuration of PCB 77, PCB 169, and hexachlorobenzene by zebra mussel (*Dreissena polymorpha*). *Ecotoxicol Environ Saf* 26:153–165.

Bunck CM, Prouty RM, Krynitsky AJ. 1987. Residues of OC pesticides and polychlorobiphenyls in starlings (*Sturnus vulgaris*) from the continental United States, 1982. *Environ Monit Assess* 8:59–75.

Burdick GE, Harris EJ, Dean HJ, Walker TM, Skea J, Colby D. 1964. The accumulation of DDT in lake trout and the effect on reproduction. *Trans Am Fish Soc* 93:127–136.

Comba ME, Norstrom RJ, Macdonald CR, Kaiser KLE. 1993. A Lake Ontario-Gulf of St. Lawrence dynamic mass budget for mirex. *Environ Sci Technol* 27:2198–2206.

Cooke AS. 1973. Shell thinning in avian eggs by environmental pollutants. *Environ Pollut* 4:85–152.

Cooper K. 1991. Effects of pesticides on wildlife. In: Hayes Jr WJ, Laws Jr ER, editors. Handbook of pesticide toxicology. San Diego CA: Academic Pr. p 463–496.

de Boer J, Stronck CJN, Traag WA, van der Meer J. 1993. Non-ortho and mono-orthosubstituted chlorobiphenyls and chlorinated dibenzo-*p*-dioxins and dibenzofurans in marine and freshwater fish and shellfish from the Netherlands. *Chemosphere* 26:1823–1842.

DeVault DS, Wifford WA, Hesselberg RJ, Nortrupt DA, Rundberg EGS, Alwan AK, Bautista C. 1986. Contaminant trends in lake trout (*Salvelinus namaycush*) from the upper Great Lakes. *Arch Environ Contam Toxicol* 15:349–356.

Draper NR, Smith H. 1981. Applied regression analysis. 2nd ed. New York: Wiley 736 p.

Elliot JE. 1984. Collecting and archiving wildlife samples in Canada. In: Lewis RA, Stein N, Lewis CW, editors. Environmental specimen banking and monitoring as related to banking. Boston MA: Martinus Nijhoff. p 45–66.

Elliot JE, Butler RW, Norstrom RJ, Whitehead PE. 1989. Environmental contaminants and reproductive success of great blue heron *Ardea herodias* in British Columbia. *Environ Pollut* 59:91–114.

Elliot JE, Martin PA, Arnold TW, Sinclair PH. 1994. OCs and reproductive success of birds in orchard and non-orchard areas of central British Columbia, Canada, 1990–1991. *Arch Environ Contam Toxicol* 26:435–443.

Ewins JP, Weseloh DVC, Norstrom RJ, Legierse K, Aumen H, Ludwig JP. 1994. Caspian terns on the Great Lakes: OC contamination, reproduction, diet and population changes, 1972–1991. Occasional Paper No. 8. Ottawa, Ontario: Canadian Wildlife Service. 30 p.

Fisher SW, Gossiaux DC, Bruner KA, Landrum PF. 1993. Investigations of the toxicokinetics of hydrophobic contaminants in zebra mussel (*Dreissena polymorpha*). In: Nalepa TF, Schloesser DW, editors. Zebra mussels: biology, impacts, and controls. Chelsea MI: Lewis. p 465–490.

Fox GA. 1993. What have biomarkers told U.S. about the effects of contaminants on the health of fish-eating birds in the Great Lakes? The theory and a literature review. *J Great Lakes Res* 19:722–736.

Fry DM, Toone CK. 1981. DDT-induced feminization of gull embryos. *Science* 213:922–924.

Giesy JP, Ludwig JP, Tillitt DE. 1994. Deformities in birds of the Great Lakes region. *Environ Sci Technol* 28:128A–135A.

Gilbertson M. 1974. Pollutants in breeding herring gulls in the lower Great Lakes. *Can Field-Nat* 88:273–280.

Gilbertson M. 1975. A Great Lakes tragedy. *Nat Can* 4:22–25.

Gilbertson M. 1983. Etiology of chick edema disease in herring gulls in the lower Great Lakes. *Chemosphere* 12:357–370.

Gilbertson M. 1989. Effects on fish and wildlife populations. In: Kimbrough RD, Jensen AA, editors. Halogenated biphenyls, terphenyls, naphthalenes, dibenzodioxins and related products. Amsterdam, The Netherlands: Elsevier. p 103–127.

Gilbertson M, Fox GA. 1977. Pollutant-associated embryonic mortality of Great Lakes herring gulls. *Environ Pollut* 12:211–216.

Gilbertson M, Hale R. 1974a. Early embryonic mortality in a herring gull colony in Lake Ontario. *Can Field-Nat* 88:354–355.

Gilbertson M, Hale R. 1974b. Characteristics of the breeding failure of a colony of herring gull in Lake Ontario. *Can Field-Nat* 88:356–357.

Gilbertson M, Reynolds LM. 1972. Hexachlorobenzene (HCB) in the eggs of common terns in Hamilton Harbor. *Bull Environ Contam Toxicol* 7:371–373.

Gilman AP, Peakall DB, Hallett DJ, Fox GA, Norstrom RJ. 1979. Herring gulls (*Larus argentatus*) as monitors of contamination in the Great Lakes. In: Animals as monitors of environmental pollutants. Washington DC: National Academy of Sciences.

Gooch JW, Matsamura F. 1987. Toxicity of chlorinated bornane (toxaphene) residues isolated from Great Lakes lake trout (*Salvelinus namaycush*). *Arch Environ Contam Toxicol* 16:349–355.

Grier JW. 1982. Ban of DDT and subsequent recovery of reproduction in bald eagles. *Science* 218:1232–1235.

Guillette Jr LJ, Gross TS, Masson GR, Matter JM, Percival HF, Woodward AR. 1994. Developmental abnormalities of the gonad and abnormal sex hormone concentrations in juvenile alligators from contaminated and control lakes in Florida. *Environ Health Perspect* 102:680–688.

Harder HW, Carter TV, Bidleman TF. 1983. Acute effects of toxaphene and its sediment-degraded products on estuarine fish. *Can J Fish Aq Sci* 40:2119–2125.

Hayes Jr WJ, Laws ER. 1991. Handbook of pesticide toxicology. Volume 1. San Diego CA: Academic Pr.

Heath RG, Prouty RM. 1967. Trial monitoring of pesticides in wings of mallards and black ducks. *Bull Environ Contam Toxicol* 2:101–110.

Hebert CE, Norstrom RJ, Simon M, Braune BM, Weseloh DV, Macdonald CR. 1994. Temporal trends and sources of PCDDs and PCDFs in the Great Lakes: herring gull monitoring, 1981–1991. *Environ Sci Technol* 28:1268–1277.

Henny CJ, Ward FR, Riddle KR, Prouty RM. 1982. Migratory peregrine falcons, *Falco perigrinus*, accumulate pesticides in Latin America during winter. *Can Field-Nat* 96:333–338.

Hickey JJ, Anderson DW. 1968. Chlorinated hydrocarbons and eggshell changes in raptorial and fish eating birds. *Science* 162:271–273.

Hoff RM, Muir DCG, Grift NP. 1992. Annual cycle of polychlorinated biphenyls and organohalogen pesticides in air in southern Ontario. Atmospheric transport and sources. *Environ Sci Technol* 26:276–283.

Järnberg U, Asplund L, de Wit C, Grafström A-K, Haglund P, Jansson B, Lexèn K, Strandell M, Olsson M, Jonsson B. 1993. Polychlorinated biphenyls and polychlorinated naphthalenes in Swedish sediment and biota: levels, patterns and time trends. *Environ Sci Technol* 27:1364–1374.

Jensen S, Johnels AG, Olsson M, Otterlind G. 1969. DDT and PCB in marine animals from Swedish waters. *Nature* 224:247–250.

Johnson RE, Carver TC, Dustman EH. 1967. Indicator species near top of food chain chosen for assessment of pesticide base levels in fish and wildlife—clams, oysters, and sediment in estuarine environment. *Pestic Monit J* 1:7–13.

Jones KC, Sanders G, Wild SR, Burnett V, Johnson AE. 1992. Evidence for decline of PCBs and PAHs in rural vegetation and air in the United Kingdom. *Nature* 356:137–140.

Kaiser KLE. 1987. The rise and fall of mirex. *Environ Sci Technol* 12:520–528.

Keith JA. 1966. Reproduction in a population of herring gulls (*Larus argentatus*) contaminated by DDT. *J Appl Ecol (Suppl.)* 3:57–70.

Kennedy SW, Lorenzen A, James CA. 1993. A rapid and sensitive cell culture bioassay system for measuring ethoxyresorufin-O-deethylase (EROD) activity in cultured hepatocytes exposed to halogenated aromatic hydrocarbons extracted from wild bird eggs. *Chemosphere* 27:367–373.

Kubiak TJ, Harris HJ, Smith LM, Schwartz TR, Stalling DL, Trick JA, Sileo L, Docherty D, Erdman TC. 1989. Microcontaminants and reproductive impairment of the Forster's tern in Green Bay, Lake Michigan—1983. *Arch Environ Contam Toxicol* 18:706–727.

Loganathan BG, Tanabe S, Tanaka H, Watanabe S, Miyazaki N, Amano M, Tatsukawa R. 1990. Comparison of OC residue levels in the striped dolphin from western North Pacific. *Mar Pollut Bull* 21:435–439.

Ludwig JP, Auman HJ, Kurita H, Ludwig ME, Campbell LM, Giesy JP, Tillitt DE, Jones PD, Yamashita N, Tanabe S, Tatsukawa R. 1993. Caspian tern reproduction in the Saginaw Bay ecosystem following a 100 year flood event. *J Great Lakes Res* 19:96–108.

Ludwig JP, Tomoff CS. 1966. Reproductive success and insecticide residues in Lake Michigan herring gulls. *Jack Pine Warbler* 44:77–85.

Mackay D, Paterson S, Eisenreich SJ, Simmons MS. 1983. Physical behavior of PCBs in the Great Lakes. Ann Arbor MI: Ann Arbor Science.

Mackay D, Sang S, Vlahos P, Gobas F, Diamond M, Dolan D. 1994. A rate constant model of chemical dynamics in a lake ecosystem: PCBs in Lake Ontario. *J Great Lakes Res* 20:625–642.

Mineau P, Fox GA, Norstrom RJ, Weseloh DV, Hallett DJ, Ellenton JA. 1984. Using the herring gull to monitor levels and effects of OC contamination in the Canadian Great Lakes. In: Nriagu JO, Simmons MS, editors. Toxic contaminants in the Great Lakes. New York: J Wiley. p 425–452.

Mora MA. 1995. Residues and trends of OC pesticide and polychlorinated biphenyls in birds from Texas—1965–1988. Washington DC: U.S. Department of the Interior, National Biological Service. Fish and Wildlife Research No. 14. 26 p.

Muir DCG, Ford CA, Stewart REA, Smith TG, Addison RF, Zinck ME, Bèland P. 1990. OC contaminants in belugas, *Delphinapterus leucas*, from Canadian waters. *Can Bull Fish Aq Sci* 224:165–190.

Muir DCG, Wagemann R, Hargrave BT, Thomas DJ, Peakall DB, Norstrom RJ. 1992. Arctic marine ecosystem contamination. *Sci Total Environ* 122:75–134.

Nalepa TF, Schloesser DW. 1993. Zebra mussels: biology, impacts, and controls. Chelsea MI: Lewis.

Norstrom RJ, Muir DCG. 1994. Chlorinated hydrocarbon contaminants in arctic marine mammals. *Sci Total Environ* 154:107–128.

Norstrom RJ, Clark TP, Kearney JP, Gilman AP. 1986. Herring gull energy requirements and body constituents in the Great Lakes. *Ardea* 74:1–23.

Norstrom RJ, Simon M, Muir DCG. 1990. Polychlorinated dibenzo-*p*-dioxins and dibenzofurans in marine mammals in the Canadian North. *Environ Pollut* 66:1–19.

Olsson M. 1986. PCBs in the Baltic environment. In: Waid JS, editor. PCBs and the environment. Volume III. Boca Raton FL: CRC Pr. p 181–208.

Olsson M, Reutergårdh L. 1986. DDT and PCB pollution trends in the Swedish aquatic environment. *Ambio* 15:103–109.

O'Shea TJ, Brownell RL. 1994. OC and metal contaminants in baleen whales: a review and evaluation of conservation implications. *Sci Total Environ* 154:179–200.

O'Shea T, Fleming III WJ, Cromartie E. 1980. DDT contamination at Wheeler National Wildlife Refuge. *Science* 209:509–510.

Peakall DB. 1988. Known effects of pollutants on fish-eating birds in the Great Lakes of North America. In: Proceedings of the World Conference on Large Lakes; 1986 May 18-21; Mackinac Island MI. Volume 1, Chronic effects of toxic contaminants in large lakes.

Peakall DB, Fox GA. 1987. Toxicological investigations of pollutant-related effects in Great Lakes gulls. *Environ Health Perspect* 71:187–193.

Peakall DB, Norstrom RJ, Rahimtula AD, Butler RD. 1986. Characterization of mixed-function oxidase systems of the nestling herring gull and its implications for bioeffects monitoring. *Environ Toxicol Chem* 5:379–385.

Pearce PA, Peakall DB, Reynolds LM. 1979. Shell thinning and residues of OCs and mercury in seabird eggs, Eastern Canada, 1970–1976. *Pestic Monit J* 13:61–68.

Pekarik C, Weseloh DV. 1995. Contaminant trends in herring gull eggs from the Great Lakes: regression analysis of long term and recent data, 1974–1993. Abstracts, 21st Annual Meeting of the International Association for Great Lakes Research;1995 May 28 - Jan 2; E. Lansing MI.

Peterson RE, Theobald HM, Kimmel GL. 1993. Developmental and reproductive toxicity of dioxins and related compounds: cross-species comparisons. *Crit Rev Toxicol* 23:283–335.

Pettit KE, Bishop CA, Weseloh DVC, Norstrom RJ. 1994. An atlas of contaminants in eggs of fish-eating colonial birds of the Great Lakes (1989–1992). 2 Vols. Technical Report Series Nos. 193 and 194. Burlington, Ontario: Ontario Region, Canadian Wildlife Service.

Price IM, Weseloh DV. 1986. Increased numbers and productivity of double-crested cormorants, *Phalacrocorax auritus*, on Lake Ontario. *Canadian Field-Naturalist* 100:474–482.

Prouty RM, Bunck CM. 1986. OC residues in adult mallard and black duck wings, 1981–1982. *Environ Monit Assess* 6:49–57.

Quinney TE, Ankney CD. 1985. Prey size selection by tree swallows. *Auk* 102:245–250.

Ribick MA, Dubay GR, Stalling DL, Petty JD, Schmitt CJ. 1982. Analysis of toxaphene residues in fish samples at part-per-billion levels. *Environ Sci Technol* 16:310–318.

Sanderson JT, Bellward GD. 1995. Hepatic microsomal ethoxyresorufin O-deethylase-inducing potency in ovo and cytosolic Ah receptor binding affinity of 2,3,7,8-tetrachlorodibenzo-*p*-dioxin: comparison of four avian species. *Toxicol Appl Pharmacol* 132:131–145.

Sanderson JT, Elliot JE, Norstrom RJ, Whitehead PE, Hart LE, Cheng KM, Bellward GD. 1994. Monitoring biological effects of polychlorinated dibenzo-*p*-dioxins, dibenzofurans, and biphenyls in great blue heron (*Ardea herodias*) in British Columbia. *J Toxicol Environ Health* 41:435–450.

Schmitt CJ, Brumbaugh WG. 1990. National contaminant biomonitoring program: concentrations of arsenic, cadmium, copper, lead, mercury, selenium, and zinc in freshwater fishes of the United States, 1976–1984. *Arch Environ Contam Toxicol* 19:731–747.

Schmitt CJ, Zajicek JL, Peterman PH. 1990. National contaminant biomonitoring program: residues of OC chemicals in freshwater fishes of the United States—1976–1984. *Arch Environ Contam Toxicol* 19:748–782.

Spitsbergen JM, Walker MK, Olson JR, Peterson RE. 1991. Pathologic alterations in early life stages of lake trout, *Salvelinus namaycush*, exposed to 2,3,7,8-tetrachlorodibenzo-*p*-dioxin as fertilized eggs. *Aquat Toxicol* 19:41–72.

Stow CA, Carpenter SR, Amrhein JF. 1994. PCB concentration trends in Lake Michigan coho (*Oncorhynchus kisutch*) and chinook salmon (*O. tshawytscha*). *Can J Fish Aq Sci* 50:1384–1390.

Tanabe S. 1988. PCB problems in the future: foresight from current knowledge. *Environ Pollut* 50:5–28.

Tanabe S, Iwata H, Tatsukawa R. 1994. Global contamination by persistent OCs and their ecotoxicological impact on marine mammals. *Sci Total Environ* 154:163–177.

Tanabe S, Kannan N, Fukushima M, Okamoto T, Wakimoto T, Tatsukawa R. 1989. Persistent OCs in Japanese coastal waters: an introspective summary from a Far East developed nation. *Mar Pollut Bull* 20:344–352.

Tanabe S, Kannan N, Subramanian A, Watanabe S, Tatsukawa R. 1987. Highly toxic coplanar PCBs: occurrence, source, persistency and toxic implications to wildlife and humans. *Environ Pollut* 47:147–163.

Tanabe S, Sung J, Choi D, Baba N, Kiyota M, Yoshida K, Tatsukawa R. 1994. Persistent OC residues in northern fur seal from the Pacific coast of Japan since 1971. *Environ Pollut* 85:305–314.

Tanabe S, Tatsukawa R. 1991. Persistent OCs in marine mammals. In: Jones KC, editor. Organic contaminants in the environment. New York: Elsevier. p 275–289.

Tillitt DE, Ankley GT, Giesy JP, Ludwig JP, Kurita H, Weseloh DV, Ross PS, Bishop CA, Sileo L, Stromborg KL, Larson J, Kubiak TJ. 1992. PCB residues and egg mortality in double crested cormorants from the Great Lakes. *Environ Toxicol Chem* 11:1281–1288.

[USDI] U.S. Department of the Interior, Fish and Wildlife Service. 1978. Determination of certain bald eagle populations as endangered or threatened. *Fed Reg* 43:6230–6233.

[USDI] U.S. Department of the Interior, Fish and Wildlife Service. 1981. Chlorinated hydrocarbons as a factor in the reproduction and survival of lake trout (*Salvelinus namaycush*) in Lake Michigan. Technical Paper 105. Washington DC: USDI. 42 p.

[USDI] U.S. Department of the Interior, Fish and Wildlife Service. 1994. Endangered threatened wildlife and plants; reclassify the bald eagle from endangered to threatened in most of the lower 48 states. *Fed Reg* 59:35584–35594.

Walker MK, Spitsbergen JM, Olson JR, Peterson RE. 1991. 2,3,7,8-Tetrachlorodibenzo-*p*-dioxin (TCDD) toxicity during early life stage development of lake trout (*Salvelinus namaycush*). *Can J Fish Aq Sci* 48: 875–883.

Weseloh DV, Custer TW, Braune BM. 1989. OC contaminants in eggs of common terns from the Canadian Great Lakes. *Environ Pollut* 59:141–160.

Wiemeyer SN, Bunck CM, Stafford CJ. 1993. Environmental contaminants in bald eagle eggs—1980–1984—and further interpretations of relationships to productivity and shell thickness. *Arch Environ Contam Toxicol* 24:213–227.

[WHO] World Health Organization. 1983. Integrated vector control. WHO Tech. Rep. Ser. No. 688. Geneva Switzerland: World Health Organization.

[WHO] World Health Organization. 1984. Malaria control as a part of primary health care. WHO Tech. Rep. Ser. No. 712. Geneva Switzerland: World Health Organization.

Wren CD. 1991. Cause-effect linkages between chemicals and populations of mink (*Mustella vison*) and otter (*Lutra canadensis*) in the Great Lakes basin. *J Toxicol Environ Health* 33:549–585.

Wuster DH, Wuster Jr CF, Strickland WN. 1965 bird mortality following DDT spray for Dutch elm disease. *Ecology* 49:488–499.

Appendix 4

Focus on Chlorine Chemistry Warranted

Wayne A. Schmidt
National Wildlife Federation, Vienna VA

Current chemical management and risk assessment methodologies provide meager defense against use and release of dangerous chemicals. Public concerns about transgenerational effects from exposure to common environmental pollutants form a sound basis for proposals to phase out certain chlorine-based chemicals and manage organochlorines (OCs) as a class. Of particular concern are effects of chemical mixtures on endocrine and immune systems. A new precautionary approach is warranted. Because of its widespread use and environmental distribution, the class of OCs is the logical starting point.

There is increasing public concern about transgenerational effects from exposure to common environmental pollutants. This has driven a search for bold new approaches to management of chemicals. OC compounds, such as dioxins and dioxin-like chemicals, have been singled out because of their widespread use and release into the environment and because many OCs have reproductive and endocrine-disrupting effects.

The purpose of this paper is to 1) clarify the nature of the proposal to phase out certain chlorine-based chemicals, 2) highlight limitations of the status quo, and 3) conclude that a precautionary approach is warranted regarding the use and release of OCs.

The Nature of the Proposal

> "Are humans and our environment in danger from persistent toxic substances now? Are future generations in danger? Based on a review of scientific studies and other recent information, we believe the answer to both questions is yes."

These conclusions of the U.S.–Canada International Joint Commission (IJC 1992) form the underpinnings for its policy recommendations on organochlorines (OCs). It is a position shared by environmental organizations, which also generally concur with the commission's recommendations on OCs.

One basis for this concern stems from emerging knowledge of the link between environmental pollutants and damage to endocrine systems, as well as multi-generational damage from such exposure (Colborn and Clement 1992). Environmentalists believe that current testing and regulatory checks are inadequate to protect against these dangers from hormone-mimicking chemicals, though some chemical manufacturers dispute this conclusion.

Improved productivity of bald eagles in the Great Lakes region is evidence that some problems from pollution are decreasing. Indeed the Great Lakes are cleaner now than they were 20 years ago. But compared to inland eagles, those eagles nesting near the Great Lakes shores have higher concentrations of PCBs and DDT in their blood and poorer reproductive success, even in recent years, than do birds feeding on inland waters, away from the shores (Bowerman 1993). In 1993, 3 eagles with crossed bills and 1 with curled claws were discovered in Michigan. There is no proof of the cause of these deformities; nevertheless, because similar deformities caused by contaminants are found in other Great Lakes birds, such as terns (Kubiak et al. 1989) and cormorants (Fox et al. 1991), the eagles appear to signal risks from continued exposure of people and wildlife to hormone-mimicking chemicals in the environment.

Are OC chemicals the sole source of such danger? Are all OCs causing equal harm? Would a phase out of chlorine-based chemicals, as proposed by the IJC, "solve" the problem? In answer to each question: "Of course not." For example, mercury is one of the most damaging and widespread pollutants and its uses must be phased out, yet it is not an organochlorine.

Yet, OC chemicals make up approximately half of the 362 chemicals confirmed to be present in the Great Lakes ecosystem (IJC 1987,1992) and about half of the chemicals with widespread distribution in the environment that are reported to have reproductive and endocrine-disrupting effects (Colborn et al. 1993). Therefore, the IJC concluded: "In practice, the mix and exact nature of these various [chlorinated] compounds cannot be precisely predicted or controlled in production processes. Thus, it is prudent, sensible and indeed necessary to treat these substances as a *class* rather than as a series of isolated, individual chemicals" (IJC 1992, emphasis added).

Contrary to claims of some critics, this policy does not contemplate banning the use of chlorine for making pharmaceuticals or disinfecting water and sewage.[1] Rather:

> The Commission's proposed approach envisions governmental consultations with industry and other interests to restrict, phase out and *eventually* ban the use of chlorine and chlorine-containing compounds from industrial feedstocks. This approach explicitly recognizes the need to take socio-economic and other consequences (including those of the available alternatives) into account in determining timetables.
>
> The federal governments prefer to focus on individual chemicals. The Commission is concerned that, despite the worldwide concern for the use and discharge of OCs and both countries' commitment to the virtual elimination of persistent toxic substances, many of which are OCs, there is reticence in both governmental and industry declarations to begin consulting on a bold alternative to the wide-scale production and use of these compounds (IJC 1994).

[1] For example, a press packet distributed by the Michigan Chemical Council in 1994 states, "If chlorine and chlorine compounds are banned, tens-of-thousands will die as municipal water supplies, public swimming pools and public eating facilities can no longer be made safe with chlorine-based disinfectants. Additional tens-of-thousands will die when the manufacture of many pharmaceuticals is ended, and when medical research on such terrible diseases as cancer and AIDS comes to a stop because of the unavailability of a basic tool of pharmaceuticals and medical research: organochloride compounds."

Limitations of the Status Quo

A chemical-by-chemical approach to chemical management will never work. Here is the dilemma: too many new chemicals and unpredictable byproducts, too little money to properly assess their effects, and too poor our understanding. For example, agriculture in the U.S. commonly uses dozens of products with some 1.1 billion pounds of pesticide active ingredients annually (Lang 1993; U.S. General Accounting Office 1993).

Based on the sciences of chemistry, toxicology, and biology, it is logical to conclude that we should be able to safely produce and use most OC chemicals. The current framework for risk assessment should be adequate to safely manage them. Many scientists have concluded that they can do just this, including most attending the 1994 SETAC Ecological Risk Assessment for Organochlorine Compounds workshop.

Yet, the devil is in the details. In practice, for example, the U.S. Toxic Substances Control Act of 1976 (TSCA) is a meager defense against use and release of dangerous chemicals. Few existing-use chemicals are assessed; among new chemicals reviewed by the U.S. Environmental Protection Agency (USEPA), only about 5% have been tested to determine their ecological toxicity to aquatic and terrestrial life (Zeeman et al. 1993). So, USEPA must base numerous decisions about potential human health and ecological dangers of new chemicals on models and risk-assessment methodologies—and do it within 90 days (Zeeman et al. 1993; Zeeman and Gilford 1993).

Some conclude that toxicologists know much about a few chemicals, a little about many, and next to nothing about most (Rodricks 1992). That has led environmentalists to presume that "so long as we use risk assessment as our chief guide for allowing chemical exposures, we can expect an unending series of unpleasant surprises as today's 'safe' dose is discovered tomorrow to be unsafe" (Montague 1994).

Others conclude that regulatory toxicologists have done an admirable job in protecting people and the environment from unreasonable risks from exposure to toxic substances. If this were not the case, they argue, cancers caused by exposure to contaminated drinking water and other adverse affects would be much worse. Whether this toxics "glass" is half full or half empty depends on one's perspective.

Likewise, divergent opinions among politicians and the public about the use of risk assessment are well described (Wentsel 1994; Lackey 1994; Shrader-Frechette 1995). The environmental community is divided on the acceptability of risk assessment as a policy tool (Silbergeld 1993; Ziegler 1993). Yet, there is no question that risk-based models will continue to be used widely to make regulatory and chemical management decisions.

The practical limitations of risk assessments are helping drive the search for alternatives to the current method of managing OCs. Risks to an adult are different from those to a developing fetus. Individual susceptibility among humans may depend on gender, age, hormonal factors, diet, preexisting diseases, genetics, lifestyle, and possibly ethnic differences (Lucier 1995). Risk assessment captures none of these subtleties, nor does current testing adequately deal with emerging concerns about potential damage from pollution to endocrine and immune systems.

Assessments performed one chemical at a time may give only a peek at a chemical's dangers. Real-world exposures are to mixtures of chemicals, which may have unpredicted effects. For example, physicians are warned that "doses of therapeutic drugs and hormones may require adjustment in patients exposed to [the OC pesticide] chlordane" and that "exposure to other pesticides or chemicals that induce microsomal enzymes may increase the toxicity of chlordane" (Agency for Toxic Substances and Disease Registry 1994). The warning is based on animal studies that show chlordane toxicity is at least additive in combination with most other pesticides.

It is unlikely that many physicians are even aware of this obscure warning about chlordane (which was banned in the U.S. in 1988 but still persists in the environment and continues to be produced for export). A 1992 survey of Wisconsin doctors performed for Physicians for Social Responsibility revealed that more than half the respondents did not know any of the health effects from exposure to mercury or PCBs—2 of the most widespread pollutants in Great Lakes-area sport fish. Despite this paucity of knowledge in its profession, the American Medical Association (AMA) insists that the risks of OCs be evaluated chemical-by-chemical (1994 AMA Resolution #503).

Often, critics of the IJC's chlorine policy are quick to agree that mistakes have been made in the past but that all the truly bad, persistent OC compounds (e.g., PCBs, DDT, chlordane) have been banned. "Today's chemicals are safe," they posit. It is a seductive argument but founded on hubris. Each generation of new chemicals has been defended with more or less the same rationale.

For example, in responding to Rachel Carson's *Silent Spring* in 1962, a chemical industry spokesman proclaimed the following in defense of DDT and other agricultural chemicals: "Actually, of course, each new agricultural chemical is thoroughly tested for safety for at least two years prior to its being released for general use. And, federal and state laws have been designed and are administered to protect the health of the consumer of farm products" (White-Stevens 1962).

Whatever the current limitations of human health risk assessment, they appear simple compared to the complexities of ecological risk assessment. Multiple endpoints must be predicted from multiple chemicals on multiple species facing multiple stresses. Consider some of the acknowledged limitations of one method recently proposed for assessing contaminant mixtures: It applies to a single life stage of a single species; it cannot account for interacting chemicals or water quality parameters, indirect mortality and community-level changes; and it includes numerous simplifying assumptions and professional judgments (Logan and Wilson 1995).

Risk assessment is a useful tool in some circumstances, although no more or less valid or subjective than alternatives, such as the phasing out of key classes of OCs. Risk assessment may even be used to predict the relative risks of alternatives to some OC products targeted for phasing out — notably polyvinyl chloride plastics.

Yet, risk assessment can become an end-unto-itself, used inadvertently or intentionally to delay obvious changes needed in public policy. This is evident, for example, in USEPA's recent (1994) "dioxin reassessment." If ever a case has been made that a class of toxic OCs should be banned by all necessary measures, it is here, and now, with dioxin and dioxin-like compounds. Predictably, the chemical industry's lobbyists are calling for more study before regulatory action. Moreover, a phalanx of scientists is criticizing USEPA's risk characterization as unproven and warranting "additional careful scientific study" and expressing concern that without clarification by USEPA "the press and public will surely misinterpret the hypothetical risks predicted in the reassessment as real" (Bradfield et al. 1995).

Precautionary Approach Warranted

A precautionary approach to the use and release of OC chemicals is warranted. The status quo has resulted in global contamination by OCs, such as PCBs and DDT. The status quo provides little assurance that current-use and new OCs are safe for humans and the environment. Science, politics, industry, and society have failed to anticipate and restrict uses and releases of environmental pollutants with capabilities to disrupt endocrine, immune, and nervous systems. There is increasing evidence that this failure is diminishing global biodiversity, human reproductive potential, and the health and well-being of our children (Colborn and Clement 1992).

It is not enough to defend the status quo by decrying the lack of data or resources to make risk assessment "work" correctly. There never will be enough data or resources to answer questions satisfactorily about the thousands of current-use and new chemicals introduced each year. Of particular concern are persistent and bioaccumulative chemicals. For these chemicals, there is no acceptable assimilative capacity. That is why the IJC concluded "[C]onventional scientific concepts of dose-response and acceptable 'risk' can no longer be defined as 'good' scientific and management bases for defining acceptable levels of pollution. They are outmoded and inappropriate ways of thinking about persistent toxics" (IJC 1994).

Therefore, a new approach addressing classes of chemicals is appropriate. OCs are the logical starting point because of their widespread use and distribution in the environment. Practical limitations likely will prevent current chemical testing and regulatory programs from ever providing adequate protection from the dangers of exposure to mixtures of environmental pollutants, particularly those causing transgenerational adverse effects.

The guiding principle for release of new substances into the environment must embody the concept of "reverse onus" advocated by the IJC: "When approval is sought for the manufacture, use or discharge of any substance which will or may enter the environment, the applicant must prove, as a general rule, that the substance is not harmful to the environment or human health" (IJC 1990).

In the final analysis, science cannot provide the answer to whether chemical management using reverse onus and based on classes of compounds, rather than the chemical-by-chemical approach used today, is best. That is a community decision that must weigh a mix of science, ethics, history, economics, law, and the collective good judgment of the public.

References

Agency for Toxic Substances and Disease Registry. 1994. Toxicological profile for chlordane. Atlanta GA: U.S. Dept. of Health and Human Services.

[AMA] American Medical Association. 1994. AMA Resolution #503. Chicago IL: AMA.

Bowerman WW. 1993. Regulation of bald eagle (*Haliaeetus leucocephalus*) productivity in the Great Lakes Basin: an ecological and toxicological approach [Ph.D. thesis]. E. Lansing MI: Michigan State Univ.

Bradfield CA, Gallo MA, Gasiewicz TA, Greenberg RS, Greenlee WF, Margolick J, Mattison DR, Munson P, Neal RA, Okey AB, Olive DL, Poland A, Rifkind AB, Rodricks JV, Rozman K, Safe SH, Starr TB, Sutter TR. 1995. USEPA dioxin reassessment (letter). *Science* 266:1628–1629.

Colborn T, Clement C, editors. 1992. Chemically induced alterations in sexual and functional development: the wildlife/human connection. Princeton NJ: Princeton Scientific.

Colborn T, vom Saal FS, Soto AM. 1993. Developmental effects of endocrine-disrupting chemicals in wildlife and humans. *Environ Health Perspect* 101:378–384.

Fox GA, Collins B, Hayakawa E, Weseloh DV, Ludwig JP, Kubiak TJ, Erdman TC. 1991. Reproductive outcomes in colonial fish-eating birds: a biomarker for developmental toxicants in Great Lakes food chains. II. Spatial variation in the occurrence and prevalence of bill defects in young double-crested cormorants in the Great Lakes, 1979–1987. *J Great Lakes Res* 17:158–167.

[IJC] International Joint Commission. 1987., 1987 Report on Great Lakes Water Quality. Windsor, Ontario: Great Lakes Water Quality Board.

[IJC] International Joint Commission. 1990. Fifth biennial report under the Great Lakes Water Quality Agreement of 1978 to the governments of the United States and Canada and the state and provincial governments of the Great Lakes basin; Part II. Washington DC and Ottawa, Ontario.

[IJC] International Joint Commission. 1992. Sixth biennial report under the Great Lakes Water Quality Agreement of 1978 to the governments of the United States and Canada and the state and provincial governments of the Great Lakes basin. Washington DC and Ottawa, Ontario.

[IJC] International Joint Commission. 1994. Seventh biennial report under the Great Lakes Water Quality Agreement of 1978 to the governments of the United States and Canada and the state and provincial governments of the Great Lakes basin. Washington DC and Ottawa, Ontario.

Kubiak TJ, Harris HJ, Smith LM, Schwartz TR, Stalling DL, Trick JA, Sileo L, Docherty DE, Erdman TC. 1989. Microcontaminants and reproductive impairment of the Forster's tern on Green Bay, Lake Michigan—1983. *Arch Environ Contam Toxicol* 18:706–727.

Lackey RT. 1994. Ecological risk assessment. *Fisheries* 19:14–18.

Lang L. 1993. Are pesticides a problem? *Environ Health Perspect* 101:578–583.

Logan DT, Wilson HT. 1995. An ecological risk assessment method for species exposed to contaminant mixtures. *Environ Toxicol Chem* 14:351–359.

Lucier GW. 1995. Not your average Joe. *Environ Health Perspect* 103:10.

Montague P. 1994. The emperor's scientific new clothes. *Rachel's Hazardous Waste News*, #393. Annapolis, MD: Environ. Research Foundation.

Rodricks JV. 1992. Calculated risks. New York: Cambridge Univ Pr.

Shrader-Frechette K. 1995. Number crunching and comparative assessment of environmental risks. *BioScience* 45:66.

Silbergeld EK. 1993. Risk assessment: the perspective and experience of U.S. environmentalists. *Environ Health Perspect.* 101:100–104.

[USEPA] U.S. Environmental Protection Agency. 1994. Health assessment document for 2,3,7,8-tetrachlorodibenzo-*p*-dioxin (TCDD) and related compounds. Volume 3 of 3, Review Draft. EPA/600/BP-92/001c.

U.S. General Accounting Office. 1993. Issues concerning pesticides used in the Great Lakes Watershed. Washington DC. GAO/RCED-93-128.

Wentsel RS. 1994. Risk assessment and environmental policy. *Environ Toxicol Chem* 13:1381.

White-Stevens R. 1962. Faddists hampering progress. *Agricultural Chemicals* September. p 28, 32.

Zeeman M, Nabholz JV, Clements RG. 1993. The development of SAR/QSAR for use under USEPA's Toxic Substances Control Act (TSCA): an introduction. In: Gorsuch JW, Dwyer FJ, Ingersoll CG, LaPoint TW. Volume 2, Environmental toxicology and risk assessment. Philadelphia PA: American Society for Testing and Materials. ASTM STP 1216. p 523–539.

Zeeman M, Gilford J. 1993. Ecological hazard evaluation and risk assessment under USEPA's Toxic Substances Control Act (TSCA): an introduction. In: Landis WG, Hughes JS, Lewis MA, editors Volume 2. Environmental toxicology and risk assessment. Philadelphia PA: American Society for Testing and Materials. ASTM STP 1179. p 7–21.

Ziegler J. 1993. Health risk assessment research: the OTA report. *Environ Health Perspect* 101:402–406.

APPENDIX 5

Species Differences in Carcinogenicity of C_1–C_3 Chlorinated Aliphatic Compounds

Trevor Green
Zeneca Central Toxicology Laboratory
Alderley Park, Macclesfield, Cheshire, UK

The chlorinated aliphatics (C_1–C_3) are major industrial chemicals that have been used in large quantities for many years. Although these chemicals have found their way into the environment, they tend not to persist or bioaccumulate, and attention has focused on their possible adverse effects on human health. Of these the principal concern has been the potential to cause human cancer, as evidenced by their carcinogenicity in laboratory animals. Only one of this group, vinyl chloride, is a known human carcinogen; the remainder produce markedly different responses among animal species, and therefore their relevance to human health is uncertain. Mechanistic studies have revealed a wide variety of carcinogenic mechanisms, which in addition to the formation of reactive electrophiles, include chronic tissue damage, peroxisome proliferation, and protein droplet nephropathy, the latter two being largely specific to rodents. Metabolism and pharmacokinetic studies have also shown marked differences between rodents, particularly mice, and humans in the in vivo fate of these chemicals. Overall, these studies conclude that in most cases the risk of cancer from chlorinated aliphatics is either specific to rodents or is considerably reduced in humans.

The smaller chlorinated aliphatics of commercial importance, principally C_1 alkanes and C_2 alkanes and alkenes, have been produced in large quantities for more than 50 years. Their commercial and industrial use (Table 1) as solvents, intermediates, and starting materials for plastics are based on the unique properties conferred on these molecules by the presence of chlorine. Consequently, their replacement by nonchlorinated substitutes has proved difficult and in some cases impossible. During their manufacture and use, chlorinated aliphatics have entered the environment in significant quantities through direct releases and by accident, and indirectly through chlorination of drinking water. In almost all cases these volatile, low-molecular-weight chemicals do not persist or bioaccumulate in the environment. Only one, 1,1,1-trichloroethane, adversely effects the ozone layer and is currently being phased out under the Montreal Protocol.

In the absence of any major environmental impact, interest in this chemical group has centered on the human health hazards arising from exposure in the workplace or from the low levels present in groundwater and drinking water. As with their chemical properties, the toxicology of chlorinated aliphatics is dominated by chlorine. The chlorine in these molecules is not normally directly reactive with biological systems. However, the normal metabolic processes of oxidation and conjugation can result in the formation of metabolites such as epoxides, aldehydes, acyl halides, and carbonium ions, which have enhanced reactivity because of the presence of chlorine.

Superficially at least, it would be reasonable to assume that closely related chemicals of this type would have common mechanisms of action and close structure-activity

Table 1 Production and use of a number of chlorinated aliphatics

Chlorinated aliphatic	Year of first commercial production	Major uses
Vinyl chloride	1933	polyvinylchloride (PVC)
Vinylidene chloride	1940	polymers and co-polymers
Trichloroethylene	1908	metal degreasing
Perchloroethylene	1925	dry cleaning
Methyl chloride	1920	methylating agent, foam blowing
Methylene chloride	1934	paint stripper, solvent
Chloroform	1922	synthesis of fluorocarbons
Carbon tetrachloride	1907	synthesis of fluorocarbons
1,2 Dichloroethane	1922	synthesis of vinyl chloride
1,1,1-Trichloroethane	1946	component cleaning

relationships. Nothing could be further from the truth. Their toxicology is characterized by marked species differences in response in laboratory animals, diverse mechanisms of action, and uncertainty about the health hazards to humans exposed to these chemicals. In the last 25 years, these variable responses in standard animal toxicology tests have attracted considerable interest from researchers seeking to identify mechanisms of action and explain species differences. As a result, knowledge of these chemicals is now probably greater than is knowledge of any other comparable group, and a much clearer understanding of their true health hazards now exists.

The common property of almost all chlorinated aliphatics, and the one that is of most concern, is their ability to cause an increase in cancer in laboratory animals exposed to very high dose levels. This is also the property that has attracted the most research interest and is the subject of this paper.

Cancer Studies

All of the chlorinated methanes and ethanes, with the exception of 1,1,1-trichloroethane, have been shown to cause cancer in rodents (Table 2) (Maltoni et al. 1974, 1984, 1986; National Cancer Institute [NCI] 1976a, 1976b, 1977, 1978; National Toxicology Program [NTP] 1977, 1985, 1986; European Centre for Ecotoxicology and Toxicology of Chemicals [ECETOC] 1985; Herren-Freund et al. 1987; Quast et al. 1988). This apparently common property, and the assumption that it will apply to humans, is attributed to the presence of chlorine and has led the drive for nonchlorinated alternatives. As with most concerns related to chlorine, the assumption that the carcinogenicity seen in rodents will occur in other species is a gross over-simplification of the true picture.

Table 2 Carcinogenicity of chlorinated aliphatics

Chlorinated aliphatic	Carcinogenicity (species and target organs)
Vinyl chloride	animal and human carcinogen
Vinylidene chloride	male Swiss mouse kidney
Trichloroethylene	mouse liver and lung, rat kidney
Perchloroethylene	mouse liver, rat kidney
Methyl chloride	inadequately tested
Methylene chloride	mouse liver and lung
Chloroform	mouse liver, rat kidney
Carbon tetrachloride	rat and mouse liver
1,2 Dichloroethane	multiple sites, rat and mouse
1,1,1-Trichloroethane	noncarcinogenic

Only one of this group of chemicals, vinyl chloride, is a known human carcinogen (Creech and Johnson 1974). Its carcinogenicity was established simultaneously in laboratory animals and in humans in 1974 (Creech and Johnson 1974; Maltoni et al. 1974). By 1986, 12 epidemiology studies had identified 120 cases of vinyl chloride-induced angiosarcoma of the liver in persons employed in vinyl chloride manufacture. In contrast to vinyl chloride, it is interesting to note that none of the other chlorinated aliphatics have been identified as carcinogens by epidemiology studies, even though such studies are numerous, the chemicals have been manufactured commercially for over 50 years, and exposures over that time are known to have been considerable (IARC 1986; Brown and Kaplan 1987; Hearne et al. 1990; Spirtas et al. 1991).

Why has the carcinogenicity predicted by the animal studies not been realized? Clearly, the power of the epidemiology studies to detect an increase in cancer, if it occurs, is always questionable. Alternatively, a more detailed analysis of the animal bioassay data suggests why animal studies may not be accurate predictors of human risk. First, the studies have been conducted at massive dose levels, as high as 3g/kg by gavage and several thousands of ppm by inhalation (NCI 1976; NTP 1986). In many, corn oil gavage was used for purely economical reasons, when inhalation was the only appropriate choice (NCI 1976). When both routes have been used, it is clear that tumor incidences are higher in the presence of corn oil (NCI 1976; Maltoni et al. 1986). A second plausible explanation can be found in the results of the studies themselves (Table 2). In many cases, cancers are seen in mice and not rats, or vice versa, following exposure to the same dose levels. Furthermore, the expected structure-activity relationships for a small group of chemicals have not been realized. For example, vinylidene chloride (1,1-dichloroethylene) is the closest analog to vinyl chloride (chloroethylene), and might be expected to have similar carcinogenic potency. Vinyl chloride was carcinogenic in all of the laboratory animals in which it was tested (rats, mice, and hamsters) and in humans (Creech and Johnson 1974; Maltoni et al. 1974) yet vinylidene chloride only induced a low incidence of kidney cancer in one sex of one mouse strain in one out of 18 cancer bioassays (Maltoni et al. 1984; ECETOC 1985). Other examples such as the chlorinated solvents (trichloroethylene, perchloroethylene, dichloromethane) increased the incidences of liver tumors in mice but not in rats (NCI 1976a, 1977; NTP 1986).

These differences, which by no means are restricted to the examples given, make extrapolation to humans with any degree of certainty very difficult. For risk assessment purposes, these chemicals are assumed to have the ability to cause cancer and a worst-case, conservative type of risk assessment is normally used. In the absence of other data, such an approach is acceptable for regulatory purposes but is scientifically less than satisfactory in view of the uncertainties involved. Those uncertainties have led to numerous mechanistic studies, seeking explanations for the species differences in order to more accurately evaluate the risks to humans from exposure to these chemicals.

Genotoxicity

A key part of any mechanistic evaluation of a carcinogen is the role of genotoxicity. Being "old" chemicals, the chlorinated aliphatics have been evaluated in a vast array of short-term tests for mutagenicity and clastogenicity (IARC 1979a, 1979b, 1986). Many of those tests were never fully validated and are not used today, yet the results remain in the literature and confuse the reader. The picture is further confounded by poor control of the materials tested. Chlorinated solvents frequently contain stabilizers, and in the past it was common for these to be reactive, and mutagenic, epoxides (Crebelli and Carere 1989). Thus, there is confusion between tests conducted with pure material, those with stabilized material, and as frequently happened, those tests where the purity of the material was not specified. Today, with a standard battery of short-term tests and pure material, the picture is somewhat clearer but remains contentious. Although the details are beyond the scope of this paper, it is true that very few of the chlorinated aliphatics give unequivocally positive results in a full range of modern in vitro and in vivo tests and that in many cases the data are of little help in determining the mechanisms involved in the development of cancer. A summary of what is believed to be the consensus view for the C_1 and C_2 chlorinated aliphatics is given in Table 3; refer to Crebelli and Carere (1989) for an example of a full and typical evaluation of the genotoxicity of one member (trichloroethylene) of this series. The uncertainty over genotoxicity has led to the search for and the discovery

Table 3 Mutagenicity of chlorinated aliphatics[1]

Chlorinated aliphatic	Mutagenicity
Vinyl chloride	0
Vinylidene chloride	+ / –
Trichloroethylene	+ / –[2]
Perchloroethylene	–
Methyl chloride	+
Methylene chloride	+
Chloroform	–
Carbon tetrachloride	–
1,2 Dichloroethane	+
1,1,1-Trichloroethane	–

[1] Summary from IARC (1979a, 1979b, 1986) and Crebelli and Carere (1989).
[2] +/– = variable results; no overall clear pattern in a wide range of tests.

of other mechanisms of carcinogenicity, which appear not to involve a direct interaction between these chemicals or their metabolites and the genome.

Metabolism and Pharmacokinetics

None of the chlorinated aliphatics are direct-acting carcinogens. They all require metabolic activation, and it is appropriate to briefly discuss the main metabolic pathways before considering the mechanisms of action. All of the members of this group are either gases or volatile liquids, and the primary route of occupational exposure is by inhalation or skin absorption. However, in nonoccupational settings, exposure can also occur through ingestion of food and drinking water. A typical example of the latter is the ingestion of chloroform present in drinking water as a result of chlorination. Once absorbed, these lipophilic chemicals partition to fat compartments and/or are metabolized by the liver to a variety of water soluble products. Exhalation of unchanged material through the lungs is also a major route of elimination (Guengerich and Watanabe 1979; Schumann et al. 1982; Anders and Pohl 1985; Henschler 1985; Dekant et al. 1986).

Metabolism occurs by one or both of 2 possible pathways. Oxidation by the cytochrome P-450 enzymes may lead to radicals (carbon tetrachloride), epoxides (chloroethylenes) or chlorinated alcohols, aldehydes, and acyl halides (e.g., dichloromethane, 1,2-dichloroethane, 1,1,1-trichloroethane) (Bonse and Henschler 1976; Guengerich and Watanabe 1979; Schumann et al. 1982; Anders and Pohl 1985; Henschler 1985; Dekant et al. 1986). Alternatively, the presence of chlorine and electron-deficient carbon atoms leads to conjugation with the tripeptide glutathione, either as a primary metabolic step (methyl chloride, dichloromethane, 1,2-dichloroethane, perchloroethylene) (Anders and Pohl 1985; Henschler 1985), or as a secondary reaction with reactive epoxides and aldehydes produced by oxidation (Henschler 1985).

Being volatile materials, the chlorinated aliphatics are not dependent on metabolism for their clearance from mammalian systems. Unmetabolized material may be eliminated through the lungs, and in some cases over 90% of the absorbed chemical is removed in this way. This also leads to some marked species differences in the extent to which these chemicals are metabolized (Brown et al. 1974; Prout et al. 1985). In those species with high metabolic rates, such as the mouse, much of the dose is metabolized before it is cleared through the lungs. In larger species such as primates and humans, the converse is true and a large percentage of the dose is excreted unchanged. The proportion of the dose exhaled unchanged is also dose-dependent. Whereas most species are able to extensively metabolize small doses, only those with high metabolic rates are able to metabolize significant proportions of the high doses used in cancer bioassays (see Figure 1 for an example). Clearly, this difference in metabolic rates has implications for the carcinogenicity of these chemicals, since cancer is associated with the metabolites rather than the parent chemicals.

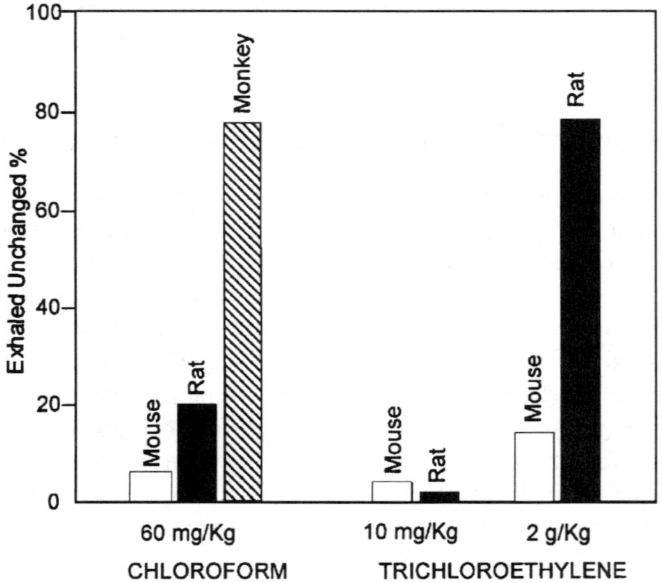

Figure 1 Species- and dose-dependent elimination of chloroform and trichloroethylene following a single oral dose (Brown et al. 1974; Prout et al. 1985)

Mechanisms

Research into mechanisms of carcinogenicity followed the realization that vinyl chloride is a human carcinogen. The outcome of that research was the discovery of a mechanism of classical genotoxicity. Vinyl chloride was metabolized to a highly reactive epoxide, which reacted with DNA to form a number of adducts that were subsequently characterized both in vitro and in vivo (Green and Hathway 1978). Following the publication of this work, it was assumed that the other chloroethylenes would be similarly mutagenic and carcinogenic through their respective epoxides (Greim et al. 1975). Today, there is no evidence to suggest that these epoxides are involved in any of the cancers caused by vinylidene chloride, trichloroethylene, or perchloroethylene (Green 1990). The few tumors seen in the male Swiss mouse kidney after exposure to vinylidene chloride are believed to result from chronic toxicity, probably caused by chloroacetyl chloride, a metabolite of vinylidene chloride (Speerschneider and Dekant 1995). Trichloroethylene and perchloroethylene both cause liver cancer in mice but not in rats (NCI 1976a; NCI 1977). Both share a common metabolite, trichloroacetic acid, which is known to induce peroxisomes, a response linked with liver cancer in rodents (Elcombe 1985; Odum et al. 1988). Further support for the existence of such a mechanism came from a cancer bioassay in which trichloroacetic acid itself was shown to induce peroxisomes and liver cancer in mice (Herren-Freund et al. 1987). Thus, from a belief that tri- and perchloroethylene caused liver cancer in mice as a result of epoxide formation and DNA damage, it is now generally accepted that these tumors arise from a nongenotoxic mechanism involving trichloroacetic acid and peroxisomes. Other nongenotoxic mechanisms exist: kidney tumors involving protein droplet nephropathy (perchloroethylene) and liver tumors resulting from chronic toxicity (carbon tetrachloride and chloroform) (Reitz 1987; Green et al. 1990). Even where there is evidence of genotoxicity (dichloromethane, 1,2-dichloroethane), there are still marked species differences in carcinogenicity as a result of differences in metabolism and metabolic rates between species (Graves et al. 1995).

Species Differences and Relevance to Humans

Certain responses in rodents are believed to have little or no relevance to humans. In these examples protein droplet nephropathy is a male rat phenomenon and the peroxisome proliferation seen with trichloroacetic acid appears not to occur in human liver cells (Elcombe 1985; Odum et al. 1988; Green et al. 1990). Even in rodents, these nongenotoxic responses have a threshold below which neither the biochemical response nor the cancer will occur (Elcombe 1985). This is illustrated with tri- and perchloroethylene in their metabolism to trichloroacetic acid. Only the mouse is able to metabolically produce sufficient quantities of this metabolite to induce peroxisome proliferation and consequently liver tumors are seen only in mice (Elcombe 1985; Odum et al. 1988). This difference in metabolic capacity between species, illustrated in Figure 2 for the metabolism of perchloroethylene to trichloroacetic acid, is a common feature of the chlorinated aliphatics and the basis of most of the species differences. A further example is shown in Figure 3 for dichloromethane. This chemical is carcinogenic in mice as a result of its conjugation with glutathione to form a reactive electrophile (Figure 4) (Graves et al. 1995). Again, the metabolic rates in the mouse far exceed those in other species to the extent that liver and lung cancer is seen only in mice (Green 1989).

Overall there is a common theme that the chlorinated aliphatics are linked with cancer in organs of those species with the highest metabolic rates. Where metabolic rates are lower, these volatile chemicals are cleared through the lungs before they are significantly metabolized, thus reducing the overall body burden of key metabolites and reducing the risks. The metabolites themselves tend to be either reactive electrophiles whose rate of detoxification is species-independent or metabolites with very long half-lives (e.g., trichloroacetic acid) where clearance is not a major factor between species. Consequently, the rate of formation of metabolites is the one key determinant associated with risk. Where rates have been determined experimentally, they are typically many-fold lower in human tissues than in rodent tissues, providing large margins of safety. In addition, a number of the mechanisms are nongenotoxic and are based on responses that are seen entirely or predominantly in rodents.

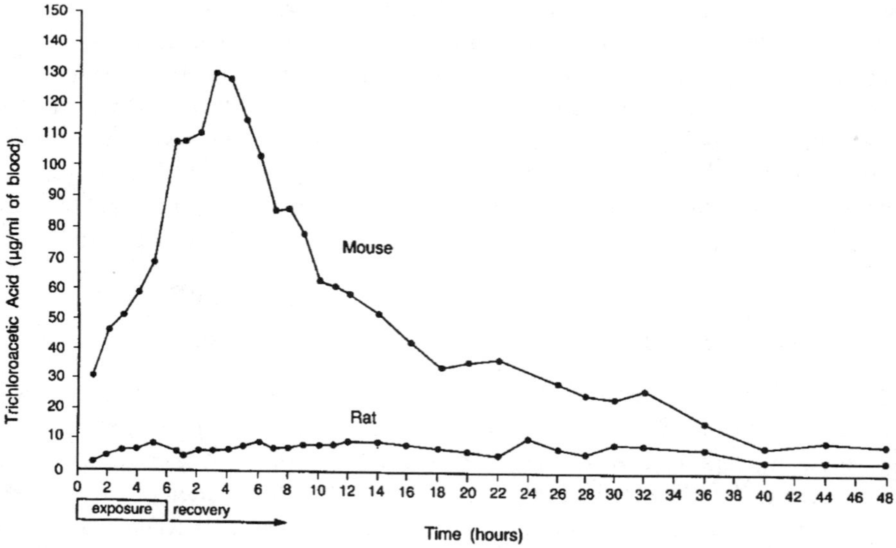

Figure 2 Blood levels of trichloroacetic acid in rats and mice exposed to 400 ppm perchloroethylene for 6 h (Odum et al. 1988)

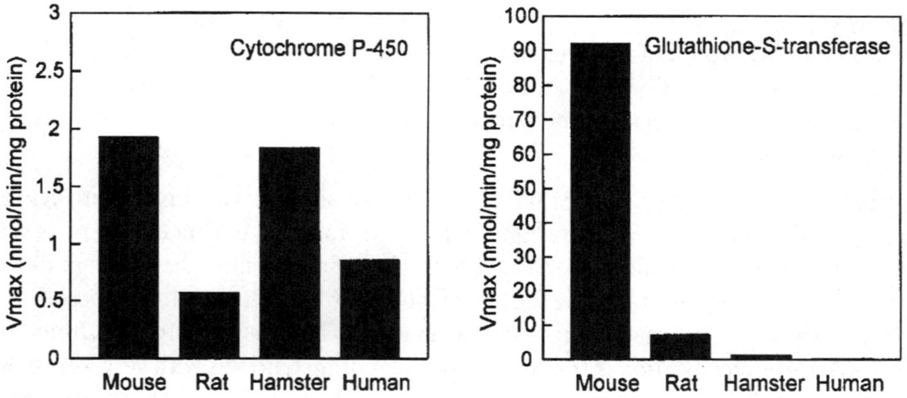

Figure 3 Metabolism of dichloromethane in liver fractions from 4 species (Green 1989)

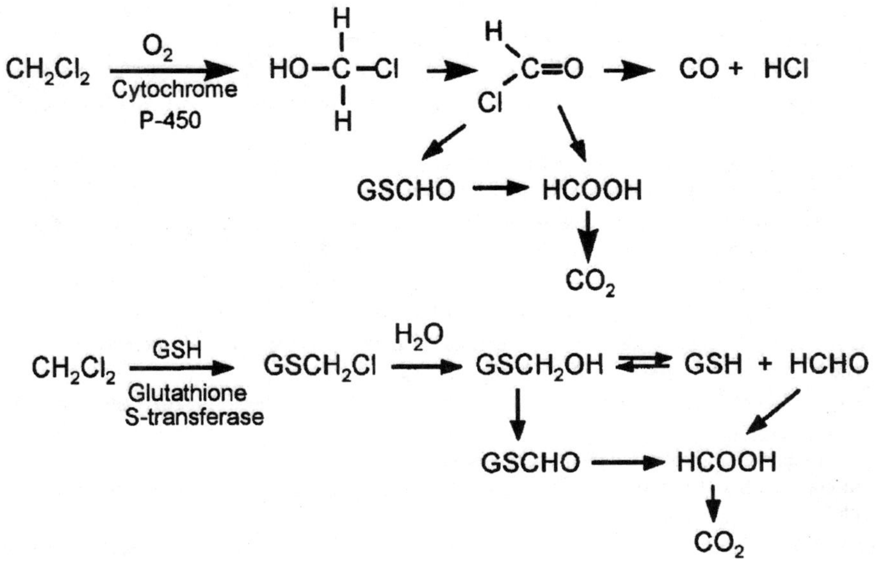

Figure 4 Dichloromethane metabolism

Conclusions

With only one or two exceptions, the C_1–C_3 chlorinated aliphatics share the common property of inducing cancer in rodents at high, maximum-tolerated dose levels. However, only one of these chemicals, vinyl chloride, has proven to be a human carcinogen. Further expectations that all of these chemicals would act through mutagenic intermediates have also not been realized and nongenotoxic mechanisms involving cell damage, protein droplet nephropathy, and peroxisomes have now been identified. Such mechanisms exhibit thresholds and in many cases show marked species specificity. Widely differing metabolic rates also contribute to the marked species differences in carcinogenicity resulting from these chemicals. Studies comparing mechanisms and metabolic rates between animals and humans have concluded that the risks to the humans exposed either occupationally or environmentally are minimal.

References

Anders MW, Pohl LR. 1985. Halogenated alkanes. In: Anders MW,editor. Bioactivation of foreign compounds. Orlando FL: Academic Pr. p 283–315.

Bonse G, Henschler D. 1976. Chemical reactivity, biotransformation and toxicity of poly-chlorinated aliphatic compounds. *CRC Crit Rev Toxicol* 4:395–409.

Brown DM, Langley PF, Smith D, Taylor DC. 1974. Metabolism of chloroform. 1. The metabolism of [14C] chloroform by different species. Xenobiotica 4:151–163.

Brown DP, Kaplan SD. 1987. Retrospective cohort mortality study of dry cleaning workers using perchloroethylene. *J Occupational Medicine* 29:535–541.

Crebelli R, Carere A. 1989. Genetic toxicology of 1,1,2-trichloroethylene. *Mutation Res* 221:11–37.

Creech JL, Johnson MN. 1974. Angiosarcoma of the liver in the manufacture of PVC. *J Occup Med* 16:150–151.

Dekant W, Schulz A, Meztlerm M, Henschler D. 1986. Absorption, elimination, and metabolism of trichloroethylene: a quantitative comparison between rats and mice. *Xenobiotica* 16:143–152.

[ECETOC] European Centre for Ecotoxicology and Toxicology of Chemicals. 1985. Vinylidene chloride. Brussels, Belgium: Avenue E. Van Nieuwenhuyse 4, B1160. JACC Report No. 5.

Elcombe CR. 1985. Species differences in carcinogenicity and peroxisome proliferation due to trichloroethylene. A biochemical human hazard assessment. *Arch Toxicol Suppl* 8:6–17.

Graves RJ, Coutts C, Green T. 1995. Methylene chloride induced DNA damage: An interspecies comparison. *Carcinogenesis* 16:1919–1926.

Green T, Hathway DE. 1978. Interactions of vinyl chloride with rat liver DNA in vivo. *Chem Biol Interact* 22:211–224.

Green T. 1989. A biological database for methylene chloride risk assessment. In Travis CC, editor. Biologically based methods for cancer risk assessment. NATO ASI Series Volume 159. New York: Plenum Pr. p 289–300.

Green T. 1990. Chloroethylenes: a mechanistic approach to human risk evaluation. *Annu Rev Pharmacol Toxicol* 30:73–89.

Green T, Odum J, Nash JA, Foster JR. 1990. Perchloroethylene induced rat kidney tumors: An investigation of the mechanisms involved and their relevance to humans. *Toxicol Appl Pharmacol* 103:77-89.

Greim H, Bonse G, Radwan Z, Reichert D, Henschler D. 1975. Mutagenicity in vitro and potential carcinogenicity of chlorinated ethylenes as a function of metabolic oxirane formation. *Biochem Pharmacol* 24:2013–2017.

Guengerich FP, Watanabe PG. 1979. The metabolism of vinyl chloride in vivo and in vitro. *Biochem Pharmacol* 28:589–596.

Hearne FT, Pifer JW, Grose F. 1990. Absence of adverse mortality effects in workers exposed to methylene chloride: An update. *J Occupational Medicine* 32:234–251.

Herren-Freund SL, Pereira MA, Khoury MD, Olson G. 1987. The carcinogenicity of trichloroethylene and its metabolites, trichloroacetic acid and dichloroacetic acid, in mouse liver. *Toxicol Appl Pharmacol* 90:183–189.

Henschler D. 1985. Halogenated alkenes and alkynes. In Anders MW, editor Bioactivation of foreign compounds. Orlando FL: Academic Pr. p 317–347.

[IARC] International Agency for Research on Cancer. 1979a. Monographs on the evaluation of the carcinogenic risk of chemicals to humans. Volume 19, Some monomers, plastics and synthetic elastomers, and acrolein. Lyon, France: IARC.

[IARC] International Agency for Research on Cancer.1979b. Monographs on the evaluation of the carcinogenic risk of chemicals to humans. Volume 20, Some halogenated hydrocarbons. Lyon, France: IARC.

[IARC] International Agency for Research on Cancer. 1986. Monographs on the evaluation of the carcinogenic risk of chemicals to humans. Volume 41, Some halogenated hydrocarbons and pesticide exposures. Lyon, France: IARC.

Maltoni C, Lefemine C, Chieco P, Carrettu D. 1974. Vinyl chloride carcinogenesis: current results and perspectives. *Med Lav* 65:421.

Maltoni C, Cotti G, Chieco P. 1984. Chronic toxicity and carcinogenicity bioassays of vinylidene chloride. *Acta Oncolog* 5:91–145.

Maltoni C, Lefermine G, Cotti G. 1986. Experimental research on trichloroethylene carcinogenesis. In: Maltoni C, Mehlman MA, editors. Volume 5, Archives of Research on Industrial Carcinogenesis. Princeton NJ: Princeton Scientific. p 1–393.

[NCI] National Cancer Institute. 1976a. Carcinogenesis bioassay of trichloroethylene. CAS No. 79-01-6. National Cancer Institute Technical Report Series No. 2. DHEW pub. No. 76-802.

[NCI] National Cancer Institute. 1976b. Report on carcinogenesis bioassay of chloroform. Bethesda MD: NCI (NTIS PB-264-018).

[NCI] National Cancer Institute. 1977. Bioassay of tetrachloroethylene for possible carcinogenicity. DHEW pub. No. 77-813.

[NCI] National Cancer Institute. 1978. Bioassay of 1,2-dichloroethane for possible carcinogenicity (Technical Report Series No. 55), DHEW Publication No. (NIH) 78-1361. Washington DC: U.S. Department of Health, Education and Welfare.

[NTP] National Toxicology Program. 1985. Technical Report on the toxicology and carcinogenesis studies of tetrachloroethylene (perchloroethylene) in F344/N rats and B6C3F1 mice (inhalation studies). DHSS-NIH Report 85-2657.

[NTP] National Toxicology Program. 1986-Jan. Technical report on the toxicology and carcinogenesis studies of dichloromethane in F344/N rats and B6C3F1 mice. NTP TR 306, Final Report.

Odum J, Green T, Foster JR, Hext PM. 1988. The role of trichloroacetic acid and peroxisome proliferation in the species differences in carcinogenicity of perchloroethylene in the mouse and rat. *Toxicol Appl Pharmacol* 92:103–112.

Prout MS, Provan WM, Green T. 1985. Species differences in response to trichloroethylene. 1. Pharmacokinetics in rats and mice. *Toxicol Appl Pharmacol* 79:389–400.

Quast JF, Calhoun LL, Frauson LE. 1988. 1,1,1-Trichloroethane formulation: a chronic inhalation toxicity and oncogenicity study in Fischer 344 rats and B6C3F1 mice. *Fund Appl Toxicol* 11:611–625.

Reitz RH. 1987. Role of cytotoxicity in the carcinogenic process. In: Butterworth BE, Slaga TJ, editors. Nongenotoxic mechanisms of carcinogenesis. Banbury Report No. 25. New York: Cold Spring Harbor Laboratory.

Schumann AM, Fox TR, Watanabe PG. 1982. A comparison of the fate of inhaled methylchloroform (1,1,1-trichloroethane) following single or repeated exposure in rats and mice. *Fundam Appl Toxicol* 2:27–32.

Speerschneider P, Dekant W. 1995. Renal tumorigenicity of 1,1-dichloroethene in mice: the role of male specific expression of cytochrome P-450 2E1 in the renal bioactivation of 1,1-dichloroethene. *Toxicol Appl Pharmacol* 130:48–56.

Spirtas R, Stewart P, Lee S. 1991. Retrospective cohort mortality study of workers at an aircraft maintenance facility I. Epidemiological results. *Br J Ind Med* 48:515–530.

Abbreviations

2,3,7,8-TCDD	2,3,7,8-tetrachlorodibenzo-*p*-dioxin
AChE	Acetylcholine esterase
ADMT	Air-dried metric ton
Ah	Aromatic hydrocarbon
AHH	Aryl hydrocarbon hydroxylase
AhR	Aryl hydrocarbon receptor
AMA	American Medical Association
AMP	Adenosine monophosphate
AOX	Adsorbable organic halogen
ARAR	Applicable, relevant, and appropriate requirement
ARM	Agricultural Runoff Model
Arnt	AhR nuclear translator protein
ATP	Adenosine triphosphate
BAF	Bioaccumulation factor
BaP	Benzo(a)pyrene
BAT	Best available technology
BCF	Bioconcentration factor
BIC	Biotic carbon
BMF	Biomagnification factor
BOD	Biological oxygen demand
BSAF	Biota-sediment accumulation factor
BUA	*Beratergreminum umweltrelevante Altstoffe*
bw	Body weight
CAA	Clean Air Act (U.S.)
CAAA	Clean Air Act Amendments (U.S.)
CAG	Carcinogen Assessment Group (USEPA)
CBR	Critical body residue
CCREM	Canadian Council of Resource and Environment Ministries
CCRP	California Comparative Risk Project (U.S.)
CEAA	Canadian Environmental Assessment Act
CemoS	Chemical Exposure Modeling System
CENR	Committee on Environment and Natural Resources (White House National Science and Technology Council)
CEPA	Canadian Environmental Protection Act

CERCLA	Comprehensive Environmental Response, Compensation, and Liability Act (U.S.)
CFC	Chlorofluorocarbon
CNS	Central nervous system
CP	Chlorinated paraffin
CPSC	Consumer Product Safety Commission (U.S.)
CRAM	Committee on Risk Assessment Methodology (NRC)
CREAMS	Chemicals, Runoff, and Erosion from Agricultural Management Systems
CWA	Clean Water Act (U.S.)
CWS	Canadian Wildlife Service
DCPA	Dimethyl tetrachloroterephthalate
DDD	Dichlorodiphenylethane
DDE	Dichlorodiphenyldichloroethylene
DDT	Dichlorodiphenyltrichloroethane
DDTr	DDT and related compounds
DES	Diethylstilbestrol
DNA	Deoxyribonucleic acid
DOC	Dissolved organic carbon
DTH	Delayed type hypersensitivity
ECETOC	European Centre for Ecotoxicology and Toxicology of Chemicals
EDC	Ethylene dichloride
EOCl	Extractable organic chlorine
EOX	Extractable organic halogen
EROD	7-Ethoxyresorufin-*O*-deethylase
ESA	Endangered Species Act (U.S.)
EU	European Union
FDA	Food and Drug Administration (U.S.)
FGETS	Food Gill Exchange of Toxic Substances
FIFRA	Federal Insecticide, Fungicide, and Rodenticide Act (U.S.)
GABA	γ–aminobutyric acid

GBMBS	Green Bay Mass Balance Study
GLNPO	Great Lakes National Program Office
GnRH	Gonadotropin-releasing hormone
GRIF	Gonadotropin release-inhibiting factor
GTH	Gonadotropin, gonadotropic hormone
HA	Active hydrogen
HAP	Hazardous air pollutant
HCB	Hexachlorobenzene
HCH	Hexachlorocyclohexane
HGMP	Herring Gull Monitoring Program
HLC	Henry's Law constant
HPG	Hypothalamus-pituitary-gonad
HPLC	High pressure liquid chromatography
HRS	Hazard ranking system
HxCB	Hexachlorobiphenyl
IAHS	International Association of Hydrological Sciences
IJC	International Joint Commission
IRLG	Interagency Regulatory Liaison Group (U.S.) (disbanded ~1980)
IRPTC	International Registry of Potentially Toxic Chemicals
ITC	Interagency Testing Committee
IUPAC	International Union of Physicists and Chemists
IUR	Inventory use review
K_{OW}	Octanol/water partition coefficient
K_{OC}	Organic carbon/water partition coefficient
LC50	Lethal concentration to 50% of a population
LD50	Lethal dose to 50% of a population
LH	Luteinizing hormone
LOAEL	Lowest-observed-adverse-effects level
MACT	Maximum available control technology
MCL	Maximum contaminant level

MCLG	Maximum contaminant level goal
MEI	Maximum exposed individual
MFO	Mixed-function oxidase or oxygenase; multifunction oxygenase
MW	Molecular weight
NADPH	Adenosine triphosphate
NAS	National Academy of Sciences (U.S.)
NBS	National Biological Service (U.S.)
NCASI	National Council of the Paper Industry for Air and Stream Improvement (U.S.)
NCBP	National Contaminant Biomonitoring Program (U.S.)
NCI	National Cancer Institute (U.S.)
NOAA	National Oceanic and Atmospheric Administration (U.S.)
NOAEC	No-observed-adverse-effects concentration
NOAEL	No-observed-adverse-effects-level
NPL	National Priority List (U.S.)
NRC	National Research Council (U.S.)
NTP	National Toxicology Program
OC	Organochlorine
OCDD	Octachloro-dibenzo-dioxin
OECD	Organization for Economic Cooperation and Development
OH-PCB	Hydroxylated PCB
o′,p′-DDT	*Ortho-para′*-DDT
OPPE	Office of Policy, Planning, and Evaluation (USEPA)
OSHA	Occupational Safety and Health Administration (U.S.)
OSTP	Office of Science and Technology Policy (U.S.)
PAH	Polycyclic aromatic hydrocarbon/Polynuclear aromatic hydrocarbon
PBTK	Physiologically based toxicokinetic model
PCA	Polychlorinated aniline
PCB	Polychlorinated biphenyl
PCBT	Polychlorinated benzyltoluene
PCBz	Polychlorinated benzene
PCDD	Polychlorinated dibenzo-*p*-dioxin

PCDE	Polychlorinated diphenyl ether
PCDF	Polychlorinated dibenzofuran
PCDH	Polychlorinated diaromatic hydrocarbon
PCDT	Polychlorinated diphenyltoluene
PCN	Polychlorinated naphthalene
PCP	Pentachlorophenol/Polychlorinated phenol
PCPP	Polychlorinated phenoxyphenol
PCT	Polychlorinated terphenyl
PDC	Particulate detrital carbon
PHOXA	Photochemical Oxidant Acid Deposition Model
pK_a	Ionization constant; degree of ionization
PMN	Premanufacturing notice
POC	Particulate organic carbon
POX	Purgeable organic halogen
ppb	Parts per billion
ppm	Parts per million
PRZM	Pesticide Root Zone Model
PTFE	Polytetrafluoroethylene
PVC	Polyvinylchloride
QALE	Quality-adjusted life expectancy
QSAR	Quantitative structure-activity relationship
R	Natural gas constant
RCRA	Resource Conservation and Recovery Act (U.S.)
RI/FS	Remedial investigation/feasibility study
RPR	Risk policy report
SAR	Structure-activity relationship
SARA	Superfund Amendments and Reauthorization Act (U.S.)
SWDA	Safe Drinking Water and Toxic Enforcement Act (U.S.)
SWRRB	Simulator for Water Resources in Rural Basins
T	Temperature
TCDD	Tetrachlorodibenzo-*p*-dioxin
TEF	Toxic equivalency factor

TEQ	Toxic equivalent
TIE	Toxicity identification and evaluation
ToCl	Total organochlorine compound
TRI	Toxic Release Inventory
TSCA	Toxic Substances Control Act (U.S.)
TSMP	Toxic Substances Management Policy (Canada)
TSS	Total suspended solid
TTR	Trans-thyretin carrier protein
UN-ECE	United Nations Economic Commission for Europe
UNEP	United Nations Environmental Program
USDI	U.S. Department of the Interior
USDOE	U.S. Department of Energy
USEPA	U.S. Environmental Protection Agency
USFWS	U.S. Fish and Wildlife Service
VCM	Vinyl chloride monomer
WHO	World Health Organization

References

Abernethy SG, Mackay D, McCarty LS. 1988. "Volume fraction" correlation for narcosis in aquatic organisms: the key role of partitioning. *Environ Toxicol Chem* 7:469–481.

Addison R, Brody PF. 1987. Transfer of organochlorine residues from blubber through the circulatory system to milk in the lactating grey seal *Halichoerus grypus*. *Can J Aquat Sci* 44:782–786.

Addison RF, Hansen PD, Pluta HJ, Willis DE. 1991. Effects of Ugilec-141, PCB substitute based on tetrachlorobenzyltoluenes, on hepatic mono-oxygenase induction in estuarine fish. *Mar Environ Res* 31:137–144.

Addison RF, Smith TG. 1974. Organochlorine residue levels in Arctic ringed seals: variation with age and sex. *Oikos* 25:335–337.

Ahlborg UG, Becking GC, Birnbaum LS, Brouwer A, Derks HJG, Feeley M, Golor G, Hanberg A, Larsen JC, Liem AKD, Safe SH, Schlatter C, Waern F, Younes M, Yrjänheikki E. 1994. Toxic equivalence factors for dioxin-like PCBs. Report on a WHO-ECEH and IPCS consultation, December 1993. *Chemosphere* 28:1049–1067.

Ahlborg UG, Brouwer A, Fingerhut MA, Jacobson JL, Jacobson SW, Kennedy SW, Kettrup AA, Koeman JH, Poiger H, Rappe C, Safe SH, Segal RF, Tuomisto I, van den Berg M. 1992. Impact of polychlorinated dibenzo-*p*-dioxins, dibenzofurans, and biphenyls on human and environmental health with special emphasis on application of the toxic equivalency factor concept. *Eur J Pharmacol* 228:179–199.

Ahokas JT, Holdway DA, Brennan SE, Goudey RW, Bibrowska HB. 1994. MFO activity in carp (*Cyprinus carpio*) exposed to treated pulp and paper mill effluent in Lake Coleman, Victoria, Australia, in relation to AOX, EOX and muscle PCDD/PCDF. *Environ Toxicol Chem* 13:41–50.

Albers PH. 1991. Oil spills and the environment: a review of chemical fate and biological effects of petroleum. In: White J, Frink L, Williams TM, Davis RW, editors. The effects of oil on wildlife. Hanover PA: Sheridan Pr. p 1–12.

Andersen ME, Clewell HJ, Gargas ML, Smith FA, Reitz RH. 1987. Physiologically based pharmacokinetics and risk assessment process for methylene chloride. *Toxicol Appl Pharmacol* 87:185–205.

Anderson DP, Zeeman M. 1995. Immunotoxicology in fish. In: Rand G, editor. Fundamentals of aquatic toxicology: effects, environmental fate, and risk assessment. Washington DC: Taylor and Francis. p 371–404.

Ankley GT, Cook PM, Carlson AR, Call DJ, Swenson JA, Corcoran HF, Hoke RA. 1992. Bioaccumulation of PCBs from sediments by oligochaetes and fishes: comparison of laboratory and field studies. *Can J Fish Aquat Sci* 49:2080–2085.

Anonymous. 1984. The correlation between the recovery rate of neurotoxic esterase activity and sensitivity to organophosphorus-induced delayed neurotoxicity. *Toxicol Appl Pharmacol* 75:350–357.

Anonymous. 1991. Toxic chemicals in the Great Lakes and associated effects (synopsis). Ottawa ON: Environment Canada, [Canadian] Department of Fisheries and Oceans, Health and Welfare Canada.

Aqua/Info. 1994. PC version of Aquire database. McLean VA: AScI Corporation.

Arkoosh MR, Casillas E, Clemons E, McCain BB, Varanasi U. 1991. Suppression of immunological memory in juvenile Chinook salmon (*Oncorhynchus tshawytscha*) from an urban estuary. *Fish Shellfish Immunol* 1:261–277.

Arkoosh MR, Clemons E, Myers M, Casillas E. 1994. Suppression of B-cell mediated immunity in juvenile Chinook salmon (*Oncorhynchus tshawytscha*) after exposure to either a polycyclic aromatic hydrocarbon or to polychlorinated biphenyls. *Immunopharmacol Immunotoxicol* 16:293–314.

Armitage PR, Doll R. 1957. A two-stage theory of carcinogenesis in relation to the age distribution of human cancer. *Br J Cancer* 11:161–169.

ASci. 1994. Aqua/Info PC version of Aquire database. McLean VA: ASci Corporation.

Asplund G, Grimvall A. 1991. Organohalogens in nature: more widespread than previously assumed. *Environ Sci Technol* 25:1346–1350.

Asplund G, Grimvall A, Jonsson S. 1994. Determination of the total and leachable amounts of organohalogens in soil. *Chemosphere* 28:1467–1475.

Atkinson R. 1985. Kinetics and mechanisms of the gas-phase reactions of hydroxyl radical with organic compounds under atmospheric conditions. *Chemical Rev* 85:69–201.

Atkinson R. 1989. Kinetics and mechanisms of the gas-phase reactions of the hydroxyl radical with organic compounds. *J Phys Chem Ref Data* Monograph No. 1.

Aulerich RJ, Ringer RK. 1977. Current status of PCB toxicity to mink and effects on their reproduction. *Arch Environ Contam Toxicol* 6:279–292.

Aulerich RJ, Ringer RK, Safronoff J. 1986. Assessment of primary vs. secondary toxicity of Aroclor 1254 to mink. *Arch Environ Contam Toxicol* 15:393–399.

Baart AC, Diederen IHSMA. 1991. Calculation of the atmospheric deposition of 29 contaminants to the Rhine catchment area. Delft NL: TNO Institute of Environmental Sciences. TNO Report R91/219.

Bacci E, Calamari D, Gaggi C, Vight M. 1990. Bioconcentration of organic chemical vapors in plant leaves: experimental measurements and concentration. *Environ Sci Technol* 24:885–889.

Bachmann A, Wallet P, Wijnen P, de Bruin W, Huntjens JLM, Roelofsen W, Zehnder AJB. 1988. Biodegradation of alpha- and beta-hexachlorocyclohexane in a soil slurry under different redox conditions. *Appl Environ Microbiol* 54:143–149.

Baldwin WS, LeBlanc GA. 1994. Identification of multiple steroid hydroxylases in Daphnia magna and their modulation by xenobiotics. *Environ Toxicol Chem* 13:1013–1021.

Baldwin WS, Millam DL, LeBlanc GA. 1995. Physiological and biochemical perturbations in Daphnia magna following exposure to the model environmental estrogen diethylstilbestrol. *Environ Toxicol Chem* 14:946–952.

Ballschmiter K. 1992. Transport and fate of organic compounds in the global environment. *Angewante Chemie* (international edition in English) 31:487–515.

Baloer J. 1994. The gender benders. Are environmental hormones emasculating wildlife? *Science News* 145:24–27.

Banerjee S, Howard PH, Rosenberg AM, Dombrowski AE, Sikka H, Tullis DL. 1984. Development of a general kinetic model for biodegradation and its application to chlorophenols and related compounds. *Environ Sci Technol* 18:416–422.

Barber MC, Suárez LA, Lassiter RR. 1991. Modelling bioaccumulation of organic pollutants in fish with an application to PCBs in Lake Ontario salmonids. *Can J Fish Aquat Sci* 48:318–337.

Barrie LA, Gregor D, Hargrave B, Lake R, Muir D, Shearer R, Tracey B, Bidleman T. 1992. Arctic contaminants: sources, occurrence and pathways. *Sci Tot Environ* 122:1–74.

Barrio-Lage G, Parsons FZ, Nassar RS, Lorenzo PA. 1986. Sequential dehalogenation of chlorinated ethenes. *Environ Sci Technol* 20:96–99.

Battersby NS. 1990. A review of biodegradation kinetics in the aquatic environment. *Chemosphere* 21:1243–1284.

Baughman GL, Paris DF, Steen WC. 1980. Quantitative expression of biotransformation rate. In: Maki AW, Dickson KL, Cairns J, editors. Biotransformation and fate of chemicals in the aquatic environment. Washington DC: American Soc for Microbiology. p 67–86.

Bauman PC, Harshbarger JC, Hartman KJ. 1990. Relationship between liver tumors and age in brown bullhead populations from two Lake Erie tributaries. *Sci Tot Environ* 94:71–87.

Bauman PC, Whittle DM. 1988. The status of selected organics in the Laurentian Great Lakes: an overview of DDT, PCBs, dioxins, furans, and aromatic hydrocarbons. *Aquat Toxicol* 11:241–257.

Baxter RM, Carey JH. 1982. Reactions of singlet oxygen in humic waters. *Freshwater Biol* 12:285–292.

Baxter RM, Carey JH. 1983. Evidence for the photochemical generation of superoxide in humic waters. *Nature* 306:575–576.

Beaver DL. 1980. Recovery of an American robin population after earlier DDT use. *J Field Ornithol* 51:220–228.

Becker DM, Sieg CH. 1987. Eggshell quality and organochlorine residues in eggs of merlins, *Falco columbarius*, in Southeastern Montana. *Can Field Naturalist* 101:369–72.

Becker M, Phillips T, Safe S. 1992. Polychlorinated diphenyl ethers—a review. *Toxicol Environ Chem* 33:189–200.

Bednarz JC. 1990. Migration counts of raptors at Hawk Mountain Pennsylvania USA as indicators of population trends 1934–1986. *Auk* 107:96–109.

Bellrose FC. 1959. Lead poisoning as a mortality factor in waterfowl populations. *Illinois Natural History Survey Bull* 27:235–288.

Bellward GD, Norstrom RJ, Whitehead PE, Elliott JE, Bandiera SM, Dworschak C, Chang T, Forbes S, Cadario B, Hart LE. 1990. Comparison of polychlorinated dibenzodioxin levels with hepatic mixed-function oxidase induction in great blue herons. *J Toxicol Environ Health* 30:33–52.

Bennett E, Mercier M. 1987. Preface, In: Berlin A, Dean J, Draper MH, Smith EMB, Spreafico F, editors. Immunotoxicology. Boston MA: Matinus Nijhoff. p IX.

Bergman A, Olsson M. 1986. Pathology of Baltic grey seal and ringed seal females with special reference to adrenocortical hyperplasia: is environmental pollution the cause of a widely distributed disease syndrome. *Finnish Game Res* 44:47–62.

Bergman A, Olsson M. 1989. Pathology of Baltic grey and ringed seal males. Report regarding animals sampled 1977–1985. In: Yablokov AV, Olsson M, editors. Influence of human activities on the Baltic ecosystem. Proceedings of the Soviet-Swedish Symposium: Effect of Toxic Substances on Dynamics of Seal Populations; 1986 Apr 14–18; Moscow, USSR. Leningrad: Gidrovmeteoizdat.

Bergman HL, Kimerle RA, Maki AW, editors. 1986. Environmental hazard assessment of effluents. New York: Pergamon.

Bernier J, Brousseau P, Tryphonas H, Krzystyniak K, Fournier M. 1995. Immunotoxicity of selected heavy metals pertinent to Great Lakes contamination. *Environ Health Perspect* 103:23–34.

Beurskens JEM, Dekker CGC, Jonkhoff J, Pompstra L. 1993. Microbial dechlorination of hexachlorobenzene in a sedimentation area of the Rhine River. *Biogeochemistry* 19:61–83.

Beurskens JEM, Dekker CGC, van den Heuvel H, Swart M, de Wolf J, Dolfing J. 1994. Dechlorination of chlorinated benzenes by an anaerobic microbial consortium that selectively mediates the thermodynamic most favourable reactions. *Environ Sci Technol* 28:701–706.

Beurskens JEM, Mol GAJ, Barreveld HL, van Munster B, Winkels HJ. 1993. Geochronology of priority pollutants in a sedimentation area of the Rhine River. *Environ Toxicol Chem* 12:1549–1566.

Bidleman TF. 1988. Atmospheric processes: wet and dry deposition of organic compounds are controlled by their vapor/particle partitioning. *Environ Sci Technol* 22:361–367.

Bidleman TF, Jantunen LM, Falconer RL, Barrie LA, Fellin P. 1995. Decline of hexachlorocyclohexane in the arctic atmosphere and reversal of air-sea gas exchange. *Geophys Res Letters* 22:219–222.

Bidleman TF, McConnell LL. 1995. A review of field experiments to determine air-water gas exchange of persistent organic pollutants. *Sci Tot Environ* 159:101–108.

Bidleman TF, Walla MD, Roura R, Carr E, Schmidt S. 1993. Organochlorine pesticides in the atmosphere of the Southern Ocean and Antarctica. *Mar Pollut Bull* 26:258–262.

Bierman VJ, DePinto JV, Young TC, Rodgers PW, Martin SC, Raghunathan RK. 1992. Development and validation of an integrated exposure model for toxic chemicals in Green Bay, Lake Michigan. Final

Report for Cooperative Agreement CR-814885. Grosse Ile MI: ERL-Duluth, Large Lakes and Rivers Research Branch. 350 p.

Bignert A, Litzén K, Odsjö T, Olsson M, Persson W, Reutergårdh W, Reutergårdh L. 1994. Time-related factors influence the concentrations of DDT, PCBs and shell parameters in eggs of Baltic Guillemot (*Uria aalge*), 1861–1989. *Environ Pollut* 89:26–36.

Bignert A, Olsson M, de Wit C, Litzen K, Rappe C, Reutergardh L. 1994. Biological variation—an important factor to consider in ecotoxicological studies based on environmental samples. *Fresenius J Analytical Chem* 348:76–85.

Binns W, Thacker EJ, James LF, Huffman WT. 1959. A congenital cyclopian-type malformation in lambs. *J Am Veterinary Med Assn* 134:180–183.

Birnbaum LS. 1994. The mechanism of dioxin toxicity: relationship to risk assessment. *Environ Health Perspect* 102:157–167.

Bjeldanes LF, Kim J-Y, Grose KR, Bartholomew JC, Bradfield CA. 1991. Ah-receptor agonists generated from indole-3-carbinol in vitro and in vivo: comparisons with 2,3,7,8-tetrachlorodibenzo-*p*-dioxin. *Proc National Academy Sci USA* 89:8185–8189.

Bjerke DL, Peterson RE. 1994. Reproductive toxicity of 2,3,7,8-tetrachlorobenzo-*p*-dioxin in male rats: different effects of in utero versus lactational exposure. *Toxicol Appl Pharmacol* 127:241–249.

Black MC, McCarthy JF. 1988. Dissolved organic macromolecules reduce the uptake of hydrophobic organic contaminants by the gills of rainbow trout (*Salmo gairdneri*). *Environ Toxicol Chem* 7:593–600.

Blectchy JD. 1983. Polychlorinated biphenyls: production, current use and possible rates of future disposal in OECD member countries. In: Barros MC, Konemann H, Visser R, editors. Proceedings of the PCB Seminar. The Hague NL: Netherlands Ministry of the Environment. p 343–365.

Blus LJ, Henny CJ, Lehart DJ, Cromartie E. 1979. Effects of heptachlor-treated cereal grains on Canada geese in the Columbia basin. In: Jarvis RL, Bartonek JC, editors. Management and biology of Pacific flyway geese: a symposium. Corvallis OR: OSU Book Stores. p 105–116.

Blus LJ, Neely Jr BS, Belisle AA, Prouty RM. 1974. Organochlorine residues in brown pelican eggs: relation to reproductive success. *Environ Pollut* 7:81–91.

Boethling RS, Howard PH, Meylan W, Stiteler W, Beauman J, Tirado N. 1994. Group contribution method for predicting probability and rate of aerobic biodegradation. *Environ Sci Technol* 28:459–465.

Bookstaff RC, Kamel F, Moore RW, Bjerke DL, Peterson RE. 1990. Altered regulation of pituitary gonadotropin-releasing hormone (GnRH) receptor number and pituitary responsiveness to GnRH in 2,3,7,8-tetrachlorodibenzo-*p*-dioxin-treated male rats. *Toxicol Appl Pharmacol* 105:78–92.

Bookstaff RC, Moore RW, Peterson RE. 1990. 2,3,7,8-tetrachlorodibenzo-*p*-dioxin increases the potency of androgens and estrogens as feedback inhibitors of luteinizing hormone secretion in male rats. *Toxicol Appl Pharmacol* 104:212–224.

Boon JP, Oostingh I, van der Meer J, Hillebrand MTJ. 1994. A model for the bioaccumulation of chlorobiphenyl congeners in marine mammals. *Europ J Pharmacol Environ Toxicol Pharmacol* 270:237–251.

Borg K, Wanntorp H, Erne K, Hanko E. 1969. Alkyl mercury poisoning in terrestrial Swedish wildlife. *Viltrevy* 6:301–379.

Borlakoglu JT, Wilkins JPG, Walker CH, Dils RR. 1990a. Polychlorinated biphenyls (PCBs) in fish-eating sea birds-I. Molecular features of PCB isomers and congeners in adipose tissue of male and female razorbills (*Alca tarda*) of British and Irish coastal waters. *Comparative Biochem Physiol* 97C:151–160.

Borlakoglu JT, Wilkins JPG, Walker CH, Dils RR. 1990b. Polychlorinated biphenyls (PCBs) in fish-eating sea birds-II. Molecular features of PCB isomers and congeners in adipose tissue of male and female puffins (*Fratercula arctica*), guillemots (*Uria aalga*), shags (*Phalacrocorax aristotelis*) and cormorants (*Phalacrocorax carbo*) of British and Irish coastal waters. *Comparative Biochem Physiol* 97C:161–171.

Borlakoglu JT, Wilkins JPG, Walker CH, Dils RR. 1990c. Polychlorinated biphenyls (PCBs) in fish eating sea birds-III. Molecular features and metabolic interpretations of PCB isomers and congeners in adipose tissue. *Comparative Biochem Physiol* 97C:173–177.

Bosveld ATC, Gradener J, Murk AJ, Brouwer A, Van Kampen M, Evers E, van den Berg M. 1995. Effects of PCDDs, PCDFs and PCBs in common tern (*Sterna hirundo*) breeding in estuarine and coastal colonies in The Netherlands and Belgium. *Environ Toxicol Chem* 14:99–106.

Boule P, Guyon C, Tissot A, Lemaire J. 1987. Specific phototransformation of xenobiotic compounds: chlorobenzenes and chlorophenols. In: Zika RG, Cooper WJ, editors. Photochemistry of environmental aquatic systems. Washington DC: American Chemical Soc (ACS). p 10–26.

Bouwman CA, Fase K, Koppe JG, Seinen W, van den Berg M. 1993. Effects of 2,3,7,8-TCDD and 2,2',4,4',5,5'-HxCB on the activity of the vitamin K cycle enzymes in rats: a novel mechanism of action. *Organohalogen Compounds* 10:13–16.

Bouwman CA, Seinen W, Koppe JG, van den Berg M. 1992. Effects of 2,3,7,8-tetrachlorodibenzo-*p*-dioxin or 2,2',4,4',5,5'-hexachlorobiphenyl on vitamin K-dependent blood coagulation in female germ free WAG/Rij-rats. *Toxicology* 72:109–120.

Bowerman WW, Giesy JP, Best DA, Kramer VJ. 1995. A review of factors affecting productivity of bald eagles in the Great Lakes Region: implications for recovery. *Environ Health Perspect* 103(Suppl 4):51–59.

Bowes GW, Jonkel CJ. 1975. Presence of polychlorinated biphenyls (PCB) in arctic and subarctic marine food chains. *J Fisheries Res Board Can* 32:2111–2123.

Bowes GW, Simoneit BR, Burlingame AL, de Lappe BW, Risebrough RW. 1973. The search for chlorinated dibenzofurans and chlorinated dibenzodioxins in wildlife populations showing elevated levels of embryonic death. *Environ Health Perspect* 5:191–198.

Boyle M. 1989. The environmental microbiology of chlorinated aromatic decomposition. *J Environ Quality* 18:395–402.

Brandt I, Lonsson C-J, Lund BO. 1992. Comparative studies on adrenocorticolytic DDT-metabolites. *Ambio* 21:602–605.

Braune BN, Norstrom RJ. 1989. Dynamics of organochlorine compounds in herring gulls: III. Tissue distribution and bioaccumulation in Lake Ontario gulls. *Environ Toxicol Chem* 8:957–968.

Briggs GG, Elliott M, Farnham WW, Janes NF, Needham PH, Pulman DA, Young SR. 1976. Insecticidal activity of the pyrethrins and related compounds. VIII. Relation of polarity with activity in pyrethroids (*Musca domestica, Phaedon cochleariae*). *Pesticide Sci* 7:236–240.

Brinkman UATh, de Kok A. 1980. Production, properties and usage. In: Kimbrough R, editor. Halogenated biphenyls, terphenyls, naphthalenes, dibenzodioxins and related products. Amsterdam NL: Elsevier Biomedical. p 1–40.

Brouwer A, van den Berg KJ. 1986. Binding of a metabolite of 3,4,3',4'-tetrachlorobiphenyl to transthyretin reduces vitamin A transport by inhibiting the formation of the protein complex, carrying both retinol and thyroxin. *Toxicol Appl Pharmacol* 85:301–312.

Brown JF, Bedard DL, Brennan MJ, Carnahan JC, Feng H, Wagner RE. 1987. Polychlorinated biphenyl dechlorination in aquatic sediments. *Science* 236:709–712.

Brown JF, Wagner RE, Bedard DL, Brennan MJ, Carnahan JC, May RJ, Tofflemire TJ. 1984. PCB transformations in upper Hudson River sediments. *Northeastern Environ Sci* 3:167–179.

Brown JF, Wagner RE, Feng H, Bedard DL, Brennan MJ, Carnahan JC, May RJ. 1987. Environmental dechlorination of PCBs. *Environ Toxicol Chem* 6:579–593.

Brownlee BG, Carey JH, MacInnis GA, Pellizzari IT. 1992. Aquatic environmental chemistry of 2-(thiocyanomethyl)-berlzothiazole and related benzothiazoles. *Environ Toxicol Chem* 11:1153–1168.

Brunström B. 1988. Sensitivity of embryos from duck, goose, herring gull, and various chicken breeds to 3,3', 4,4'-tetrachlorobiphenyl. *Poultry Sci* 67:52–57.

Brunström B, Håkansson H, Lundberg K. 1991. Effects of technical PCB and fractions thereof on ethoxyresurofin-O-deethylase activity, vitamin A levels and lymphoid development in the mink (*Mustela vison*). *Pharmacol Toxicol* 69:421–426.

Bryant CW, Amy GL, Neil R, Ahmad S. 1988. Partitioning of organic chlorine between bulk water and benthal interstitial water through a draft mill aerated lagoon. *Water Sci Technol* 20:73–79.

Bulger WH, Kupfer O. 1985. Estrogenic activity of pesticides and other xenobiotics on the uterus and male reproductive tract. In: Thomas J, Korach KS, McLachlan JA, editors. Endocrine toxicology. New York: Raven Pr. p 1–32.

Bunce NJ, Kumar Y, Brownlee BG. 1978. An assessment of the impact of solar degradation of polychlorinated biphenyls in the aquatic environment. *Chemosphere* 7:155–164.

Bunce NJ, Landers JP, Langshaw J-A, Nakai JS. 1989. An assessment of the importance of direct solar degradation of some simple chlorinated benzenes and biphenyls in the vapour phase. *Environ Sci Technol* 23:213–218.

Bunce NJ, Nakai JS, Yawching M. 1991. A model for estimating the rate of chemical transformation of a VOC in the troposphere by two pathways: photolysis by sunlight and hydroxyl radical attack. *Chemosphere* 22:305–315.

Burnison BK, Hodson PV, Nuttley D, Efler S. 1994. Isolation and characterization of a MFO-inducing substance. Presented at the 29th Central Canadian Symposium on Water Pollution Research, Canada Centre for Inland Waters; 1994 Feb 9–10; Burlington ON.

Buser HR, Bosshardt H-P, Rappe C. 1978a. Identification of polychlorinated dibenzo-*p*-dioxin isomers found in fly ash. *Chemosphere* 7:165–172.

Buser HR, Bosshardt H-P, Rappe C. 1978b. Identification of polychlorinated dibenzofuran isomers in fly ash and PCB pyrolysis. *Chemosphere* 7:419–429.

Buser HR, Muller MD. 1986. Methylthio metabolites of polychlorobiphenyls identified in sediment samples from two lakes in Switzerland. *Environ Sci Technol* 20:730–735.

[CAAA] Clean Air Act Amendments. 1990. Public Law 101-549, 104 Stat. 2399, November 15, 1990.

Cairns Jr J, Dickson KL, Maki AW, editors. 1978. Estimating the hazard of chemical substances to aquatic life. Philadelphia PA: American Soc for Testing and Materials (ASTM). Tech Publ 657. 278 p.

Calamari D, Bacci E, Focardi S, Gaggi C, Morosini M, Vighi M. 1991. Role of plant biomass in the global environmental partitioning of chlorinated hydrocarbons. *Environ Sci Technol* 25:1489–1495.

Calamari D, Tremolada P, DiGuardo A, Vighi M. 1994. Chlorinated hydrocarbons in pine needles in Europe: fingerprint for the past and recent use. *Environ Sci Technol* 28:429–434.

Carey JH. 1994. Transformation processes of contaminants in rivers. Hydrological, chemical and biological processes of transformation and transport of contaminants in aquatic environments. Proceedings of the Rostov-on-Don Symposium; 1993 May. Rostov, Russia: International Association of Hydrological Sciences (IAHS) Publ Nr 219. p 41–50.

Carlberg GE, Martinsen K, Kringstad A, Gjessing E, Grande M, Källqvist T, Skåre JU. 1986. Influence of aquatic humus on the bioavailability of chlorinated micropollutants in Atlantic salmon. *Arch Environ Contam Toxicol* 15:543–548.

Caron RJ, Reeve DW. 1991. Environmental photolysis of chlorinated organic matter discharged in kraft pulp bleaching effluents. Preprint from author of paper presented at the 1991 Pulp and Paper Industry Environment Conference; 1991Oct 8–10; Quebec ON.

Carsel RF, Smith CN, Mulkey LA, Dean JD, Jowise P. 1984. User's manual for the pesticide root zone model (PRZM). Release 1. Athens GA: U.S. Environmental Protection Agency (USEPA). EPA-600/3-84-109. 216 p.

Carson R. 1962. Silent spring. New York: Houghton Mifflin.

Casida JE, Maddrell SPH. 1971. Diuretic hormone release on poisoning Rhodnius with insecticide chemicals. *Pesticide Biochem Physiol* 1:71–83.

Casillas E, Misitano DA, Johnson LL, Rhodes LD, Collier TK, Stein JE, McCain BB, Varanasi U. 1991. Inducibility of spawning and reproductive success of female English sole (*Parophrys vetulus*) from urban and nonurban areas of Puget Sound, Washington. *Mar Environ Res* 31:99–122.

Castro CE. 1993. Biodehalogenation: the kinetics and rates of microbial cleavage of carbon-halogen bonds. *Environ Toxicol Chem* 12:1609–1618.

[CCREM] Canadian Council of Resource and Environment Ministries. 1987. Canadian water quality guidelines, task force on water quality guidelines of the Canadian Council of Resource and Environment Ministers. Ottawa: Water Quality Branch, Inland Waters Directorate, Environment Canada.

[CEAA] Canadian Environmental Assessment Act. 1992. Third Session, Thirty-fourth Parliament of Canada, 40-41 Elizabeth II, 1991–1992. p 1–54.

[CEPA] Canadian Environmental Protection Act. 1990. Priority Substances Report No. 1: Polychlorinated Dibenzodioxins and Polychlorinated Dibenzofurans. Ottawa ON: CEPA, Ministry of Supply and Services. En40-215/1E.

[CEPA] Canadian Environmental Protection Act. 1997. Environmental assessments of priority substances under the Canadian Environmental Protection Act. Guidance Manual Version 1.0, March 1997. Ottawa ON: Chemical Evaluation Division, Commercial Chemicals Evaluation Branch, Environment Canada. EPS/2/CC/3E.

Chapdelaine G, LaPorte P, Nettleship DN. 1987. Population, productivity and DDT contamination trends of Northern gannets (*Sula bassarus*) at Bonaventure Island, Quebec, 1967–1984. *Can J Zool* 65:2922–2926.

Cheung MO, Gilbert EF, Peterson RE. 1981. Cardiovascular teratogenicity of 2,3,7,8-tetrachlorodibenzo-*p*-dioxin in the chick embryo. *Toxicol Appl Pharmacol* 61:197–204.

Chu I, Villeneuve DC, Yagminas A, LeCavalier P, Håkansson H, Ahlborg UG, Valli VE, Kennedy SW, Bergman Å, Seegal RF, Feeley M. 1995. Toxicity of PCB 77 (3,3,4,4-tetrachlorobiphenyl) and PCB 118 (2,3,4,4,5-pentachlorobiphenyl) in the rat following subchronic dietary exposure. *Fundamental Appl Toxicol* 26:282–292.

Chu I, Villeneuve DC, Yagminas A, LeCavalier P, Poon R, Feeley M, Kennedy SW, Seegal RF, Håkansson H, Ahlborg UG, Valli VE. 1994. Subchronic toxicity of 3,3',4,4',5-pentachlorobiphenyl in the rat. Part I: Clinical, biochemical, hematological and histopathological changes. *Toxicol Appl Pharmacol* 22:457–468.

Clark CM, Bunck E, Cromartie E, LaVal RK. 1983. Year and age effects of residues of dieldrin and heptachlor in dead gray bats, Franklin County, Missouri - 1976, 1977, and 1978. *Environ Toxicol Chem* 2:387–393.

Clark P, Ogasawara A, Smith GJ, Ohlendorf HM. 1989. Selenium accumulation by raccoons exposed to irrigation drainwater at Kesterson National Wildlife Refuge, California, 1986. *Arch Environ Contam Toxicol* 18:787–794.

Clark TP, Norstrom RJ, Fox GA, Won HY. 1987. Dynamics of organochlorine compounds in herring gulls (*Larus argentatus*): II. A two-compartment model and data for ten compounds. *Environ Toxicol Chem* 6:547–559.

Coats JR, Metcalf RL, Kapoor IP. 1977. Effective DDT analogues with altered aliphatic moieties. Isobutanes and chloropropanes. *J Agric Food Chem* 25:859–868.

Coburn DR, Treichler R. 1946. Experiments on toxicity of DDT to wildlife. *J Wildlife Management* 10:208–216.

Connell DW. 1988. Bioaccumulation behaviour of persistent organic chemicals with aquatic organisms. *Rev Environ Contam Toxicol* 101:117–154.

Connell DW, Shüürmann G. 1988. Evaluation of various molecular parameters as predictors of bioconcentration in fish. *Ecotoxicol Environ Safety* 15:324–335.

Connolly JP. 1991. Application of a food chain model to PCB contamination of the lobster and winter flounder food chains in New Bedford harbour. *Environ Sci Technol* 25:760–769.

Connolly JP, Parkerton T. 1994. Development and application of a model of PCBs in Green Bay walleye and brown trout. Final Technical Report to USEPA, Large Lakes and Rivers Research Branch, Grosse Isle, MI.

Connolly JP, Pedersen CJ. 1988. A thermodynamic based evaluation of organic chemical accumulation in aquatic organisms. *Environ Sci Technol* 22:99–103.

Connolly JP, Tonelli R. 1985. Modelling kepone in the Striped Bass food chain of the James River estuary. Estuarine and Coastal Shelf Science 20:349–366.

Connolly RB, Anderson ME. 1991. Biologically based pharmacodynamic models: tools for toxicological research and risk assessment. *Annu Rev Pharmacol Toxicol* 31:503–523.

Cook PM. 1994. Risk assessment for dioxins, furans and PCBs in Lake Ontario sediments to lake trout reproduction. Paper presented at the SETAC Workshop on Environmental Eisk Assessment for Organochlorine Chemicals; 1994 Jul 24–29; Alliston ON.

Cooper WJ, Lean DRS, Carey JH. 1989. Spatial and temporal patterns of hydrogen peroxide in lakes. *Can J Fish Aquat Sci* 46:1227–1231.

Cooper WJ, Zika RG, Petasne RG, Fischer AM. 1989. Sunlight-induced photochemistry of humic substances in natural waters: major reactive species. In: Suffet IH, MacCarthy P, editors. Aquatic humic substances: influence on fate and treatment of pollutants. ACS Advances in Chemistry Series 219. Washington DC: American Chemical Soc (ACS). p 333–362.

Cooper WJ, Zika RG, Petasne RG, Plaue JMC. 1988. The photochemical formation of hydrogen peroxide in lakes. *Environ Sci Technol* 22:1156–1160.

Corely RA, Mendrala AL, Smith FA, Staats DA, Gargas ML, Connolly RB, Anderson ME, Reitz RH. 1990. Development of a physiologically based pharmacokinetic model for chloroform. *Toxicol Appl Pharmacol* 103:512–527.

Cotham WE, Bidleman TF. 1991. Estimating the atmospheric deposition of organic contaminants to the Arctic. *Chemosphere* 22:165–188.

Cottam C, Higgins E. 1946. DDT: its effect on fish and wildlife. U.S. Fish and Wildlife Service Circular 11. 14 p.

Couch LK. 1946. Effects of DDT on wildlife in a Mississippi River bottom woodland. *Audubon Bull* 58:1–7.

Couture LA, Abbott BD, Birnbaum LS. 1990. A critical review of the developmental toxicity and teratogenicity of 2,3,7,8-tetrachlorodibenzo-*p*-dioxin: recent advances toward understanding the mechanism. *Teratology* 42:619–627.

Crossland NO, Mitchell GC, Dorn PB. 1992. Use of outdoor artificial streams to determine threshold toxicity concentrations for a petrochemical effluent. *Environ Toxicol Chem* 11:49–59.

Crossland NO, Wolff CJM. 1985. Fate and biological effects of pentachlorophenol in outdoor ponds. *Environ Toxicol Chem* 4:73–86.

Crump KS, Guess HA, Deal LL. 1977. Confidence intervals and test of hypothesis concerning dose-response relations inferred from animal carcinogenicity data. *Biometrics* 33:437–451.

Crump KS, Watson WW. 1979. GLOBAL 79: a FORTRAN program to extrapolate dichotomous animal carcinogenicity data to low dose.Rushton LA: National Institute of Environmental Health Sciences. Contract I-ES-2123.

Cullen MC, Connell DW. 1994. Pesticide bioaccumulation in cattle. *Ecotoxicol Environ Safety* 28:221–231.

Cunningham KM, Goldberg MC, Weiner ER. 1988. Mechanisms for aqueous photolysis of adsorbed benzoate, oxalate and succinate on iron oxyhydroxide (goethite) surfaces. *Environ Sci Technol* 22:1090–1097.

Curlin LC, Bommaraju TV. 1993. Alkali and chlorine products. In: Volume 1, Kirk-Othmer encyclopedia of chemical technology. 4th ed. New York: J Wiley. p 938–1025.

Dahlman O, Mrck R, Ljungquist P, Reimann A, Johansson C, Born H, Grimvall A. 1993. Chlorinated structural elements in high-molecular-weight organic matter from unpolluted waters and bleached-kraft mill effluents. *Environ Sci Technol* 27:1616–1620.

Davidson JWF, Sumner DD, Parker JD. 1982. Chloroform: a review of its metabolism, mutagenic and carcinogenic potential. *Drug Chem Toxicol* 5:1–87.

Delzell E, Doull J, Giesy JP, Mackay D, Munro IC, Williams GM. 1994. An interpretive review of chlorinated compounds in the environment. *Regulatory Toxicol Pharmacol* 20: Special Supplement, No. 1. Part 2 of 2 parts.

Demayo A, Taylor M, Hodson PV. 1982. Lead in the aquatic environment. *CRC Critical Rev Environ Control* 12:257–305.

DePinto JV, Raghunathan R, Bierman Jr VJ, Rodgers PW, Young TC, Martin SC. 1993. Analysis of organic carbon sediment-water exchange in Green Bay, Lake Michigan. *Water Sci Technol* 28:149–159.

DePinto JV, Raghunathan RK, Sierzenga PM, Zhang X, Bierman Jr VJ, Rodgers PW, Young TC. 1994. Recalibration of GBTOX: an integrated exposure model for toxic chemicals in Green Bay, Lake Michigan. Grosse Ile MI: ERL-Duluth, Large Lakes and Rivers Research Branch. Technical Report submitted as part of Cooperative Agreement CR-818560. 139 p.

Descotes J. 1988. Immunotoxicity of chemicals. In: Descotes J, editor. Immunotoxicology of drugs and chemicals. Amsterdam NL: Elsevier.

Devine MD, Duke SO, Fedtke C. 1993. *Physiology of Herbicide Action.* Englewood Cliffs NJ: PTR Prentice Hall. p 129.

De Voogt P, Brinkman UA. 1989. Production, properties and usage of polychlorinated biphenyls. In: Kimbrough R, Jensen S, editors. Halogenated biphenyls, terphenyls, naphthalenes, dibenzodioxins and related products. Amsterdam NL: Elsevier. p 3–48.

De Vries MB. 1987. IMPAQT: a mathematical model for simulation of the fate of micropollutants in aquatic systems. Model description. Delft NL: Delft Hydraulics. Report Nr T250.

De Vries MB, Goossens H. 1987. UPTAQE: a mathematical model for simulation of the accumulation of micro-pollutants in aquatic food chains. Model description. Delft NL: Delft Hydraulics. Report Nr T250.

Di Domenico A, Silano V, Viviano G, Zapponi G. 1980. Accidental release of 2,3,7,8-tetrachlorodibenzo-*p*-dioxin (TCDD) at Seveso, Italy. *Ecotoxicol Environ Safety* 4:298–320.

Dickson KL, Maki AW, Cairns Jr J, editors. 1979. Analyzing the hazard evaluation process. Bethesda MD: American Fisheries Soc. 159 p.

DiToro DM, Zarba CS, Hansen DJ, Barrey WJ, Swartz RC, Cowan CE, Pavlou SP, Allen HE, Thomas NA, Paquin PR. 1991. Technical basis for establishing sediment quality criteria for nonionic organic chemicals using equilibrium partitioning. *Environ Toxicol Chem* 10:1541–1583.

Draper WM, Crosby DG. 1981. Hydrogen peroxide and hydroxyl radical: intermediates in indirect photolysis reactions in water. *J Agric Food Chem* 29:699–702.

Draper WM, Crosby DG. 1983a. The photochemical generation of hydrogen peroxide in natural waters. *Arch Environ Contam Toxicol* 12:121–126.

Draper WM, Crosby DG. 1983b. Photochemical generation of superoxide radical anion in water. *J Agric Food Chem* 32:734–737.

Drummond RA, Russom CL. 1990. Behavioural toxicity syndromes: a promising tool for assessing toxicity mechanisms in juvenile fathead minnows. *Environ Toxicol Chem* 9:37–46.

Dung MH, O'Keefe PW. 1994. Comparative rates of photolysis of PCDFs in organic solvents and in aqueous solutions. *Environ Sci Technol* 28:549–554.

Dunier M. 1994. Modulators of fish immune response models for environmental toxicology-biomarkers immunostimultors. In: Stolen JS, Fletcher TC, editors. Modulators of fish immune responses. Fair Haven NY: SOS. p 123–139.

Ehman J, Ballschmiter K. 1989. Isomer specific determination of tetrachlorobenzyltoluenes (TCBT) in the technical mixture Ugilec 141 by capillary gas chromatography. *Fresensius J Anal Chem* 332:904–911.

Eisen HJ, Hannah RR, Legraverend C, Okey AB, Nebert DW. 1983. The Ah receptor: controlling factor in the induction of drug-metabolizing enzymes by certain chemical carcinogens and other environmental pollutants. In: Volume X, Biochemical actions of hormones. Academic Pr. p 227–258.

Eisenreich SJ, Strachan WMJ. 1992. Estimating atmospheric deposition of toxic substances to the Great Lakes. an update. Burlington ON: National Water Research Institute, Environment Canada. 59 p.

Eisler R. 1986. Diazinon hazards to fish, wildlife, and invertebrates: a synoptic review. Washington DC: Biological Report, Contaminant Hazard Reviews. Washington DC: U.S. Dept. of Interior Report No. 9.

Eisler R, Jacknow J. 1985. Toxaphene hazards to fish, wildlife, and invertebrates: a synoptic review. Washington DC: Biological Report, Contaminant Hazard Review. U.S. Dept. of Interior Report No. 4.

Eitzer BD, Hites RA. 1989. Polychlorinated dibenzo-*p*-dioxins and dibenzofurans in the ambient atmosphere of Bloomington, Indiana. *Environ Sci Technol* 23:1389–1395.

Ellington JJ, Stancil Jr FE, Payne WD. 1987a. Measurement of hydrolysis rate constants for evaluation of hazardous waste land disposal. Volume 2, Data on 54 chemicals. Washington DC: U.S. Environmental Protection Agency (USEPA). EPA/600/S3-87/019 (PB 87-227 344/AS).

Ellington JJ, Stancil Jr FE, Payne WD. 1987b. Measurement of hydrolysis rate constants for evaluation of hazardous waste land disposal. Volume 1, Data on 32 chemicals. Washington DC: U.S. Environmental Protection Agency (USEPA). EPA/600/S3-86/043 (PB 87-140 349/AS).

Ellington JJ, Stancil Jr FE, Payne WD. 1988. Measurement of hydrolysis rate constants for evaluation of hazardous waste land disposal. Volume 3, Data on 70 chemicals. Washington DC: U.S. Environmental Protection Agency (USEPA). EPA/600/S3-88/028 (PB 88-234 042/AS).

Elliott M. 1976. Chemistry, biochemistry and insecticidal action of natural and synthetic pyrethroids. *Pesticide Sci* 7:223–224.

Elliott M. 1989. The pyrethroids: early discovery, recent advances and the future. *Pesticide Sci* 27:337–351.

Elliott M, Farnham AW, Janes NF, Needham PH, Pulman DA. 1974. Synthetic insecticide with a new order of activity. *Nature* 248 (5450):710–711.

Endicott DD, Cook PM. 1994. Modelling the partitioning and bioaccumulation of TCDD and other hydrophobic organic chemicals in Lake Ontario. *Chemosphere* 28:75–87.

Endicott DD, Richardson WL, Kandt DJ. 1992. MICHTOX: a mass balance and bioaccumulation model for toxic chemicals in Lake Michigan. Grosse Ile MI: USEPA, ERL-Duluth, Large Lakes Research Station.

Endicott DD, Velleux M, Freeman K. 1992. Development and calibration of a mass balance model for estimating the export of in-place pollutants from the Lower Fox River to Green Bay. Draft Report. Grosse Ile MI: USEPA Large Lakes and Rivers Research Branch, ERL-Duluth.

Enell M, Wennberg L. 1991. Distribution of halogenated organic compounds (AOX): Swedish transport to surrounding sea areas and mass balance studies in five drainage systems. *Water Sci Technol* 24:385–395.

[EU] European Union. 1993a. Council Regulation 67/93 of 7 January 1993 on the evaluation and control of the risks of new substances. Official Journal of the European Communities L 84/1. Brussels, Belgium.

[EU] European Union. 1993b. Council Regulation 793/93 of 23 March 1993 on the evaluation and control of the risks of existing substances. Official Journal of the European Communities L 84/1. Brussels, Belgium.

[EU] European Union. 1993c. Technical guidance documents in support of the risk assessment directive (93/67/EEC) for new substances notified in accordance with the requirements of council directive 67/548/EEC. Brussels, Belgium.

[EU] European Union. 1994. Technical guidance documents in support of the risk assessment of existing chemicals in accordance with the requirements of council directive (93/793/EEC). Draft, March 1994. Brussels, Belgium.

Fabian P. 1986. Halogenated hydrocarbons in the atmosphere. In: Hutzinger O, editor. Volume 4a, Environmental chemistry. Berlin, Germany: Springer Verlag. p 23–52.

Falconer RL, Bidleman TF. 1994. Vapor pressures and predicted particle/gas distributions of PCB congeners as functions of temperature and ortho-chlorine substitution. *Atmospher Environ* 28:547–554.

Faust BC. 1994. Photochemistry of clouds, fogs, and aerosols. *Environ Sci Technol* 28:217A–222A.

Ferguson J. 1939. The use of chemical potentials as indices of toxicity. *Proc Royal Soc London* 127B:387–404.

Fiedler H. 1996. Sources of PCDD/PCDF and impact on the environment. *Chemosphere* 32:55–64.

Fiedler H, Timms CW, Hutzinger O. 1990. Dioxins: sources of environmental load and human exposure. *J Toxicol Environ Chem* 29:157–234.

[FIFRA] Federal Insecticide, Fungicide, and Rodenticide Act. 1947. Ch 125, 61 Stat. 163 (7 U.S.C. § 136 et seq.), June 25, 1947.

Fisher JP, Fitzsimons JD, Coombs Jr GF, Spitsbergen JM. 1996. Naturally occurring thiamine deficiency causing reproductive failure in Finger Lakes Atlantic salmon and Great Lakes lake trout. *Trans Am Fisheries Soc* 125:167–178.

Fisher JP, Iamonte T, Little EE, DeLonay A. 1995. Pathological and behavioural manifestations of the "Cayuga Syndrome," a thiamine deficiency in larval landlocked Atlantic salmon. *J Aquat Animal Health* 7:269–283.

Fisher JP, Spitsbergen JM, Getchell R, Symula J, Skea JE, Babenzein M, Chiotti T. 1995. Reproductive failure in landlocked Atlantic salmon from New York Finger Lakes. I. Investigations into the etiology and epidemiology of "Cayuga Syndrome." *J Aquat Animal Health* 7:81–94.

Fitzsimons JD. 1995. The effect of B-vitamins on a swim-up syndrome in Lake Ontario lake trout. *J Great Lakes Res* 21:286–289

Flickinger EL. 1979. Effects of aldrin exposure on snow geese in Texas rice fields. *J Wildlife Management* 43:994–1001.

Flickinger EL, King KA, Stout WF, Mohn MM. 1980. Wildlife hazards from Furadan 3G applications to rice in Texas. *J Wildlife Management* 44:190–197

Forget G. 1991. Pesticides and the third world. *J Toxicol Environ Health* 32:11–31.

Fox GA. 1991. Practical causal inference for ecoepidemiologists. *J Toxicol Environ Health* 33:359–373.

Fox GA, Collins B, Hayakawa E, Weseloh DV, Ludwig JP, Kubiak TJ, Erdman TC. 1991. Reproductive outcomes in colonial fish-eating birds: a biomarker for developmental toxicants in Great Lakes food chains. II. Spatial variation in the occurrence and prevalence of bill defects in young double-crested cormorants in the Great Lakes, 1979–1987. *J Great Lakes Res* 17:158–167.

[FQPA] Food Quality Protection Act. 1996. Public Law 104-170, Sec. 1, 110 Stat. 1489 (7 U.S.C. § 136).

Frank H. 1991. Airborne chlorocarbons, photooxidants and forest decline. *Ambio* 20:13–18.

Frank H, Scholl H, Sutinen S, Norokorpi Y. 1992. Trichloroacetic acid in conifer needles in Finland. *Annalen Botanica Fennica* 29:263–267.

Fries GF, Paustenbach DJ. 1990. Evaluation of potential transmission of 2,3,7,8-tetrachlorodibenzo-*p*-dioxin contaminated incinerator emissions to humans via foods. *Toxicol Environ Health* 29:1–43.

Friesen KJ, Muir DCG, Webster GBR. 1990. Evidence of sensitized photolysis of polychlorinated dibenzo-*p*-dioxins in natural waters under sunlight conditions. *Environ Sci Technol* 24:1739–1744.

Fry DM. 1995. Reproductive effects in birds exposed to pesticides and industrial chemicals. *Environ Health Perspect* (Suppl 7) 103:165–171.

Fry DM, Toone CK. 1981. DDT-induced feminization of gull embryos. *Science* 213:922–924.

Fuerst EP, Norman MA. 1991. Interactions of herbicides with photosynthetic electron transport. *Weed Sci* 39:458-464.

Fürst P. 1987. Determination of the polychlorinated biphenyl substitute Ugilec (tetrachlorobenzyltoluenes) in fish. *J Chromatography* 405:311–317.

Fürst P, Krause GHM, Hein D, Delschen T. 1992. Influence of PCDD/PCDF levels in grass and soil on the contamination of cow's milk. *Organohalogen Compounds* 8:333–336.

Gagnon MM, Bussieres D, Dodson JJ, Hodson PV. 1995. White sucker (*Catostomus commersoni*) growth, maturation and reproduction in pulp mill contaminated and reference rivers. *Environ Toxicol Chem* 14:317–327.

Gagnon MM, Dodson JJ, Hodson PV. 1994. Seasonal effects of bleached kraft mill effluent on reproductive parameters of white sucker (*Catostomus commersoni*) populations of the St. Maurice River, Quebec, Canada. *Can J Fish Aquat Sci* 51:337–347.

Gallagher K, Hale RC, Greaves J, Bush EO, Stilwell DA. 1993. Accumulation of polychlorinated terphenyls in aquatic biota of an estuarine creek. *Ecotoxicol Environ Safety* 26:302–312.

Gantzer CJ, Kollig HP, Rittman BE, Lewis DL. 1988. Predicting the rate of trace-organic compound removal by natural biofilms. *Water Res* 22:191–200.

Gergov M, Priha M, Talka E, Valttila O, Kangas A, Kukkonen K. 1988. Chlorinated organic compounds in effluent treatment at kraft mills. *Tappi J* 77:175–184.

Geyer HJ, Politzki P, Freitag D. 1984. Prediction of ecotoxicological behavior of chemicals: relationship between n-octanol/water partition coefficient and bioaccumulation of organic chemicals by algae *Chlorella*. *Chemosphere* 13:269–284.

Geyer HJ, Scheunert I, Bruggemann R, Steinberg C, Korte F, Kettrup Z. 1991. QSAR for organic chemical bioconcentration in Daphnia, algae and mussels. *Sci Tot Environ* 109/110:387–394.

Geyer HJ, Scheunert I, Rappe K, Kettrup A, Korte J, Greim H, Rozman K. 1990. Correlation between acute toxicity of 2,3,7,8-tetrachlorodibenzo-*p*-dioxin (TCDD) and total body fat content in mammals. Toxicology 65:97–107.

Ghosh S, Thomas P. 1995. Antagonistic effects of xenobiotics on steroid-induced final maturation of Atlantic croaker oocytes in vitro. *Mar Environ Res* 39:159–163.

Giesy JP, Bowerman WW, Mora MA, Jones PD, Verbrugge DA, Othoudt RA, Newsted JL, Vandervoort C, Summer CL, Aulerich RJ, Bursian SJ, Ludwig JP, Ludwig ME, Dawson GA, Kubiak TJ, Best DA, Tillitt DE. 1994. Contaminants in fishes from Great Lakes-influenced sections and above dams of three Michigan Rivers: III. Implications for health of eagles. *Arch Environ Contam Toxicol* 29:309–321.

Giesy JP, Ludwig JP, Tillitt DE. 1994. Embryolethality and deformities in colonial, fish-eating, water birds of the Great Lakes region: assessing causality. *Environ Sci Technol* 28:128A–135A.

Giesy JP, Newsted J, Garling DL. 1986. Relationships between chlorinated hydrocarbon concentrations and rearing mortality of chinook salmon (*Oncorhynchus tschawytscha*) eggs from Lake Michigan. *J Great Lakes Res* 12:82–98.

Giesy JP, Verbrugge DA, Othout R, Bowerman WW, Mora MA, Jones PD, Newsted JL, Vandervoort C, Heaton SN, Aulerich RJ, Bursian SJ, Ludwig JP, Ludwig M, Dawson GA, Kubiak TJ, Best DA, Tillitt DE. 1994. Contaminants in fishes from Great Lakes-influenced sections and above dams on three Michigan Rivers: II. Implications for the health of mink. *Arch Environ Toxicol Chem* 27:213–223.

Gilbertson M. 1983. Etiology of chick edema disease in herring gulls in the lower Great Lakes. *Chemosphere* 12:357–370.

Gilbertson M, Kubiak T, Ludwig J, Fox G. 1991. Great Lakes embryo mortality, edema, and deformities syndrome (GLEMEDS) in colonial fish-eating birds: similarity to chick-edema disease. *J Toxicol Environ Health* 33:455–520.

Gilbertson M, Morris RD, Hunter RA. 1976. Abnormal chicks and PCB residue levels in eggs of colonial birds on the lower Great Lakes (1971–1973). *Auk* 93:434–442.

Gilbertson M, Schneider RS. 1991. International joint commission workshop on cause-effect linkages: preface. *J Toxicol Environ Health* 33:v–x.

Gilbertson M, Schneider RS. 1993. Preface to the special section on cause-effect linkages. Causality: The missing link between science and policy. *J Great Lakes Res* 19:720–721.

Gilman AP, Peakall DB, Hallett DJ, Fox GA, Norstrom RJ. 1979. Herring gulls (*Larus Argentatus*) as monitors of contamination in the Great Lakes. Animals as monitors of environmental pollution, symposium on pathobiology of environmental pollution: animal models and wildlife as monitors 1977: University of Connecticut. Washington DC: National Academy of Science. p 280–289.

Gleichmann E, Kimber I, Purchase IFH. 1989. Immunotoxicology: suppressive and stimulatory effects of drugs and environmental chemicals on the immune system. *Arch Toxicol* 63:257–273.

[GLNPO] Great Lakes National Program Office, U.S. Environmental Protection Agency (USEPA). 1992. Green Bay/Fox River Mass Balance Study Preliminary Management Summary. Chicago IL: GLNPO. 24 p.

Gobas FAPC. 1993. A model for predicting the bioaccumulation of hydrophobic organic chemicals in aquatic food-webs: application in Lake Ontario. *Ecol Model* 69:1–17.

Gobas FAPC, Bedard DC, Ciborowski JJH, Haffner GD. 1989. Bioaccumulation of chlorinated hydrocarbons by the mayfly *Hexagenia limbata* in Lake St. Clair. *J Great Lakes Res* 15:581–588.

Gobas FAPC, Mcneil EJ, Lovett-Doust L, Haffner GD. 1991. Bioconcentration of chlorinated aromatic hydrocarbons in aquatic macrophytes. *Environ Sci Technol* 25:924–929.

Gobas FAPC, Muir DCG, Mackay D. 1988. Dynamics of dietary bioaccumulation and faecal elimination of hydrophobic organic chemicals. *Chemosphere* 17:943–962.

Gobas FAPC, Zhang X, Wells R. 1994. Gastrointestinal magnification: the mechanisms of biomagnification and food chain accumulation or organic chemicals. *Environ Sci Technol* 27:2855–2863.

Gohre K, Miller GC. 1983. Singlet oxygen generation on soil surfaces. *J Agric Food Chem* 31:1104–1108.

Gohre K, Scholl R, Miller GC. 1986. Singlet oxygen reactions on irradiated soil surfaces. *Environ Sci Technol* 20:934–938.

Goldberg ED. 1975. Synthetic organohalides in the sea. *Proc Royal Soc London* 189:277–289.

Gorge G, Nagel R. 1990 Toxicity of lindane, atrazine, and deltamethrin to early life stages of zebrafish (*Brachydanio rerio*). *Ecotoxicol Environ Safety* 20:246–255.

Graham JD, Wiener JB, editors. 1997. Risk versus risk: tradeoffs in protecting health and the environment. Cambridge MA: Harvard Univ Pr. 337 p.

Graney RL, Maciorowski A, Solomon KR, Nelson H, Laskowski D, Baker JL, editors. 1994. Report of the aquatic risk assessment and mitigation dialogue group. Pensacola FL: Soc of Environmental Toxicology and Chemistry (SETAC) Foundation. 220 pp.

Gray LE, Kelce WR, Monosson E, Ostby JS, Birnbaum LS. 1996. TCDD alters sex differentiation in male rats and hamsters: reduced ejaculated and epididymal sperm numbers and sex accessory gland weights in offspring with normal androgenic status. *Toxicol Appl Pharmacol* 131:108–118.

Gribble GW. 1992. Naturally occurring organohalogen compounds. *J Natural Products Res* 55:1353–1395.

Gribble GW. 1994. The natural production of chlorinated compounds. *Environ Sci Technol* 28:310A–319A.

Grimvall A, Borén H, Jonsson S, Karlsson S, Sävenhed R. 1991. Organohalogens of natural and industrial origin in large recipients of bleach-plant effluents. *Water Sci Technol* 24:373–383.

Grimvall A, Laniewski K, Borén H, Jonsson S. 1994. Organohalogens of natural or unknown origin in surface water and precipitation. *J Toxicol Environ Chem* 46:183–196.

Grue CE, Fleming WJF, Busby DG, Hill EF. 1983. Assessing hazards of organophosphate pesticides to wildlife. *Trans North Am Wildlife Nature Res Conf* 48:200–220.

Guenzi WD, Beard WE. 1967. Anaerobic degradation of DDT to DDD in soil. *Science* 156:1116–1117.

Guiney PD, Melancon Jr MJ, Lech JJ, Peterson RE. 1979. Effects of egg and sperm maturation and spawning on the distribution and elimination of a polychlorinated biphenyl in rainbow trout (*Salmo gairdneri*). *Toxicol Appl Pharmacol* 47:261–272.

Håkansson H, Johansson L, Ahlborg UG, Poiger H. 1990. Vitamin A storage in rats subchronically exposed to PCDDs/PCDFs. *Chemosphere* 20:1147–1150.

Håkansson H, Johansson L, Manzoor E, Ahlborg UG. 1991. Effects of 2,3,7,8-tetrachlorodibenzo-*p*-dioxin (TCDD) on the vitamin A status of Hartley guinea pigs, Sprague-Dawley rats, C57Bl/6 mice, DBA/2 mice, and Golden Syrian hamsters. *J Nutrition Sci Vitaminol* 37:117–138.

Håkansson H, Manzoor E, Ahlborg UG. 1992. Effects of technical PCB preparations and fractions thereof on vitamin A levels in the mink (*Mustela vison*). *Ambio* 21:588–590.

Håkansson H, Sundin P, Andersson T, Brunstroem B, Dencker L, Engvall M, Ewald G, Gilek M, Holm G, Honkasalo S, Idenstam-Almquist J, Jonsson P, Kautsky N, Lundberg G, Lund-Kvernheim A, Martinsen K, Norrgren L, Personen M, Rundgren M, Stalberg M, Tarkpea M, Wesen C. 1991. In vivo and in vitro toxicity of fractionated fish lipids, with particular reference to their content of chlorinated organic compounds. *Pharmacol Toxicol* 69:459–471.

Hansen DJ. 1989. Status of the development of water quality criteria and advisories. In. Water quality standards for the 21st century. Washington DC: U.S. Environmental Protection Agency (USEPA).

Hansen PD, Westernhagen HV, Rosenthal H. 1985. Chlorinated hydrocarbons and hatching success in Baltic herring spring spawners. *Mar Environ Res* 15:59–76.

Haraguchi K, Athanasiadou M, Bergman A, Hovander L, Jensen S. 1992. PCB and PCB methyl sulfones in selected groups of seals from Swedish waters. *Ambio* 21:546–549.

Harder HW, Carter TV, Bidleman TF. 1983. Acute effects of toxaphene and its sediment-degraded products on estuarine fish. *Can J Fish Aquat Sci* 40:2119–2125.

Harner T, Mackay D. 1995. Measurement of octanol-air partition coefficients for chlorobenzenes, PCBs and DDT. *Environ Sci Technol* 29:1599–1606.

Harper DB. 1993. The contribution of natural halogenation process to the atmospheric halomethane burden. Proceedings of International Conference on Naturally-Produced Organohalogens: Formation, Occurrence, Characteristics, and Environmental Significance; 1993 Sep 14–17; Delft NL.

Harper N, Howie L, Connor K, Arellano L, Craig A, Dickerson R, Safe SH. 1993. Immunosuppressive and monooxygenase induction activities of highly chlorinated diphenyl ether congeners in C57BL/6 and DBA/2 mice. *Fundamental Appl Toxicol* 20:496–502.

Harrad SJ, Jones KC. 1992. A source inventory and budget for chlorinated dioxins and furans in the United Kingdom environment. *Sci Tot Environ* 126:89–107.

Harris HJ, Erdman TC, Ankley GT, Lodge KB. 1993. Measures of reproductive success and polychlorinated biphenyl residues in eggs and chicks of Forster's terns on Green Bay, Lake Michigan, Wisconsin - 1988. *Arch Environ Contam Toxicol* 25:304–314.

Harshbarger JC, Cullen LJ, Calabresese MJ, Spero PM, Bauman PC, Parland WK. 1984. Epidermal, hepatocellular and cholangiocellular carcinomas in brown bullheads, (*Ictalurus nebulosus*), from industrially polluted Black River. *Ohio Mar Environ Res* 14:535–536.

Hawker DW, Connell DW. 1989. A simple water-octanol partition system for bioconcentration investigations. *Environ Sci Technol* 23:961–965.

Hayes H, Risebrough RW. 1972. Pollutant concentrations in abnormal terns from Long Island Sound. *Auk* 89:19–35.

Heaton SN, Aulerich RJ, Bursian SJ, Giesy JP, Render JA, Tillitt DE, Kubiak TJ. 1991. Reproductive effects of feeding Saginaw Bay source fish to ranch mink. Abstracts of the papers given at the Cause-Effects Linkages II Symposium; 1991 Sep 27–28. Lansing MI: Michigan Audubon Soc. p 24–25.

Hebert VR, Miller GC. 1990. Depth dependence of direct and indirect photolysis on soils surfaces. *J Agric Food Chem* 38:913–918.

Heindel JJ, Chapin RE, Gulati DK, George JD, Price CJ, Marr MC, Myers CB, Barnes LH, Fail PA, Grizzel TB, Schwetz BA, Yang RH. 1994. Assessment of the reproductive and developmental toxicity of pesticide/fertilizer mixtures based on confirmed pesticide contamination in California and Iowa groundwater. *Fundamental Appl Toxicol* 22:605–621.

Helander B. 1994. Productivity in relation to residue levels of DDE in the eggs of the white-tailed sea eagles (*Haliaeetus albicilla*) in Sweden. In: Meyburg BU, Chancellor RD, editors. Raptor conservation today. Berlin, Germany and London UK: WWGBP/Pica Pr. p 735–738.

Helander B, Bignert A. 1992. Harbour seal (*Phoca vitulina*) at the Swedish Baltic coast: population trends and reproduction. *Ambio* 5:261–263.

Helle E, Olsson M, Jensen S. 1976a. DDT and PCB levels and reproduction in ringed seal from the Bothnian Bay. *Ambio* 5:188–189.

Helle E, Olsson M, Jensen S. 1976b. PCB levels correlated with pathological changes in seal uteri. *Ambio* 5:261–263.

Hemming H, Flodström S, Wärngård L, Bergman Å, Kronevi T, Nordgren I, Ahlborg UG. 1993. Relative tumour promoting activity of polychlorinated biphenyls (3,4,5,3',4'-, 2,3,4,3',4'- and '-CB) in rat liver. *Angewante Zoologie* 248:163–174.

Hermens JLM, Opperhuizen A, editors. 1991. QSAR in environmental toxicology. IV. Amsterdam NL: Elsevier.

Hickie BE, McCarty LS, Dixon DG. 1995. A residue-based toxicokinetic model for pulse-exposure toxicity in aquatic systems. *Environ Toxicol Chem* 14:2187–2197.

Hileman B. 1992. Effects of organochlorines on marine animals to be investigated. *Chemical and Engineering News*, 9 Mar 1992. p 23–24.

Hinton WS, Lane AM. 1991. Characteristics of municipal solid waste incinerator fly ash promoting the formation of polychlorinated dioxins. *Chemosphere* 22:473–483.

Hodin F, Boren H, Grimvall A, Karlsson S. 1991. Formation of chlorophenols and related compounds in natural and technical chlorination processes. *Water Sci Technol* 24:402–410.

Hodson PV. 1988. The effect of metal metabolism on uptake, disposition and toxicity in fish. *Aquat Toxicol* 11:3–18.

Hodson PV. 1990. Indicators of ecosystem health at the species level and the example of selenium effects on fish. *Environ Monitor Assess* 15:241–254.

Hodson PV, Carey JH, Munkittrick KR, Servos MR. 1994. Canada and Sweden - contrasting regulations for chlorine discharge from the pulp and paper industry. Paper presented at the SETAC Workshop on Environmental Risk Assessment for Organochlorine Chemicals; 1994 Jul 24–29; Alliston ON.

Hodson PV, McWhirter M, Ralph K, Gray B, Thivierge D, Carey J, Van Der Kraak G, Whittle DM, Levesque M-C. 1992. Effects of bleached kraft mill effluent on fish in the St. Maurice River, Quebec. *Environ Toxicol Chem* 11:1635–1651.

Hodson PV, Whittle DM, Wong PTS, Borgmann U, Thomas RL, Chau YK, Nriagu JO, Hallett DJ. 1984. Lead contamination of the Great Lakes and its potential effects on aquatic biota. In: Nriagu JO, Simmons MS, editors. Toxic contaminants in the Great Lakes. New York: J Wiley. p 335–370.

Hoff RM. 1994. An error budget for the determination of the atmospheric mass loading of toxic chemicals in the Great Lakes. *J Great Lakes Res* 20:229–239.

Hoff RM, Bidleman TF, Eisenreich SJ. 1993. Estimation of PCC loadings from the atmosphere to the Great Lakes. *Chemosphere* 27:2047–2055.

Hoff RM, Muir DCG, Grift NP. 1992. Annual cycle of polychlorinated biphenyls and organohalogenated pesticides in air in Southern Ontario. I. Air concentration data. *Environ Sci Technol* 26:266–275.

Hoffman DJ, Rattner BA, Hall RJ. 1990. Wildlife toxicology. *Environ Sci Technol* 24:276–283.

Hoffman DJ, Rattner BA, Sileo L, Docherty D, Kubiak TJ. 1987. Embryotoxicity, teratogenicity, and aryl hydrocarbon hydroxylase activity in Forster's tern on Green Bay, Lake Michigan. *Environ Res* 42:176–184.

Hoigne J, Faust BC, Haag WR, Scully Jr FE, Zepp RG. 1989. Aquatic humic substances as sources and sinks of photochemically produced transient reactants. In: Suffet IH, MacCarthy P, editors. Aquatic humic substances: influence on fate and treatment of pollutants. ACS Advances in Chemistry Series, 219. Washington DC: American Chemical Soc (ACS). p 363–381.

Holdway DA, Dixon DG. 1986. Effects of methoxychlor exposure of flagfish eggs (*Jordanella floridae*) on hatchability, juvenile methoxychlor tolerance and whole-body levels of tryptophan, serotonin and 5-hydroxyindoleacetic acid. *Water Res* 20:893–897.

Holdway DA, Kempe EJ, Dixon DG. 1987. Acute toxicity of methoxychlor to larval white sucker (*Catostomus commersoni*) as modified by age, food availability, and egg pre-exposure. *Can J Fish Aquat Sci* 44:227–232.

Holsen TM, Noll KE. 1992. Dry deposition of atmospheric particles: application of current models to ambient data. *Environ Sci Technol* 26:1807–1815.

Holsen, TM, Noll KE, Liu S-P, Lee W-J. 1991. Dry deposition of polychlorinated biphenyls in urban areas. *Environ Sci Technol* 25:1075–1081.

Hornbuckle KC, Jeremiason JD, Sweet CW, Eisenreich SJ. 1994. Seasonal variation in the air-water exchange of PCBs in Lake Superior. *Environ Sci Technol* 28:1491–1501.

Horowitz A, Suflita JM, Tiedje JM. 1983. Reductive dehalogenations of halobenzoates by anaerobic lake sediment microorganisms. *Appl Environ Microbiol* 45:1459–1465.

Howard PA, Boethling RS, Jarvis WF, Michalenko EM. 1991. Handbook of environmental degradation rates. Boca Raton FL: CRC/Lewis.

Howard PA, Michalenko EM, Jarvis WF, Basu DK, Sage GW, Meylan WM, Beauman JA, Gray DA. 1991. Handbook of environmental fate and exposure data for organic chemicals. Volume III, Pesticides. Boca Raton FL: CRC /Lewis.

Howard PA, Boethling RS, Stiteler WM, Meylan WM, Hueber AE, Beauman JA, Larosche ME. 1992. Predictive model for aerobic biodegradability developed from a file of evaluated biodegradation data. *Environ Toxicol Chem* 11:593–603.

Howe GE, Marking LL, Bills TD, Boogard MA, Mayer Jr FL. 1994. Effects of water temperature on the toxicity of 4-nitrophenol and 2-dinitrophenol to developing rainbow trout (*Onchorynchus mykiss*). *Environ Toxicol Chem* 13:79–84.

Howe GE, Marking LL, Bills TD, Rach JJ, Mayer Jr FL. 1994. Effects of water temperature and pH on toxicity of terbufos, trichlorfon, 4-nitrophenol, and 2-dinitrophenol to the amphipod (*Gammarus pseudolimnaeus)* and rainbow trout (*Onchorynchus mykiss*). *Environ Toxicol Chem* 13:51–66.

Howie L, Dickerson R, Davis D, Safe S. 1990. Immunosuppressive and monooxygenase induction activities of polychlorinated diphenyl ether congeners in C57BL/6N mice: quantitative structure-activity relationships. *Toxicol Appl Pharmacol* 105:254–263.

Huang ES, Nelson FR. 1986. Anti-estrogenic action of chlordecone in rat pituitary gonadotrophs in vitro. *Toxicol Appl Pharmacol* 82:62–29.

Hülster A, Marschner H. 1992. Transfer of PCDD/PCDF from contaminated soils to food and fodder crop plants. *Organohalogen Compounds* 8:351–354.

Hwang H-M, Hodson RE, Lee RF. 1987. Photolysis of phenol and chlorophenols in estuarine water. In: Zika RG, Cooper WJ, editors. Photochemistry of environmental aquatic systems. Washington DC: American Chemical Soc (ACS). p 27–43.

[IARC] International Agency for Research on Cancer. 1979a. Chloroform. *IARC monographs on the evaluation of carcinogenic risk of chemicals to humans* 20:401–427.

[IARC] International Agency for Research on Cancer. 1979b. Vinyl chloride. *IARC monographs on the evaluation of carcinogenic risk of chemicals to humans* 19:337–438.

[IARC] International Agency for Research on Cancer. 1979c. Carbon tetrachloride. *IARC monographs on the evaluation of carcinogenic risk of chemicals to humans* 20:371–399.

[IJC] International Joint Commission. 1994. Seventh biennial report under the Great Lakes Water Quality Agreement of 1978 to the governments of the United States and Canada and the state and provincial governments of the Great Lakes basin. Washington DC and Ottawa ON. p 7.

[IRPTC] International Registry of Potentially Toxic Chemicals. 1988. Production and legal status of major chlorinated pesticides and industrial chemicals. Geneva, Switzerland: IRPTC.

Ivens IA, Loser E, Rinke M, Schmidt U, Mohr U. 1993. Subchronic toxicity of 2,3,7,8-tetrabromodibenzo-*p*-dioxin in rats. *Toxicology* 83:181–201.

Iwata H, Tanabe S, Sakai N, Nishimura A, Tatsukawa R. 1994. Geographical distribution of persistent organochlorines in air, water and sediments from Asia and Oceania, and their implications for global redistribution from lower latitudes. *Environ Pollut* 85:15–33.

Iwata H, Tanabe S, Sakai N, Tatsukawa R. 1993. Distribution of persistent organochlorines in the oceanic air and surface seawater and the role of ocean on their global transport and fate. *Environ Sci Technol* 27:1080–1098.

Jansson B, Andersson R, Asplund L, Litzen K, Nylund K, Sellström U, Uvemo U-B, Wahlberg C, Wideqvist U, Odsjö T, Olsson M. 1993. Chlorinated and brominated persistent organic compounds in biological samples from the environment. *Environ Toxicol Chem* 12:1163–1174.

Jantunen LJ, Bidleman TF. 1995. Reversal of the air-water gas exchange of hexachlorocyclohexanes in the Bering and Chukchi seas: 1993 vs. 1988. *Environ Sci Technol* 29:1081–1089.

Jeffers PM, Wolfe NL. 1989. Hydrolysis of carbon tetrachloride. *Science* 246:1638.

Jeffers PM, Ward LM, Wojtowitch LM, Wolfe NL. 1989. Homogeneous hydrolysis rate constants for selected chlorinated methanes, ethanes, ethenes and propanes. *Environ Sci Technol* 23:965–969.

Jensen AA, Jørgensen KF. 1983. Polychlorinated terphenyls (PCTs) use, levels and biological effects. *Sci Tot Environ* 27:231–250.

Jensen S. 1972. The PCB story. *Ambio* 1:123–131.

Jensen S, Eriksson G, Kylin H. 1992. Atmospheric pollution by persistent organic compounds: monitoring with pine needles. *Chemosphere* 24:229–245.

Jeremiason JD, Hornbuckle KC, Eisenreich SJ. 1994. PCBs in Lake Superior. 1978–92: Decreases in water concentration reflect losses by volatilization. *Environ Sci Technol* 28:903–914.

Johnson LL, Casillas E, Collier TK, McCain BB, Varanasi U. 1988. Contaminant effects on ovarian development in English sole (*Parophrys vetulus*) from Puget Sound, Washington. *Can J Fish Aquat Sci* 45:2133–2146.

Johnson LL, Stehr CM, Olson OP, Myers MS, Pierce SM, Wigren CA, McCain BB, Varanasi U. 1993. Chemical contaminants and hepatic lesions in winter flounder (*Pleuronectes americanus*) from the northeast coast of the United States. *Environ Sci Technol* 27:2759–2771.

Jokela J, Elomaa E, Salkinoja-Salonen M. 1992. Adsorbable organic halogens (AOX) in drinking water and aquatic environment in Finland. *J Water Supply Res Technol, Aqua* 41:4–11.

Jokela J, Laine M, Ek M, MS Salkinoja-Salonen. 1993. The effect of biological treatment on halogenated organics in bleached kraft pulp mill effluents studied by molecular weight distribution analyses. *Environ Sci Technol* 27:547–557.

Jokela JK, Salkinoja-Salonen M. 1992. Molecular weight distribution of organic halogens in bleached kraft pulp mill effluents. *Environ Sci Technol* 26:1190–1197.

Jones PD, Hannah DJ, Buckland SJ, Day PJ, Leathem SV, Porter LJ, Auman HJ, Sanderson T, Summer C, Ludwig JP, Colborn TL, Giesy JP. 1996. Persistent synthetic chlorinated hydrocarbons in albatross tissue samples from Midway Atoll. *Environ Toxicol Chem* 15:1793–1800.

Jones PD, Hannah DJ, Buckland SJ, Day PJ, Leathem SV, Porter LJ, Giesy JP, Auman HJ, Ludwig JP, Summer CL. 1995. PCBs and 2,3,7,8-substituted polychlorinated dibenzo-*p*-dioxins and dibenzofurans in albatross tissues from Midway Island in the central Pacific Ocean. Organohalogen Compounds 20:137–140.

Jonsson C-J, Lund B-O, Brandt I. 1993. Adrenocorticolytic DDT metabolites: studies in mink (*Mustella vison)* and otter (*Lutra lutra*). *Ecotoxicology* 2:41–53.

Jury WA, Spencer WF, Farmer WJ. 1983. Behavior assessment model for trace organics in soil. I. Model description. *J Environ Qual* 12:558–564.

Kaiser KLE. 1986. QSAR in environmental toxicology. II. Dordrecht NL: Reidel.

Kannan K, Tanabe S, Quynh HT, Huc ND, Tatsukawa R. 1992. Residue pattern and dietary intake of persistent organochlorine compounds in foodstuffs from Vietnam. *Arch Environ Contam Toxicol* 22:367–374.

Kannan K, Tanabe S, Ramesh A, Subramanian A, Tatsukawa R. 1992. Persistent organochlorine residues in foodstuffs from India and their implications on human dietary exposure. *J Agric Food Chem* 40:518–524.

Kannan K, Tanabe S, Williams RJ, Tatsukawa R. 1994. Persistent organochlorine residues in foodstuffs from Australia, Papua New Guinea and the Solomon Islands: contamination levels and human dietary exposure. *Sci Tot Environ* 153:29–49.

Kapoor IP, Metcalf RL, Hirwe AS, Coats JR, Khalsa MS. 1973. Structure activity correlations of biodegradability of DDT analogs. *J Agric Food Chem* 21:310–315.

Kapoor IP, Metcalf RL, Hirwe AS, Lu PY, Coats JR, Nystrom RF. 1972. Comparative metabolism of DDT, methylchlor, and ethoxychlor in mouse, insects, and in a model ecosystem. *J Agric Food Chem* 20:1–6.

Kavlock RJ, Daston GP, DeRosa D, Fenner-Crisp P, Gray LE, Kaattari S, Lucier G, Luster M, Mac MJ, Maczka C, Miller R, Moore J, Rolland R, Scott G, Sheehan DM, Sinks T, Tilson HA. 1996. Research needs for the risk assessment of health and environmental effects of endocrine disruptors: a report of the US EPA sponsored workshop. *Env Hlth Perspect* 104 (Suppl. 4):715–740.

Kavlock RJ, Ankley GT. 1996. A perspective on the risk assessment process for endocrine-disruptive effects on wildlife and human health. *Risk Anal* 16:731–739.

Kawano M, Inoue T, Wada T, Hidaka H, Tatsukawa R. 1988. Bioconcentration and residue patterns of chlordane compounds in marine animals: invertebrates, fish, mammals, and seabirds. *Environ Sci Technol* 22:792–797.

Keeler RF. 1984. Teratogens in plants. *J Animal Sci* 58:1029–1039.

Keene WC, Maben JR, Pzenny AAP, Galloway JN. 1993. Measurement technique for inorganic chlorine gases in the marine boundary layer. *Environ Sci Technol* 27:866–874.

Kendall RJ, Dickerson R, Giesy JP, Suk W, editors. 1998. Principles and processes for evaluating endocrine disruption in wildlife. Pensacola FL: SETAC. 491 p.

Kerkvliet NI. 1995. Immunological effects of chlorinated dibenzo-*p*-dioxins. *Environ Health Perspect* 103:47–53.

Kingsley MCS. 1993. Effects of aquatic toxicity on the belugas of the St. Lawrence Estuary. In: Baddaloo EG, Moore JW, editors. Proceedings of the Nineteenth Annual Aquatic Toxicity Workshop, Edmonton, Alberta. Canadian Technical Reports of Fisheries and Aquatic Science Nr 1942. Ottawa: Minister of Supply and Services Canada 1993. p 186–200.

Kirk-Othmer. 1993. Kirk-Othmer encyclopaedia of chemical technology. 4th ed. New York: J Wiley.

Kitunen V, Valo R, Salkinoja-Salonen M. 1985. Analysis of chlorinated phenols, phenoxyphenols and dibenzofurans around wood preserving facilities. *Int J Environ Analytical Chem* 20:13–28.

Kitunen V, Valo R, Salkinoja-Salonen M. 1987. Contamination of soil around wood preserving facilities by polychlorinated aromatic compounds. *Environ Sci Technol* 21:96–101.

Klecka GM. 1985. Biodegradation. In: Neely WB, Blau GE, editors. Volume 1, Environmental exposure from chemicals. Boca Raton FL: CRC. p 109–155.

Kleeman JM, Moore RW, Peterson RE. 1990. Inhibition of testicular steroidogensis in 2,3,7,8-tetrachlorodibenzo-*p*-dioxin-treated rats; evidence that the key lesion occurs prior to or during pregnenolone formation. *Toxicol Appl Pharmacol* 106:112–125.

Kleopfer RD, Easley DM, Haas BB, Delhl TG, Jackson DE, Wurrey CJ. 1985. Anaerobic degradation of trichloroethylene in soil. *Environ Sci Technol* 19:277–280.

Kloepper-Sams PJ, Benton E. 1994. Exposure of fish to biologically treated bleached-kraft effluent. 2. Induction of hepatic cytochrome P4501A in mountain whitefish *Prosopium williamsoni* and other species. *Environ Toxicol Chem* 13:1483–1496.

Kloepper-Sams PJ, Swanson SM, Marchant T, Schryer R, Owens JW. 1994. Exposure of fish to biologically treated bleached-kraft effluent. 1. Biochemical, physiological and pathological assessment of rocky mountain whitefish (*Prosopium williamsoni*) and longnose sucker (*Catostomus catostomus*). *Environ Toxicol Chem* 13:1469–1482.

Knisel WG, editor. 1980. CREAMS, a field scale model for chemicals, runoff, and erosion from agricultural management systems. Washington DC: Science and Education Admin, U.S.Dept of Agriculature (USDA). Conservation Research Report No. 26.

Knuth BA. 1990. Risk communication: a new dimension in sport fisheries management. *North Am J Fisheries Management* 10:374–381.

Kociba RJ, Keyes DG, Beyer JE, Carreon RM, Wade CE, Dittenber DA, Kalnins RP, Frauson LE, Park CN, Barnard SD, Hummel RA, Humiston CG. 1978. Results of a two-year chronic toxicity and oncogenicity study of 2,3,7,8-tetrachlorodibenzo-*p*-dioxin in rats. *Toxicol Appl Pharmacol* 46:279–303.

Koester CJ, Hites RA. 1992a. Wet and dry deposition of chlorinated dioxins and furans. *Environ Sci Technol* 26:1375–1382.

Koester CJ, Hites RA. 1992b. Photodegradation of PCDDs and PCDFs adsorbed to fly ash. *Environ Sci Technol* 26:502–507.

Koistinen J, Paasivirta, Lahtiperä M. 1993. Bioaccumulation of dioxins, coplanar PCBs, PCDEs, HxCNs, R-PCBs, R-PCPHs and R-PCBBs in fish from a pulp mill receipt watercourse. *Chemosphere* 27:149–156.

Kollig HP. 1988. Criteria for evaluating the reliability of literature data on environmental process constants. *Toxicol Environ Chem* 17:287–311.

Korach KS, McLachlan JA. 1995. Techniques for detection of estrogenicity. *Environ Health Perspect* 103 Supplement 7:5–8.

Krzystyniak K, Tryphonas H, Fournier M. 1995. Evaluation of chemical-induced immunotoxicity. *Environ Health Perspect* 103:17–22.

Kubiak TJ, Harris HJ, Smith LM, Schwartz TR, Stalling DL, Trick JA, Sileo L, Docherty DE, Erdmann TC. 1989. Microcontaminants and reproductive impairment of the Forster's Tern on Green Bay, Lake Michigan. *Arch Environ Contam Toxicol* 18:706–727

Kuenemann P, Battersby NS. 1992. Use of the super 14C-most-probable-number method to enumerate active heterotrophic bacteria in aqueous samples. *Chemosphere* 25:591–603.

Kuhn EP, Suflita JM. 1989a. Dehalogenation of pesticides by anaerobic microorganisms in soil and groundwater: a review. In: Sawhney BL, Brown K, editors. Special Publication 22. Madison WI: Soil Science Soc of America. p 111–180.

Kuhn EP, Suflita JM. 1989b. Sequential reductive dehalogenation of chloroanilines by microorganisms from a methanogenic aquifer. *Environ Sci Technol* 23:848–852.

Kupfer D, Bulger WH. 1980. Estrogenic properties of DDT and its analogs. In: McLachlan JA, editor. Volume 5, Estrogens in the environment. Amsterdam: Elsevier North Holland. p 239–263.

Kurz J, Ballschmiter K. 1994. Relationship between structure and retention of polychlorinated diphenylethers (PCDE) in HRGC in comparison with other groups of halogenated aromatic compounds. *Fresenius J Analytical Chem* 349:533–537.

Kurz J, Ballschmiter K. 1995. Isomer-specific determination of 79 polychlorinated diphenylethers (PCDE) in cod liver oils, chlorophenols and fly ash. *Fresenius J Analytical Chem* 351:98–109.

Kylin H, Grimvall E, Ostman C. 1994. Environmental monitoring of PCBs using pine needles as passive samplers. *Environ Sci Technol* 28:1320–1324.

Laine M, Jokela J, Salkinoja-Salonen M. 1991. Removal of different molecular sizes of organic halogen compounds during biological treatment of bleached pulp mill effluents. In: Södergren A, editor. Environmental fate and effects of bleached pulp mill effluents. SEPA Report 4031. Stockholm, Sweden. p 113–120.

Lake JL., Rubinstein NI, Lee II H, Lake CA, Heltshe J, Pavignano S. 1990. Equilibrium partitioning and bioaccumulation of sediment associated contaminants by infaunal organisms. *Environ Toxicol Chem* 9:1095–1106.

Lamb JC, Balcomb R, Bens CM, Cooper RL, Gorsuch JW, Matthiessen P, Peden-Adams MM, Voit EO. 1998. Hazard identification and epidemiology. In: Kendall RJ, Dickerson R, Giesy JP, Suk W, editors. Principles and processes for evaluating endocrine disruption in wildlife. Pensacola FL: SETAC. p 17–38.

Landahl JT, Johnson LL. 1993. Contaminant exposure and population growth of English sole in Puget Sound: the need for better early life-history data. *Am Fisheries Soc Symposium* 14:117–123.

Landner L, Grimvall A, Håkansson L, Sangfors O, Waltersson E. 1994. Chlorine and chlorine compounds in Sweden. Swedish Chemicals Inspection Report 16. ISSN 0284-1185. Stockholm, Sweden. 269 p.

Landrum PF. 1989. Bioavailability and toxicokinetics of polycyclic aromatic hydrocarbons sorbed to sediments for the amphipod *Pontoporeia hoyi*. *Environ Sci Technol* 23:588–595.

Langford CH, Carey JH. 1987. Photocatalysis by inorganic components of natural water systems. In: Zika RG, Cooper WJ, editors. Photochemistry of environmental aquatic systems. Washington DC: American Chemical Soc (ACS). p 225–239.

Laniewski K, Boren H, Grimvall A, Jonsson S, von Sydow L. 1995. Chemical characterisation of adsorbable organic halogens (AOX) in precipitation. In: Grimvall A, de Leer EBW, editors. Naturally produced organohalogens. Dordrecht NL: Kluwer Academic. p 113–129.

Lans MC, Klasson-Wehler E, Willemsen M, Meussen E, Safe S, Brouwer A. 1993. Structure dependent, competitive interaction of hydroxy-chlorobiphenyls, dibenzo-*p*-dioxins and dibenzofurans with human transthyretin. *Chem Biol Interact* 88:7–21.

Larson RJ. 1980. Role of biodegradation kinetics in predicting environmental fate. In: Maki AW, Dickson KL, Cairns J, editors. Biotransformation and fate of chemicals in the aquatic environment. Washington DC: American Soc for Microbiology. p 67–86.

Larsson P, Järnmark C, Södergren A. 1992. PCBs and chlorinated pesticides in the atmosphere and aquatic organisms of Ross Island, Antarctica. *Mar Pollut Bull* 25:281–287.

Le Boeuf BJ, Bonnell ML. 1971. DDT in California sea lions. *Nature* 234:108–110

Lee G, Ellersieck MR, Mayer FL, Krause GF. 1995. Predicting chronic lethality of chemicals to fishes from acute toxicity data: multifactor probit analysis. *Environ Toxicol Chem* 14:345–349.

Lee GF, Hughes RA, Vieth GD. 1977. Evidence for partial degradation of toxaphene in the aquatic environment. *Water Air Soil Pollut* 8:479–484.

Lee RF, Hinga K, Almquist G. 1982. In: Grice GD, Reeve MR, editors. Marine mesocosms: biological and chemical research in experimental ecosystems. New York: Springer Verlag. p 123–135.

Leister DL, Baker JE. 1994. Atmospheric deposition of organic contaminants to Chesapeake Bay. *Atmospher Environ* 28:1499–1520.

Leversee GJ, Landrum PF, Geisy JP, Fanin T. 1983. Humic acids reduce bioaccumulation of some polycyclic aromatic hydrocarbons. *Can J Fish Aquat Sci* 40:(Suppl. 2)63–69.

Lewis DL, Gattie DK. 1988. Prediction of substrate removal rates of attached microorganisms and of relative contributions of attached and suspended communities at field sites. *Appl Environ Microbiol* 54:434–440.

Lewis DL, Holm HW. 1981. Rates of transformation of methyl parathion and diethyl phthalate by aufwuchs microorganisms. *Appl Environ Microbiol* 42:698–703.

Ligocki MP, Leuenberger C, Pankow JF. 1985a. Trace organics in rain. II. Gas scavenging of neutral organic compounds. *Atmospher Environ* 19:1609–1617.

Ligocki MP, Leuenberger C, Pankow JF. 1985b. Trace organics in rain. III Particle scavenging of neutral organic compounds. *Atmospher Environ* 19:1619–1626.

Lindberg P, Schei PJ, Wikman M. 1988. The peregrine falcon in Fennoscandia. In: Cade TJ, Endersson JH, Thelander CG, White CM, editors. Peregrine falcon populations. Their management and recovery. Boise ID: Pergrine Fund. p 159–172.

Lindström K, Österberg F. 1984. Characterization of the high molecular mass chlorinated matter in spent bleach liquors (SBL). Part I. Alkaline SBL. *Holzforschung* 38:201–212.

Lipnick RL. 1993. Baseline toxicity QSAR models: a means to assess mechanism of toxicity for aquatic organisms and mammals. In: Gorsuch JW, Dwyer FJ, Ingersoll CG, LaPoint TW, editors. Volume 2, Environmental toxicology and risk assessment. Philadelphia PA: American Soc for Testing and Materials (ASTM). STP 1216. p 610–619.

Lock EA. 1993. Responses of the kidney to toxic compounds. In: Ballantyne B, Marrs T, Turner P, editors. General and applied toxicology. New York: Stockton Pr. p 507–536.

Lockhart WL, Wagemann R, Tracey B, Sutherland D, Thomas DJ. 1992. Presence and implications of chemical contaminants in the fresh waters of the Canadian Arctic. *Sci Tot Environ* 122:165–245.

Ludwig JP. 1984. Decline, resurgence and population dynamics of Michigan and Great Lakes double-crested cormorants. *Jack-Pine Warbler* 2:91–102.

Ludwig JP, Auman HJ, Kurita-Matsuba H, Ludwig ME, Campbell LM, Giesy JP, Tillitt DE, Jones P, Yamashita N, Tanabe S, Tatsukawa R. 1993. Caspian tern reproduction in the Saginaw Bay ecosystem following a 100-year flood event. *J Great Lakes Res* 19:96–108.

Ludwig JP, Giesy JP, Summer CL, Bowerman W, Heaton S, Aulerich R, Bursian S, Auman HJ, Jones PD, Williams LL, Tillitt DE, Gilbertson M. 1993. A comparison of water quality criteria in the Great Lakes basin based on human or wildlife health. *J Great Lakes Res* 19:789–807.

Lund BO, Bergman A, Brandt I. 1988. Metabolic activation and toxicity of a DDT-metabolite, 3-methylsulphonyl-DDE, in the adrenal *zona fasiculata* in mice. *Temp Biol Interactions* 65:25–40.

Lundholm E. 1987. Thinning of eggshells in birds by DDE: mode of action on the eggshell gland. *Comparative Biochem Physiol* 88C:1–22

Luster MI, Munson AE, Thomas PT, Holsaple MP, Fenters JD, White KL, Lauer LD, Germolec DR, Rosenthal GJ, Dean JH. 1988. Development of a testing battery to assess chemical-induced immunotoxicity: National Toxicology Program's guidelines for immunotoxicity evaluation in mice. *Fundamental Appl Toxicol* 10:2–19.

Mabey W, Mill T. 1978. Critical review of hydrolysis of organic compounds in water under environmental conditions. *J Phys Chem Ref Data* 7:383–415.

Mably TA, Bjerke DL, Moore RW, Gendron-Fitzpatrick A, Peterson RE. 1992. In utero and lactational exposure of male rats to 2,3,7,8-tetrachlorodibenzo-*p*-dioxin. III. Effects on spermatogenesis and reproductive capability. *Toxicol Appl Pharmacol* 114:118–126.

Mably TA, Moore RW, Goy RW, Peterson RE. 1992. In utero and lactational exposure of male rats to 2,3,7,8-tetrachlorodibenzo-*p*-dioxin. II. Effects on sexual behavior and the regulation of luteinizing hormone secretion in adulthood. *Toxicol Appl Pharmacol* 114:108–117.

Mably TA, Moore RW, Peterson RE. 1992. In utero and lactational exposure of male rats to 2,3,7,8-tetrachlorodibenzo-*p*-dioxin. I. Effects on androgenic status. *Toxicol Appl Pharmacol* 114:97–107.

Macalady DL, Tratnyek PG, Wolfe NL. 1989. Influence of natural organic matter on the abiotic hydrolysis of organic contaminants in aqueous systems. *ACS Advances in Chemistry Series* 219:323–332.

MacLatchy DL, Van Der Kraak GJ. 1995. The phytoestrogen b-sitosterol alters the reproductive endocrine status of goldfish. *Toxicol Appl Pharmacol* 134:305–312.

MacLatchy DL, Peters L, Nickle J, Van Der Kraak GJ. 1997. Exposure to b-sitosterol the endocrine status of goldfish differently than 17b-estradiol. *Environ Toxicol Chem* 16:1895–1904.

MacCubbin AE, Black JJ, Dunn BP. 1990. 32P-postlabeling detection of DNA adducts in fish from chemically contaminated waterways. *Sci Tot Environ* 94:89–104.

Maccubbin AE, Chidambaram S, Black JJ. 1988. Detection of metabolites of aromatic hydrocarbons in the bile of brown bullheads (*Ictalurus nebulosus*). *J Great Lakes Res* 14:101–108.

Mackay D. 1989. Modeling the long-term behavior of an organic contaminant in a large lake: application to PCBs in Lake Ontario. *J Great Lakes Res* 15:283–297.

Mackay D. 1991. Multimedia environmental models: the fugacity approach. Chelsea MI: Lewis. 257 p.

Mackay D, Shiu WY, Ma KC. 1992a. Illustrated handbook of physical-chemical properties and environmental fate for organic chemicals. I. Monoaromatic hydrocarbons, chlorobenzenes and PCBs. Ann Arbor MI: Lewis. 697 p.

Mackay D, Shiu WY, Ma KC. 1992b. Illustrated handbook of physical-chemical properties and environmental fate for organic chemicals. II. Polyaromatic hydrocarbons, polychlorinated dioxins and dibenzofurans. Ann Arbor MI: Lewis. 597 p.

Mackay D, Shiu WY, Ma KC. 1992c. Illustrated handbook of physical-chemical properties and environmental fate for organic chemicals. III. Volatile organic chemicals. Ann Arbor MI: Lewis. 916 p.

Mackay D, Wania F. 1995. Transport of contaminants to the Arctic: partitioning, processes and models. *Sci Tot Environ* 160/161:25–38.

Maddrell SHP, Casida JE. 1971. Mechanism of insecticide-induced diuresis in Rhodnius (*Rhodnius prolixus*). *Nature* 231(5297):55–56.

Maddrell SHP, Reynolds SE. 1972. Release of hormones in insects after poisoning with insecticides (*Rhodnius prolixus*). *Nature* 236 (5347):404–406.

Mailhot H. 1987. Prediction of algal bioaccumulation and uptake rate of organic compounds by ten physicochemical properties. *Environ Sci Technol* 21:1009–1013.

Martin SC, Hinz SC, Rodgers PW, Bierman Jr VJ, DePinto JV, Young TC. 1995. Calibration of a hydraulic transport model for Green Bay, Lake Michigan. *J Great Lakes Res* 21:599–609.

Martineau D, Beland P, Desjardins C, Lagace A. 1987. Levels of organochlorine chemicals in tissues of Beluga whales (*Delphinapterus leucas*) from the St. Lawrence estuary, Quebec, Canada. *Arch Environ Contam Toxicol* 16:137–147.

Martineau D, De Guise S, Fournier M, Shugart L, Girard C, Lagace A, Beland P. 1994. Pathology and toxicology of Beluga whales from the St. Lawrence estuary, Quebec, Canada. Past, present and future. *Sci Tot Environ* 154:201–215.

Martineau D, Lagace A, Beland P, Higgins R, Armstrong D, Shugart LR. 1988. Pathology of stranded Beluga whales (*Delphinapterus leucas*) from the St. Lawrence estuary, Quebec. *J Comparative Pathol* 98:287–311.

Mayer FL, Adams WJ, Finley MT, Michael PR, Mehrle PM, Saeger VW. 1981. Phosphate ester hydraulic fluids: an aquatic environmental assessment of Pydrauls 50E and 115E. In: Branson DR, Dickson KL, editors. Philadelphia PA: American Soc for Testing and Materials (ASTM). Tech Publ 737. p 103–123.

Mayer FL, Ellersieck MR. 1986 Manual of acute toxicity and database for 410 chemicals and 66 species of freshwater animals. Washington DC: U.S. Department of the Interior. Service Resource Publication 160.

Mayer FL, Krause GF, Ellersieck DR, Buckler MR, Lee G. 1994. Predicting chronic lethality of chemicals to fishes from acute toxicity test data: concepts and linear regression analysis. *Environ Toxicol Chem* 13:671–678.

Mayer FL, Mehrle PM, Crutcher PL. 1978. Interactions of toxaphene and vitamin C in channel catfish. *Trans Am Fish Soc* 107:326–333.

McBean EA, Rovers FA. 1992. Estimation of the probability of exceedance of a contaminant concentration. *Ground Water Monitoring Rev* 12:115–119.

McCarty LS, Mackay D. 1993. Enhancing ecotoxicological modeling and assessment: body residues and modes of toxic action. *Environ Sci Technol* 27:1719–1728.

McCarty LS, Mackay D, Smith AD, Ozburn GW, Dixon DG. 1992. Residue-based interpretation of toxicity and bioconcentration QSARs from aquatic bioassays: neutral narcotic organics. *Environ Toxicol Chem* 11:917–930.

McConnell LL, Cotham WE, Bidleman TF. 1993. Gas exchange of hexachlorocyclohexanes in the Great Lakes. *Environ Sci Technol* 27:1304–1311.

McCrady JK, Maggard SP. 1993. Uptake and photodegradation of 2,3,7,8-tetrachlorodibenzo-*p*-dioxin sorbed to grass foliage. *Environ Sci Technol* 27:343–350.

McLachlan JA. 1993. Functional toxicology: a new approach to detect biologically active xenobiotics. *Environ Health Perspect* 101:386–387.

McLachlan MS. 1994. Model of the fate of hydrophobic contaminants in cows. *Environ Sci Technol* 28:2407–2414.

McLachlan MS. 1996. Bioaccumulation of hydrophobic chemicals in agricultural food chains. *Environ Sci Technol* 30:252–259.

McMaster ME, van Der Kraak GJ, Portt CB, Munkittrick KR, Sibley PK, Smith IR, Dixon DG. 1991. Changes in hepatic mixed-function oxygenase (MFO) activity, plasma steroid levels and age at maturity of a white sucker (*Catostomus commersoni*) population exposed to bleached kraft mill effluent. *Aquat Toxicol* 21:199–218.

Mellin TN, Busch RD, Wang CC. 1983. Postsynaptic inhibition of invertebrate neuromuscular transmission by avermectin B1a. *Neuropharmacology* 22:89–96.

Meylan WM, Howard PH. 1991. Bond contribution method for estimating Henry's law constants. *Environ Toxicol Chem* 10:1283–1293.

Meylan WM, Howard PH. 1993. Computer estimation of the atmospheric gas-phase reaction rate of organic compounds with hydroxyl radicals and ozone. *Chemosphere* 12:2293–2299.

Meylan WM, Howard PH. 1995. Atom/fragment contribution method for estimating octanol/water partition coefficients. *J Pharm Sci* 84:83–92.

Meylan WM, Howard PH, Boethling RS. 1992. Molecular topology/fragment contribution method for predicting soil coefficients. *Environ Sci Technol* 26:1560–1567.

Middeldorp P, Briglia M, Kitunen V, Ralo V, Salkinoja-Salonen M. 1992. Biodegradation and transformation of polychlorinated phenols in soil. In: Behrebs D, Weisner J, editors. Mikrobiol Reinig Boeden, Beitr. Dechema-Fachgespraechs Umweltshcutz, 9th. Frankfurt/Main, Germany: Dechema. p 360–363.

Mikesell MD, Boyd SA. 1985. Reductive dechlorination of the pesticides 2,4-D, 2,4,5-T and pentachlorophenol in anaerobic sludges. *J Environ Qual* 14:337–340.

Mikesell MD, Boyd SA. 1986. Complete reductive dechlorination and mineralization of pentachlorophenol by anaerobic microorganisms. *Appl Environ Microbiol* 52:861–865.

Mill T, Mabey W. 1988. Hydrolysis of organic chemicals. In: Hutzinger O, editor. Volume 2, The handbook of environmental chemistry. Berlin, Germany: Springer-Verlag.

Millar K. 1993. Degradation of the high molecular weight fraction of bleached kraft mill effluent by biological and photochemical processes [masters thesis]. Guelph ON: Univ Guelph.

Miller GC, Hebert VR, Miille MJ, Mitzel R, Zepp RG. 1989. Photolysis of octachlorodibenzo-*p*-dioxin on soils. Production of 2,3,7,8-TCDD. *Chemosphere* 18:1265–1274.

Miller GC, Zepp RG. 1979. Photoreactivity of aquatic pollutants sorbed on suspended sediments. *Environ Sci Technol* 13:860–863.

Miller MA. 1993. Maternal transfer of organochlorine compounds in salmonines to their eggs. *Can J Fish Aquat Sci* 50:1405–1413.

Mineau P, Fox GA, Norstrom RJ, Weseloh DV, Hallett DJ, Ellenton JA. 1984. Using the herring gull to monitor levels and effects of organochlorine contamination in the Canadian Great Lakes. In: Nriagu JO, Simmons MS, editors. Toxic contaminants in the Great Lakes. New York: J Wiley.

Mohn WA, Tiedje JM. 1992. Microbial reductive dehalogenation. *Microbiol Rev* 56:482–507.

Moolgavkar SH, Knudson Jr AG. 1981. Mutation and cancer: a model for human carcinogenesis. *J National Cancer Inst* 66:1037–1052.

Moolgavkar SH, Venzon DJ. 1979. Two-event models for carcinogenesis: incidence curves for childhood and adult tumors. *Math Biosci* 47:55–77.

Moore RW, Jefcoate CR, Peterson RE. 1991. 2,3,7,8-tetrachlorodibenzo-*p*-dioxin inhibits steroidogenesis in the rat testis by inhibiting the mobilization of cholesterol to cytochrome P450scc. *Toxicol Appl Pharmacol* 109:85–97.

Moore RW, Potter CL, Theobald HM, Robinson JA, Peterson RE. 1985. Androgenic deficiency in male rats treated with 2,3,7,8-tetrachlorodibenzo-*p*-dioxin. *Toxicol Appl Pharmacol* 79:99–111.

Mora MA. 1991. Organochlorines and breeding success in cattle egrets from the Mexicali Valley, Baja California, Mexico. *Colonial Waterbirds* 14:127–132.

Morgan MJ, Kiceniuk JW. 1991. Recovery of foraging behaviour of Atlantic salmon exposed to a simulated commercial application of fenitrothion. *Environ Chem Toxicol* 10:961–965.

Morrison ML, Slack RD, Shanley E. 1978. Declines in environmental pollutants in olivaceus cormorant (*Phalacrocorax olivaceus*) eggs from Texas USA 1970–1977. *Wilson Bull* 90:639–42.

Morrissey RE, Schwetz BA. 1989. Reproductive and developmental toxicity in animals. In: Kimbrough RD, Jensen AA, editors. Halogenated biphenyls, terphenyls, naphthalenes, dibenzodioxins, and related products. 2nd ed. Amsterdam NL: Elsevier. p 195–225.

Mortensen P, Bergman A, Bignert A, Hansen HJ, Harkonen T, Olsson M. 1992. Prevalence of skull lesions in harbour seals (*Phoca vitulina*) in Swedish and Danish museum collections 1835–1988. *Ambio* 21:520–524.

Mortimer MR, Hughes JM. 1991. Effects of organophosphate pollution on genetic structure in two species of estuarine crabs. *Mar Pollut Bull* 22:352–359.

Mössner S, Spraker TR, Becker PR, Ballschmiter K. 1992. Ratios of enantiomers of alpha-HCH and determination of alpha-, beta- and gamma-HCH isomers in brain and other tissues of neonatal northern fur seals (*Callorhinus ursinus*). *Chemosphere* 24:1171–1180.

Mount DI. 1989. Methods for aquatic toxicity identification evaluations: Phase III toxicity confirmation procedures. Duluth MN: USEPA. EPA-600/3-88-036.

Mowbray DL. 1988. Pesticide use in the South Pacific. UNEP Regional Seas Reports and Studies No. 89 and South Pacific Regional Environmental Programme, Topic Review No. 26. Rome, Italy: UNEP. 275 p.

Mudambi AR, Hassett JP. 1988. Photochemical activity of mirex associated with dissolved organic matter. *Chemosphere* 17:1133–1146.

Mudambi AR, Hassett JP, McDowell WH, Scrudato RJ. 1992. Mirex and photomirex relationships in Lake Ontario. *J Great Lakes Res* 18:405–414.

Muir DCG, Knoll A, Yarechewski AL, Webster GRB. 1986. Bioconcentration of 1,3,6,8-tetrachlorodibenzodioxin and octachlorodibenzo-dioxin by fish. *Environ Toxicol Chem* 5:261–272.

Muir DCG, Segstro M, Welbourne PM, Toom D, Eisenreich SJ, Macdonald CR, Whelpdale D. 1993. Patterns of accumulation of airborne organochlorine contaminants in lichens from the upper Great Lakes region of Ontario. *Environ Sci Technol* 27:1201–1210.

Muir DCG, Wagemann R, Hargrave BT, Thomas DJ, Peakall DB, Norstrom RJ. 1992. Arctic marine ecosystem contamination. *Sci Tot Environ* 122:75–134.

Muir DCG, Yarechewski AL, Metner DS, Lockhart WL, Webster GRB, Friesen KJ. 1990. Dietary accumulation and sustained hepatic mixed function oxidase enzyme induction by 2,3,7,8-pentachlorobenzofuran in rainbow trout. *Environ Toxicol Chem* 9:1463–1472.

Mukherjee AB. 1990. The use of chlorinated paraffins and their possible effects in the environment. Publications of the water and environment administration: Series A66. Helsinki, Finland: National Board of Waters and the Environment. 53 p.

Müller JF, Hawker DW, Connell DW. 1994. Calculation of bioconcentration factors of persistent hydrophobic compounds in the air/vegetation system. *Chemosphere* 29:623–640.

Mullins JA, Carsel RF, Scarbrough JE, Ivery AM. 1993. PRZM-2 A model for predicting pesticide fate in the crop root zone and unsaturated soil zones: program and user's manual for Release 2.0. Athens GA: U.S. Environmental Protection Agency. EPA/600/R-93/046.

Munkittrick KR, Dixon DG. 1988. Evidence for a maternal yolk factor associated with increased tolerance and resistance of feral white sucker *Catostomus commersoni* to waterborne copper. *Ecotoxicol Environ Safety* 15:7–20.

Munkittrick KR, Dixon DG. 1989. Effects of natural exposure to copper and zinc on egg size and larval copper tolerance in white sucker (*Catostomus commersoni*). *Ecotoxicol Environ Safety* 18:15–26.

Munkittrick KR, McMaster ME, Portt CB, Van Der Kraak GJ, Smith IR, Dixon DG. 1992. Changes in maturity, plasma sex steroid levels, hepatic mixed function oxygenase activity, and the presence of external lesions in lake whitefish (*Coregonus clupeaformis*) exposed to bleached kraft mill effluent. *Can J Fish Aquat Sci* 49:1560–1569.

Munkittrick KR, Servos MR, Gorman K, Blunt B, McMaster ME, Van Der Kraak GJ. 1994. Characteristics of EROD induction associated with exposure to pulp mill effluent. 4th Symposium on Environmental Toxicology and Risk Assessment, American Soc for Testing and Materials (ASTM); 1994 Apr 11–13; Montreal, Quebec.

Munkittrick KR, Van Der Kraak GJ, McMaster ME, Portt CB. 1992. Response of hepatic MFO activity and plasma sex steroids to secondary treatment of bleached kraft pulp mill effluent and mill shutdown. *Environ Toxicol Chem* 11:1427–1439.

Munkittrick KR, Van Der Kraak GJ, McMaster ME, Portt CB, van den Heuvel MR, Servos MR. 1994. Survey of receiving-water environmental impacts associated with discharges from pulp mills. 2. Gonad size, liver size, hepatic EROD activity and plasma sex steroid levels in white sucker. *Environ Toxicol Chem* 13:1089–1101.

Murk AJ, Bosveld ATC, Barua A, van den Berg M, Brouwer A. 1994. Effects of polychlorinated aromatic hydrocarbons (PHAHs) on biochemical parameters in chicks of the common tern (*Sterna hirundo*). *Aquat Toxicol* 30:91–115.

Murk AJ, van den Berg JHJ, Koeman JH, Brouwer A. 1991. The toxicity of tetrachlorobenzyltoluenes (Ugilec 141) and polychlorinated biphenyls (Aroclor 1254 and PCB-77) compared in Ah-responsive and Ah-nonresponsive mice. *Environ Pollut* 72:46–67.

Murphy NBK, Kaufmann DD, Fries GF. 1979. Degradation of pentachlorophenol (PCP) in aerobic and anaerobic soil. *J Environ Sci Health* B14:1–14.

Murray FJ, Smith FA, Nitschke KO, Huniston CG, Kociba RJ, Schwetz BA. 1979. Three-generation reproduction study of rats given 2,3,7,8-tetrachlorodibenzo-*p*-dioxin (TCDD) in the diet. *Toxicol Appl Pharmacol* 50:241–252.

Myers MS, Rhodes LD, McCain BB. 1987. Pathologic anatomy and patterns of occurrence of hepatic neoplasms, putative preneoplastic lesions, and other idiopathic hepatic conditions in English sole (*Parophrys vetulus*) from Puget Sound, Washington, U.S.A. *J Nat Cancer Inst* 78:333–363.

Myers MS, Stehr CM, Olson OP, Johnson LL, McCain BB, Chan SL, Varanasi U. 1994. Relationships between toxicopathic hepatic lesions and exposure to chemical contaminants in English sole (*Pleuronectes vetulus*), starry flounder (*Platichthys stellatus*), and white croaker (*Genyonemus lineatus*) from selected marine sties on the Pacific coast, USA. *Environ Health Perspect* 102:200–215.

Nabholz JV, Miller P, Zeeman M. 1993. Environmental risk assessment of new chemicals under the toxic substances control act TSCA Section Five. In: Landis WG, Hughes JS, Lewis MA, editors. Environmental toxicology and risk assessment. Philadelphia PA: American Soc for Testing and Materials (ASTM). STP 1179. p 40–55.

[NAS] National Academy of Sciences. 1983. Risk assessment in the federal government: managing the process. Washington DC: NAS.

Neidleman SL, Geigert J. 1986. Biohalogenation: principles, basic roles, and applications. New York: Halsted Pr.

Neilson AH, Allard AS, Hynning PA, Remberger M, Landner L. 1983. Bacterial methylation of chlorinated phenols and guaiacols: formation of veFatroles from guaiacols and high-molecular weight chlorinated lignin. *Appl Environ Microbiol* 45:774–783.

Neilson AH, Allard AS, Hynning P-A, Remberger M, Viktor T. 1990. The environmental fate of chlorophenolic constituents of bleachery effluents. *Tappi J* 239–247.

Nelson JA. 1974. Effects of dichlorodiphenyltrichloroethane (DDT) analogs and polychlorinated biphenyl (PCP) mixtures on 17ß-[3H] estradiol binding to rat uterine receptor. *Biochem Pharmacol* 23:447–451.

Newton L, Bogan J. 1978. The role of different organochlorine compounds in the breeding of British sparrowhawks. *J Appl Ecol* 15:105–116.

Nichols JW, McKim JM, Andersen ME, Gargas ML, Clewell HJ, Erickson RJ. 1990. A physiologically based toxicokinetic model for the uptake and disposition of waterborne organic chemicals in fish. *Toxicol Appl Pharmacol* 106:433–447.

Niimi AJ. 1983. Biological and toxicological effects of environmental contaminants in fish and their eggs. *Can J Fish Aquat Sci* 40:306–312.

Niimi AJ. 1986. Biological half-lives of chlorinated diphenyl ethers in rainbow trout (*Salmo gairdneri*). *Aquat Toxicol* 9:105–116.

Niimi AJ, Metcalfe CD, Huestis SY. 1994. Chlorinated diphenylethers in Great Lakes fish and their environmental implication. *Environ Toxicol Chem* 13:1133–1138.

Niimi AJ, Oliver BG. 1988. Influence of molecular weight and molecular volume on dietary absorption efficiency of chemicals by fishes. *Can J Fish Aquat Sci* 45:222–227.

Norman R, Begon M, Bowers RG. 1994. The population dynamics of microparasites and vertebrate hosts: the importance of immunity and recovery. *Theoretical Population Biol* 46:96–119.

Norokorpi Y, Frank H. 1993. Effect of stand density on damage to birch (*Betula pubescens*) caused by phytotoxic air pollutants. *Annalen Botanica Fennica* 30:181–187.

Norstrom RJ. 1988. Patterns and trends of PCB contamination in Canadian wildlife. In: Crine J-P, editor. Hazards, decontamination and replacement of PCB. New York: Plenum. p 85–100.

Norstrom RJ, Simon M, Muir CDG. 1990. Chlorinated dioxins and furans in marine mammals from the Canadian arctic. *Environ Pollut* 66:1–20.

Norstrom RJ, Muir DCG, Ford CA, Simon M, Macdonald CR, Béland P. 1992. Indication of P450 monooxygenase activities in beluga (*Delphinapterus leucas*) and narwhal (*Monodon monoceros*). *Mar Environ Res* 34:278–272.

Norstrom RJ, Muir DCG. 1994. Chlorinated hydrocarbon contaminants in arctic marine mammals. *Sci Tot Environ* 154:107–128.

[NRC] National Research Council. 1993. Issues in risk assessment. Washington DC: National Academy Pr.

[NTP] National Toxicology Program. 1980. Bioassay of a mixture of 1,2,3,6,7,8-hexachlorodibenzo-*p*-dioxin and 1,2,3,7,8,9-hexachlorodibenzo-*p*-dioxin for possible carcinogenicity (gavage study). Research Triangle Park NC: U.S. Department of Health and Human Services (DHHS), Public Health Service (PHS). Technical Report, Series No 198.

[NTP] National Toxicology Program. 1993. Pesticide/fertilizer mixtures administered in drinking water to F344/N rats and B6C3F1 mice. Research Triangle Park NC: Public Health Service (PHS), National Institutes of Health (NIH). Toxicology Report Series 36. NIH 93-3385.

Odsjo T. 1993. The Swedish Environmental Specimen Bank with reference to the National Contaminant Monitoring Programme in Sweden. *Sci Tot Environ* 139/140:147–156.

Olsson M. 1985. The environmental specimen bank at the Swedish Museum of Natural History. In: Wise SA, Zeisler R, editors. International review of environmental specimen banking. Washington DC: U.S. Department of Commerce. National Bureau of Standards. Special Publication 706, Section 5. p 26–33.

Olsson M, Andersson O, Bergman A, Blomkvist G, Frank A, Rappe C. 1992. Contaminants and diseases in seals from Swedish waters. *Ambio* 21:561–562.

Olsson M, Karlsson B, Ahnland E. 1994a. Diseases and environmental contaminants in seals from the Baltic and Swedish west coast. *Sci Tot Environ* 154:217–227.

Olsson M, Karlsson B, Ahnland E. 1994b. Vad hander med salarna ? Naturvardsverket, Rapport 4254, 1–72. Stockholm: The Swedish Environmental Protection Agency. Report 4254 (English summary). p 56–57.

Opperhuizen A, Sijm DTHM. 1990. Bioaccumulation and biotransformation of polychlorinated dibenzo-*p*-dioxins and dibenzofurans in fish. *Environ Toxicol Chem* 9:175–186.

Opperhuizen A, v.d. Velde EW, Gobas FAPC, Liem DAK, v.d. Steen JMD. 1985. Relationship between bioconcentration in fish and steric factors of hydrophobic chemicals. *Chemosphere* 14:1871–1986.

Opperhuizen A, Voors PI. 1987. Bioconcentration kinetics of 2,4,5 tri and 3,3',4,4' tetrachlorobiphenyl and 2,4,5 tri and 3,3',4,4' tetrachlorodiphenylether in fish. *Chemosphere* 16:2379–2388.

Österberg F, Lindström K. 1985a. Characterization of the high molecular mass chlorinated matter in spent bleach liquors. (SBL). Part II. Acidic SBL. *Holzforschung* 39:149–158.

Österberg F, Lindström K. 1985b. Characterization of the high molecular mass chlorinated matter in spent bleach liquors. (SBL). Part III. Mass spectrometric interpretation of aromatic degradation products of SBL. *Org Mass Spectroscopy* 20:515–524.

Owens JW, Swanson SM, Birkholtz DA. 1994. Bioaccumulation of 2,3,7,8-tetrachlorodibenzo-*p*-dioxin, 2,3,7,8-tetrachlorodibenzofuran and extractable organic chlorine at a bleached-kraft mills site in a Northern Canadian river system. *Environ Toxicol Chem* 13:343–354.

Panday AC. 1988. Impact of endosulfan (thiodan) EC 35 on behavior and dynamics of oocyte development in the teleostean fish, *Colisa (Trichogaster) fasciatus*. *Ecotoxicol Environ Safety* 15:221.

Pankow JF. 1987. Review and comparative analysis of the theories on partitioning between the gas and aerosol particulate phases in the atmosphere. *Atmospher Environ* 21:2275–2283.

Pankow JF, Bidleman TF. 1992. Interdependence of the slopes and intercepts from log-log correlations of measured gas-particle partitioning and vapor pressure. 1. Theory and analysis of available data. *Atmospher Environ* 26A:1071–1080.

Paris DF, Rogers JE. 1986. Kinetic concepts for measuring microbial rate constants: effects of nutrients on rate constants. *Appl Environ Microbiol* 51:221–225.

Paris DF, Steen WC, Baughman GL, Barnett JT. 1981. Second-order model to predict microbial degradation of organic compounds in natural waters. *Appl Environ Microbiol* 41:603–609.

Parkerton TF, Connolly JP, Thomann RV, Uchrin CG. 1993. Do aquatic effects of human health end points govern the development of sediment-quality criteria for nonionic organic chemicals? *Environ Toxicol Chem* 12:507–523.

Parrott JL, Hodson PV, Servos MR, Huestis SL, Dixon DG. 1995. Relative potencies of polychlorinated dibenzo-*p*-dioxins and dibenzofurans for inducing mixed function oxygenase activity in rainbow trout. *Environ Toxicol Chem* 14:1041–1050.

Paterson S, Mackay D, Bacci E, Calamari D. 1991. Correlation of the equilibrium and kinetics of leaf-air exchange of hydrophobic organic chemicals. *Environ Sci Technol* 25:866–871.

Paterson S, Mackay D, McFarlane C. 1994. A model of organic chemical uptake by plants from soil and the atmosphere. *Environ Sci Technol* 28:2259–2266.

Peakall DB, Peakall ML. 1973. Effect of a chlorinated biphenyl on the reproduction of artificially and naturally incubated dove eggs. *J Appl Ecol* 10:863–868.

Pearson CR. 1982. C1 and C2 halocarbons. In: Hutzinger O, editor. Volume 3B, Environmental chemistry. Berlin, Germany: Springer Verlag. p 89–116.

Peijnenburg WJGM, Hart MJT, den Hollander HA, van de Meent D, Verboom HH, Wolfe NL. 1992. QSARs for predicting reductive transformation rate constants of halogenated aromatic hydrocarbons in anoxic sediment systems. *Environ Toxicol Chem* 11:301–314.

Pellinen J, Salkinoja-Salonen M. 1985. High performance size exclusion chromatography (HPSEC) of lignin and its derivatives. *J Chromatography* 328:299–308.

Perdue EM, Wolfe NL. 1983. Prediction of buffer catalysis in field and laboratory studies of pollutant hydrolysis reactions. *Environ Sci Technol* 17:635–642.

Peterle TJ. 1969. DDT in Arctic snow. *Nature* 224:620–620.

Peterson RE, Theobald HM, Kimmel GL. 1993. Developmental and reproductive toxicity of dioxins and related compounds: cross-species comparisons. *CRC Crit Rev Toxicol* 23:283–335.

Peterson RE, Walker MK. 1992. Effects on aquatic and terrestrial organisms. Current views on the impact of dioxins and furans on human health and the environment. Session V: Ecotoxicological aspects. Proceedings of the Toxicology Forum; 1992 Nov 9–11; Berlin, Germany. p 383–399.

Peterson RH. 1994. Toxicity of PCDDs, PCDFs and PCBs to juvenile salmonids. Paper presented at the SETAC Workshop on Environmental Risk Assessment for Organochlorine Chemicals; 1994 Jul 24–29; Alliston ON.

Petreas MX, Goldman LR, Hayward DG, Chang RR, Flattery JJ, Weismüller T, Stephens RD. 1991. Biotransfer and bioaccumulation of PCDD/PCDFs from soil: controlled exposure studies of chickens. *Chemosphere* 23:1731–1741.

PHOXA. 1989. Photochemical oxidants acid deposition model application: development emission data base. Delft NL: TNO Institute of Environmental Sciences. PHOXA Report No 1., TNO Report 88-120.

Pignatello JJ, Martinsen MM, Steiert JG, Carlson RE, Crawford RL. 1983. Biodegradation and photolysis of pentachlorophenol in artificial freshwater streams. *Appl Environ Microbiol* 46:1024–1031.

Pitot HC, Goldsworthy T, Campbell HA, Poland A. 1980. Quantitative evaluation of promotion by 2,3,7,8-tetrachlorodibenzo-*p*-dioxin of hepatocarcinogenesis from diethylnitrosamine. *Cancer Res* 40:3616–3620.

Poland A, Knutson JC. 1982. 2,3,7,8-tetrachlorodibenzo-*p*-dioxin and related halogenated aromatic hydrocarbons: examination of the mechanism of toxicity. *Annu Rev Pharmacol Toxicol* 22:517–554.

Poland A, Teitelbaum P, Glover E. 1989. [125I]2-iodo-3,7,8-trichlorodibenzo-*p*-dioxin-binding species in mouse liver induced by agonists for the Ah receptor: characterization and identification. *Mol Pharmacol* 36:113–120.

Pong SS, DeHaven R, Wang CC. 1982. A comparative study of avermectin B1a and other modulators of the gamma-aminobutyric acid receptor-chloride ion channel complex. *J Neuroscience* 2:966–971.

Pontasch KW, Niederlehner BR, Cairns Jr J. 1988. Comparisons of single species, microcosm, and field responses to a complex effluent. *Environ Toxicol Chem* 8:521–532.

Posner JC. 1991. Evaluation of sorbents for He collection and analysis of trace levels of airborne vapors: bis(2-chloroethyl)sulfide (Mustard), a case study. *Chemosphere* 22:461–472.

Potts GR. 1977. Population dynamics of the grey partridge: overall effects of herbicides and insecticides on chick survival rates. *Int Congress Game Biol* 13:203–211.

Price IM, Weseloh DV. 1986. Increased numbers and productivity of double-crested cormorants (*Phalacrocorax auritus*) on Lake Ontario. *Can Field Naturalist* 100:474–482.

Pzenny AAP, Keene WC, Jacob DJ, Fan S, Maben JR, Zetwo MP, Springer-Young M, Galloway N. 1993. Evidence of inorganic chlorine gases other than hydrogen chloride in marine surface air. *Geophysical Res Letters* 20:699–702.

Quensen III JF, Tiedje JM, Boyd SA. 1988. Reductive dechlorination of polychlorinated biphenyls by anaerobic microorganisms from sediments. *Science* 242:752–754.

Raloff J. 1996. The pesticide shuffle. *Science News* 149:174–175.

Ramesh A, Tanabe S, Kannan K, Subramanian AN, Kumaran PL, Tatsukawa R. 1992. Characteristic trend of persistent organochlorine contamination in wildlife from a tropical agricultural watershed, South India. *Arch Environ Contam Toxicol* 23:26–36.

Rands MRW. 1985. Pesticide use on cereals and the survival of grew partridge chicks: a field experiment. *J Appl Ecol* 22:49–54.

Ratcliffe DA. 1967. Decrease in eggshell weight in certain birds of prey. *Nature* 215:208–210.

Ratcliffe DA. 1970. Changes attributable to pesticides in egg breakage frequency and eggshell thickness in some British birds. *J Appl Ecol* 7:67–115.

[RCRA] Resource Conservation and Recovery Act. 1976. Public Law 94-580, 90 Stat. 2795 (42 U.S.C. § 6901 et seq.), October 21, 1976.

Reijnders PJH. 1976. The harbour seal (*Phoca vitulina*) population in the Dutch Wadden Sea: size and composition. *Netherlands J Sea Res* 10:223–235.

Reijnders PJH. 1980. Organochlorine and heavy metal residues in harbour seals from the Wadden Sea and their possible effects on reproduction. *Netherlands J Sea Res* 14:30–65.

Reijnders PJH. 1986. Reproductive failure in common seals feeding on fish from polluted coastal waters. *Nature* 324:456–457.

Reischl A, Reissinger M, Hutzinger O. 1987. Occurrence and distribution of atmospheric organic micropollutants in conifer needles. *Chemosphere* 16:2647–2652.

Reitz RH, Mendrala AL, Park CN, Andersen ME, Guengerich FP. 1988. Incorporation of in vitro enzyme data into the physiologically-based pharmacokinetic (PB-PK) model for methylene chloride: implications for risk assessment. *Toxicol Letters* 43:97–116.

Remberger M, Allard A-S, Nielson AH. 1986. Biotransformations of chloroguaiacols, chlorocatechols and chloroveratroles in sediments. *Appl Environ Microbiol* 51:552–558.

Ritter L, Solomon KR, Forget J, Stemeroff M, O'Leary C. 1995. Persistent organic pollutants: an assessment report on aldrin, chlordane, DDT, dieldrin, dioxins and furans, endrin, heptachlor, hexachlorobenzene, mirex, polychlorinated biphenyls, and toxaphene. IPCS Interorganization Programme for the Sound Management of Chemicals; 1995 Nov; Rome, Italy. Rome: UNEP, ILO, FAO, UNIDO, OECD.

Robbins CS, Springer PF, Webster CG. 1951. Effects of five-year DDT application on breeding bird population. *J Wildlife Management* 15:213–216.

Roberts AL, Jeffers PM, Wolfe NL, Gschwend PM. 1993. Structure-reactivity relationships in dehydrohalogenation reactions of polychlorinated and polybrominated alkanes. *Crit Rev Environ Sci Technol* 23:1–39.

Robinson RD. 1994. Evaluation and development of laboratory protocols for estimating reproductive impacts of pulp mill effluent on fish [PhD thesis]. Guelph ON: Univ Guelph.

Robinson RD, Carey JH, Solomon KR, Smith IR, Servos MR, Munkittrick KR. 1994. Survey of receiving-water environmental impacts associated with discharges from pulp mills. 1. Mill characteristics, receiving-water chemical profiles and lab toxicity tests. *Environ Toxicol Chem* 13:1075–1088.

Rochkind ML, Blackburn JW, Sayler GS. 1986. Microbial decomposition of chlorinated aromatic compounds. Washington DC: U.S. Environmental Protection Agency (USEPA). EPA/600/2-86/090.

Rogan WJ, Gladen BC, Hung K-L, Koong S-L, Shih LY, Taylor JS, Wu Y-C, Yang D, Ragan NB, Hsu C-C. 1988. Congenital poisoning by polychlorinated biphenyls and their contaminants in Taiwan. *Science* 241:334–338.

Ronnefahrt B. 1987. Nachweis und Bestimmung des PCB-Ersatprodukte Ugilec in Wasserproben und Fischen aus der Lippe. *Deutsches Lebensmittel-Rundschau* 83:214–218.

Rosene Jr W. 1965. Effects of field applications of heptachlor on bobwhite quail and other wild animals. *J Wildlife Management* 29:554–580.

Ross PS, de Swart RL, Reijnders PJH, Van Loveren H, Vos JG, Osterhaus ADME. 1995. Contaminant-related suppression of delayed-type hypersensitivity and antibody responses in harbour seals fed herring from the Baltic Sea. *Environ Health Perspect* 103:162–167.

Rozman K, Rozman T, Greim H. 1981. Enhanced fecal elimination of stored hexachlorobenzene from rats and rhesus monkeys by hexadecane or mineral oil. *Toxicology* 22:33–44.

Safe S. 1984. Polychlorinated biphenyls (PCBs) and polybrominated biphenyls (PBBs): biochemistry, toxicology, and mechanism of action. *Crit Rev Toxicol* 13:319–395.

Safe S. 1989. Polychlorinated biphenyls (PCBs): mutagenicity and carcinogenicity. *Mutation Res* 220:31–47.

Safe S. 1990. Polychlorinated biphenyls (PCBs), dibenzo-*p*-dioxins (PCDDs), dibenzofurans (PCDFs), and related compounds: environmental and mechanistic considerations which support the development of toxic equivalency factors (TEFs). *Crit Rev Toxicol* 21:51–88.

Salkinoja-Salonen MS, Jokela JK. 1991. Measurement of organic halogen compounds in urine as an indicator of exposure. *Scand J Workplace Environ Health* 17:75–78.

Sandermann H, Arjmand M, Gennity I, Winkler R. 1990. Animal bioavailability of defined xenobiotic lignin metabolites. *J Agric Food Chem* 38:1877–1880.

Sandermann H, Dieper D-H, Winkler R. 1993. Mineralization of lignin-bound and free xenobiotics by the white-rot fungus, *Phanerochaete chrysosporium*. In: Kennedy JF, Phillips GO, Williams PA. Cellulosics: pulp, fibre and environmental aspects. New York: Ellis. p 499–503.

[SARA] Superfund Amendments and Reauthorization Act. 1986. Public Law 99-499, 100 Stat. 1613, October 17, 1986. (Amended the Comprehensive Environmental Response, Compensation, and Liability Act [CERCLA], Public Law 96-510, 94 Stat. 2767 [42 U.S.C. § 9601 et seq.), December 11, 1980.

Saski A, Salkinoja-Salonen M, Vahatalo A, Salonen K. 1991. Degradation and environmental fate of bleached pulp mill effluents studied in freshwater. In: Sodergren A, editor. Environmental fate and effects of bleached pulp mill effluents. Stockholm: Swedish Environmental Protection Agency. SEPA 4031. p 183–189.

Saski E, Jokela JK, Salkinoja-Salonen MS. 1996. Biodegradability of different size classes of bleached kraft pulp mill effluent organic halogens during wastewater treatment and in lake environments. In: Servos MR, Munkittrick KR, Carey JH, van Der Kraak GJ, editors. Environmental fate and effects of pulp and paper mill effluents. Delray Beach FL: St. Lucie Pr. p 179–193.

Schantz SL, Bowman RE. 1989. Learning in monkeys exposed perinatally to 2,3,7,8-tetrachlorodibenzo-*p*-dioxin (TCDD). *Neurotoxicol Teratol* 11:13–19.

Scheil S, Baugarten G, Beiter R, Schwartz S, Wagner J-O, Mathies M, Trapp S. 1994. CemoS - Eine objekt-orientierte Software zur Expositionsmodellierung. *Ecoinforma* 7:391–404.

Schellenberg K, Leuenberger C, Schwanwnbach RP. 1984. Sorption of chlorinated phenols by natural sediments and aquifer materials. *Environ Sci Technol* 18:652–657.

Schiewe MH, Weber DD, Myers MS, Jacques FJ, Reichert WL, Krone CA, Malins DC, McCain BB, Chan S-L, Varanasi U. 1991. Induction of foci of cellular alteration and other hepatic lesions in English sole (*Parophrys vetulus*) exposed to an extract of an urban marine sediment. *Can J Fish Aquat Sci* 48:1750–1760.

Schwarzenbach RP, Giger W, Schaffner C, Wanner O. 1985. Groundwater contamination by volatile halogenated alkanes: abiotic formation of volatile sulfur compounds under anaerobic conditions. *Environ Sci Technol* 19:322–327.

Sedlak DL, Andren AW. 1991. Aqueous-phase oxidation of polychlorinated biphenyls by hydroxyl radicals. *Environ Sci Technol* 25:1419–1427.

Seefeld MS, Corbett SW, Keesey RE, Peterson RE. 1984. Characterization of the wasting syndrome in rats treated with 2,3,7,8-tetrachlorodibenzo-*p*-dioxin. *Toxicol Appl Pharmacol* 73:311–322.

Selkirk JK, Soward SM. 1993. Compendium of abstracts from long-term cancer studies reported by the National Toxicology Program of the National Institute of Environmental Health Sciences from 1976 to 1992. *Environ Health Perspect* 101(Suppl. 1):1–295.

Servos MR, Huestis SY, Whittle DM, Van Der Kraak GJ, Munkittrick KR. 1994. Survey of receiving water environmental impacts associated with discharges from pulp mills. III Polychlorinated dioxins and furans in muscle and liver of white sucker (Catostomus commersoni). *Environ Toxicol Chem* 13:1103–1115.

Servos MR, Muir DCG, Webster GRB. 1989. Effect of particulate and dissolved organic carbon on the bioavailability of polychlorinated dioxins. *Aquat Toxicol* 14:169–184.

Shaw PD, Hager LP. 1959. An enzymatic chlorination reaction. *J Am Chem Soc* 81:1011–1012.

Shea D. 1988. Developing national sediment quality criteria. *Environ Sci Technol* 22:1256–1261.

Sijm DTHM, Opperhuizen A. 1988. Biotransformation, bioaccumulation, and lethality of 2,8-dichlorodibenzo-*p*-dioxin: a proposal to explain the biotic fate of PCDDs and PCDFs. *Chemosphere* 17:83–89.

Sijm DTHM, Opperhuizen A. 1989. Biotransformation of organic chemicals by fish: a review of enzyme activities and reactions. In: Hutzinger O. Volume 2, Handbook of environmental chemistry. Part E: Reactions and Processes. Heidelberg, Germany: Springer-Verlag. p 163–235.

Sijm DTHM, Part P, Opperhuizen A. 1993. The influence of temperature on the uptake rate constants of hydrophobic compounds determined by the isolated perfused gills of rainbow trout (*Oncorhynchus mykiss*). *Aquat Toxicol* 25:1–14.

Sijm DTHM, Seinen W, Opperhuizen A. 1992. Life-cycle study on biomagnification in fish. *Environ Sci Technol* 26:2162–2174.

Sijm DTHM, Sinnige TL. 1995. Experimental octanol/water partition coefficients of chlorinated paraffins. *Chemosphere* 31:4427–4435.

Sijm DTHM, Wever H, Opperhuizen A. 1993. Congener-specific biotransformation and bioaccumulation of PCDDs and PCDFs from fly ash in fish. *Environ Toxicol Chem* 12:1895–1907.

Simonich SL, Hites RA. 1994. Vegetation-atmosphere partitioning of polycyclic aromatic hydrocarbons. *Environ Sci Technol* 28:939–943.

Singhal RL, Vijayvargiya R, Shukla GS. 1985. Toxic effects of cadmium and lead on reproductive function. In: Thomas JA, Korach KS, McLachlan JA, editors. Endocrine toxicology. New York: Raven Pr. p 149–179.

Skoglund RS, Swackhamer DL. 1992. In: Baker LA, editor. Environmental chemistry of lakes and reservoirs. Washington DC: American Chemical Soc (ACS).

Skurlatov YI, Zepp RG, Baughman GL. 1983. Photolysis rates of (2,4,5-trichlorophenoxy)acetic acid and 4-amino-3,5,6-trichloropicolinic acid in natural waters. *J Agric Food Chem* 31:1065–1071.

Slooff W, Bont PFH, Janus JA, Annema JA. 1992. Exploratory report chlorinated paraffins. Bilthoven NL: National Institute of Public Health and Environmental Protection. Report Nr 710401016. 48 p.

Smialowicz RJ, Riddle MM, Williams WC, Dilberto JJ. 1994. Effects of 2,3,7,8-tetrachlorodibenzo-*p*-dioxin (TCDD) on humoral immunity and lymphocyte subpopulations: differences between mice and rats. *Toxicol Appl Pharmacol* 124:248–256

Smith JH. 1986. Role of renal metabolism in chloroform nephrotoxicity. *Comments Toxicol* 1:125–144.

Smith JH, Maita K, Sleight SD, Hook JB. 1983. Mechanism of chloroform toxicity. I. Time course of chloroform in male and female mice. *Toxicol Appl Pharmacol* 70:467–479.

Smrchek J, Morcock R, Clements R, Rabert W. 1993. Assessing ecological hazard under TSCA: methods and evaluation of data. In: Landis WG, Hughes JS, Lewis MA, editor. Environmental toxicology and risk assessment. Philadelphia PA: American Soc for Testing and Materials (ASTM). p 22–39.

Snoo GR, Canters KJ, de Jong FMW, Cuperus R. 1994. Integral hazard assessment of side effects of pesticides in the Netherlands: a proposal. *Environ Toxicol Chem* 13:1331–1340.

Solomon KR, Baker DB, Richards P, Dixon KR, Klaine SJ, La Point TW, Kendall RJ, Giddings JM, Giesy JP, Hall Jr LW, Weisskopf CP, Williams M. 1996. Ecological risk assessment of atrazine in North American surface waters. *Environ Toxicol Chem* 15:31–76.

Solomon KR, Bergman H, Mackay D, McKague B. 1994. A review and assessment of the ecological risks associated with the use of chlorine dioxide for the bleaching of pulp. Proceedings of the 1994 International Pulp Bleaching Conference; 1994 Jun 13–16; Vancouver BC. Montreal PQ: Canadian Pulp and Paper Assn.

Soto AM, Chung KL, Sonnenschein C. 1994. The pesticides endosulfan, toxaphene and dieldrin have effects on human estrogen-sensitive cells. *Environ Health Perspect* 102:380–383.

Spencer WF, Claith MM, Jury WA, Zhang L-Z. 1988. Volatilization of organic chemicals from soil as related to their Henry's Law Constants. *J Environ Qual* 17:504–509.

Spitsbergen JM, Walker MK, Olson JR, Peterson RE. 1991. Pathologic alterations in early life stages of lake trout, *Salvelinus namaycush*, exposed to 2,3,7,8-tetrachlorodibenzo-*p*-dioxin as fertilized eggs. *Aquat Toxicol* 19:41–72.

Spitzer PREA. 1978. Productivity of ospreys in Connecticut: Long Island increases as DDE residues decline. *Science* 202:333–335.

Stein JE, Collier TK, Reichert WL, Casillas E, Varanasi U. 1992. Bioindicators of contaminant exposure and sublethal effects: studies with benthic fish in Puget Sound, WA. *Environ Toxicol Chem* 11:701–714.

Steuer J, Jaeger S, Hammond R, Patterson D, Bhattacharyya J. 1994. Deterministic modeling of PCB transport in the Lower Fox River between Lake Winnebago and DePere Wisconsin: data synthesis and model development and validation. Madison WI: Wisconsin Dept. Natural Resources.

Stickel WH. 1975. Some effects of pollutants in terrestrial ecosystems. In: McIntyre AD, Mills FF, editors. Ecological research. New York: Plenum. p 25–74.

Stolen JS, Fletcher TC, editors. 1994. Modulators of fish immune responses. Volume 1, Models for environmental toxicology/biomarkers, immunostimulators. Fair Haven NY: SOS. 252 p.

Stone WB, Gradoni PB. 1985. Wildlife mortality related to use of the pesticide diazinon. *Northeast Environ Sci* 4:30–38.

Strachan WMJ, Eriksson G, Kylin H, Jensen S. 1993. Organochlorine compounds in pine needles: methods and trends. *Environ Toxicol Chem* 13:443–451.

Strachan WMJ, Eisenreich SJ. 1988. Mass balancing of toxic chemicals in the Great Lakes: the role of atmospheric deposition. Report of the Workshop on Estimation of Atmospheric Loadings of Toxic

Chemicals to the Great Lakes Basin; 1986 Oct; Scarborough ON. Windsor ON: International Joint Commission (IJC). 113 p.

Suedel BC, Boraczek JA, Peddicord RK, Clifford PA, Dillon TM. 1994. Trophic transfer and biomagnification potential of contaminants in aquatic ecosystems. *Rev Environ Contam Toxicol* 136:21–89.

Suflita JM, Horowitz A, Shelton DR, Tiedje JM. 1982. Dehalogenation: a novel pathway for the anaerobic biodegradation of haloaromatic compounds. *Science* 218:1115–1117.

Suflita JM, Stout J, Tiedje JM. 1984. Dechlorination of 2,4,5-trichlorophenoxy acetic acid by anaerobic microorganisms. *J Agric Food Chem* 32:218–221.

Sun K, Krause GF, Ellersieck MR, Mayer FL, Basu AP. 1994. Predicting chronic toxicity based on the theory of accelerated life testing. EPA 600/R-94/058. Gulf Breeze FL: U.S. Environmental Protection Agency (USEPA).

Suter G. 1993. Ecological risk assessment. Ann Arbor MI: Lewis.

Svanberg O. 1983. Chlorinated paraffins. A review of environmental behaviour and effects. Report No. 1614, National Swedish Environmental Protection Board, Solna, Sweden. 52 p.

Svenson A, Kaj L. 1989. Photochemical conversion of chlorinated phenolic substances in aquatic media as studied by AOX and microtox tests. *Sci Tot Environ* 78:89–98.

Swackhamer DL, Hites RA. 1988. Occurrence and bioaccumulation of organochlorine compounds in fishes from Siskiwit Lake, Isle Royale, Lake Superior. *Environ Sci Technol* 22:543–548.

Swackhamer DL, Skoglund RS. 1992. Bioaccumulation of PCBs by algae: kinetics versus equilibrium. *Environ Toxicol Chem* 12:831–838.

Swanson SM, Schryer R, Shelast B, Holley K, Berbekar I, Kloepper-Sams P, Owens JW, Steeves L, Birkholz D, Marchant T. 1993. Wapiti Smoky River ecosystem study. Grande Prairie, Alberta, Canada: Weyerhaeuser. 175 p.

Swindoll CM, Aelion CM, Dobbins DC, Jiang O, Long SC, Pfaender FK. 1988. Aerobic biodegradation of natural and xenobiotic organic compounds by subsurface microbial communities. *Environ Toxicol Chem* 7:291–299.

Swindoll CM, Aelion CM, Pfaender FK. 1988. Influence of inorganic and organic nutrients on aerobic biodegradation and on the adaptation response of subsurface microbial communities. *Appl Environ Microbiol* 54:212–217.

Takeoka H, Ramesh A, Iwata H, Tanabe S, Subramanian AN, Mohan D, Magendran A, Tatsukawa R. 1991. Fate of the insecticide HCH in the tropical coastal area of South India. *Mar Pollut Bull* 22:290–297.

Tanabe S, Gondaira F, Subramanian AN, Ramesh A, Mohan D, Kumaran P, Venugopalan VK, Tatsukawa R. 1990. Specific pattern of persistent organochlorine residues in human breast milk from South India. *J Agric Food Chem* 38:899–903.

Tanabe S, Kannan K, Tabucanon MS, Siriwong C, Ambe Y, Tatsukawa R. 1991. Organochlorine pesticide and polychlorobiphenyl residues in foodstuffs from Bangkok, Thailand. *Environ Pollut* 72:191–203.

Tanabe S, Mori T, Tatsukawa R, Miyazaki N. 1983. Global pollution of marine mammals by PCBs, DDTs and HCHs (BHCs). *Chemosphere* 12:1269–1275.

Tanabe S, Ramesh A, Sakashita D, Iwata H, Tatsukawa R, Mohan D, Subramanian AN. 1991. Fate of HCH (BHC) in tropical paddy field: application test in South India. *Int J Environ Analytical Chem* 45:45–53.

Tanabe S, Tatsukawa R, Kawano M, Hidaka H. 1982. Global distribution and atmospheric transport of chlorinated hydrocarbons HCH (BHC) isomers and DDT compounds in the Western Pacific, Eastern Indian and Antarctic Oceans. *J Oceanographic Soc Japan* 38:137–148.

Tanabe S, Watanabe S, Kan H, Tatsukawa R. 1988. Capacity and mode of PCB metabolism in small cetaceans. *Mar Mammal Sci* 4:103–124.

Tatsukawa R, Tanabe S. 1990. Fate and bioaccumulation of persistent organochlorine compounds in the marine environment. In: Baumgartner DJ, Duedall IW, editors. Volume 6, Oceanic process in marine pollution. Malabar FL: Krieger. p 39–52.

Tejning S. 1967. Mercury in pheasants (*Phasianus colchicus* L.) deriving from seed grain dressed with methyl and ethyl mercury compounds. *Oikos* 18:334–344.

Thomann RV. 1981. Equilibrium model of fate of microcontaminants in diverse aquatic food chains. *Can J Fish Aquat Sci* 38:280–296.

Thomann RV. 1989. Bioaccumulation model of organic chemical distribution in aquatic food chains. *Environ Sci Technol* 23:699–707.

Thomann RV, Connolly JP. 1984. Model of PCB in the Lake Michigan lake trout food chain. *Environ Sci Technol* 18:65–71.

Thomann RV, Connolly JP, Parkerton T. 1992. An equilibrium model of organic chemical accumulation in aquatic food webs with sediment interaction. *Environ Toxicol Chem* 11:615–629.

Thomann RV, DiToro DM. 1983. Physico-chemical model of toxic substances in the Great Lakes. *J Great Lakes Res* 9:474–496.

Thomas DJ, Hamilton MC. 1988. Organochlorine residues in biota of the Baffin Island region. Sidney, BC: SeaKem Oceanography Ltd. Report prepared for Indian and Northern Affairs Canada, Ottawa.

Thomas DJ, Tracey B, Marshall H, Norstrom RJ. 1992. Arctic terrestrial ecosystem contamination. *Sci Tot Environ* 122:135–164.

Thomas P. 1995. Pesticide-induced immunotoxicity: are Great Lakes residents at risk? *Environ Health Perspect* 103:55–61.

Thomas PT. 1989. Effects of Arochlor 1254 and cadmium on reproductive endocrine function and ovarian growth in Atlantic croaker. *Mar Environ Res* 28:499–503.

Thomas PT. 1990. Teleost model for studying the effects of chemicals on female reproductive endocrine function. *J Experimental Zool Suppl* 4:126–128.

Thomas PT. 1993. Effects of cadmium on gonadotropin secretion from Atlantic croaker pituitaries incubated in vitro. *Mar Environ Res* 35:141–145.

Thomas PT. 1995. Pesticide-induced immunotoxicity in experimental animals and man. *Environ Health Perspect* 103:55–61

Thomas PT, Smith J. 1993. Binding of xenobiotics to the estrogen receptor of spotted seatrout. *Mar Environ Res* 35:147–151.

Tillitt DE, Ankley GT, Giesy JP, Ludwig JP, Kurita-Matsuba H, Weseloh DV, Ross PS, Bishop CA, Sileo L, Stromberg K, Larson J, Kubiak TJ. 1992. Polychlorinated biphenyl residues and egg mortality in double-crested cormorants from the Great Lakes. *Environ Toxicol Chem* 11:1281–1288.

Tremblay LT, Van Der Kraak GJ. 1998. Use of a series of in vitro and in vivo bioassays developed in rainbow trout to evaluate the actions of b-sitosterol. *Aquatic Toxicol* (Accepted January 1998).

Tremblay LT, Van Der Kraak GJ. 1999. Comparison between the effects of the phytosterol b-sitosterol and pulp and paper mill effluents on sexually immature rainbow trout. *Environ Toxicol Chem* (accepted June 1998).

Trizio D, Basketter DA, Botham PA, Graepel PH, Lambre C, Magda SJ, Pal TM, Riley AJ, Ronneberger H, Van Sittert NJ, Bontinck WJ. 1988. Identification of immunotoxic effects of chemicals and assessment of their relevance to man. *Food Chem Toxic* 26:527–539.

Tryphonas H. 1995. Great Lakes health effects: immunotoxicity of PCBs (Aroclors). *Environ Health Perspect* 103:33–46.

Tryphonas H, Fournier M, Lacroix P, Filipo D, McGuire D, Fernie P, Hayword S, Bryce F, Arnold D. 1998. Effects of Great Lakes contaminants on the immune system of two-generation Sprague-Dawley rats. Part B: Functional considerations. *Regulatory Toxicol Pharmacol* (submitted).

Tryphonas H, Fournier M, McGuire D, Lacroix F, Fernie P, Hayword S, Stapley R, Bryce F, Arnold D. 1998. Effects of Great Lakes contaminants on the immune system of two-generation Sprague-Dawley rats. Part A: Effects on peripheral blood and spleen cellularity. *Regulatory Toxicol Pharmacol* (submitted).

Tryphonas H, Hayward S, O'Grady L, Arnold JCK, Bryce F, Zawidzka ZZ. 1989. Immunotoxicity studies of PCB (Aroclor 1254) in adult Rhesus monkey (*Macaca mulatta*): preliminary report. *Int J Immunopharmacol* 11:199–206.

Tryphonas H, Luster MI, White Jr KL, Naylor PH, Erdos R, Burlesson GR, Germolec D, Hodgen M, Hayward S, Arnold DL. 1991. Effects of PCB (Aroclor 1254) on nonspecific immune parameters in Rhesus (*Macaca mulatta)* monkey. *Int J Immunopharmacol* 13:639–648.

[TSCA] Toxic Substances Control Act. 1976. Public Law 94-469, 90 Stat. 2003 (15 U.S.C. § 2601 et seq.), October 11, 1976.

Turner RT, Eliel LP. 1978. o,p'-DDT as an estrogen: an evaluation of its ability to compete with 3-H-estradiol for nuclear estrogen receptor sites in quail oviduct. *Bull Environ Contam Toxicol* 19:139–142.

UK Department of the Environment. 1994. Environmental hazard assessment: chlorinated paraffins. Garston, Watford UK: UK Department of the Environment, WD2 7JR. Report TSD/19-1994. 47 p.

[UN-ECE] United Nations Economic Commission for Europe. 1994. State of knowledge report of the UN-ECE Task Force on Persistent Organic Pollutants. 96 p.

[UNEP] United Nations Environmental Program. 1987. Montreal Protocol on the substances that deplete the ozone layer: final act; 1987 Sep; Nairobi, Kenya. Rome, Italy: UNEP.

[UNEP] United Nations Environmental Program. 1990. Adjustments and Amendments to the Montreal Protocol: 3rd meeting of the parties; 1990 Oct; London UK. Rome, Italy: UNEP.

[UNEP] United Nations Environmental Program. 1992. Adjustments and amendments to the Montreal Protocol: 4th meeting of the parties; 1992; Copenhagen, Denmark. Rome, Italy: UNEP.

U.S. Department of the Interior. 1980. Handbook of acute toxicity of chemicals to fish and invertebrates. Washington DC: U.S. Dept of Interior. Service Resource Publication 137.

[USEPA] U.S. Environmental Protection Agency. 1985. SWRRB Manual. Pesticide runoff simulation. simulator for water resources in rural basins. 2nd ed. Washington DC: USEPA Office of Pesticide Programs.

[USEPA] U.S. Environmental Protection Agency. 1986. Guidelines for carcinogen risk assessment. *Federal Register* 51(185):33992–34003.

[USEPA] U.S. Environmental Protection Agency. 1987. Unfinished business: a comparative assessment of environmental problems. Washington DC: USEPA.

[USEPA] U.S. Environmental Protection Agency. 1989. Interim procedures for estimating risks associated with exposures to mixtures of chlorinated dibenzo-*p*-dioxins and dibenzofurans (CDDS and CDFs) and 1989 update. Washington DC: Risk Assessment Forum. EPA/626/3-89/016.

[USEPA] U.S. Environmental Protection Agency. 1990a. OPTS IUR. *Federal Register* 55:39586.

[USEPA] U.S. Environmental Protection Agency. 1990b. Reducing risks: setting priorities and strategies for environmental protection. Washington DC: USEPA. SAB-EC-90-021.

[USEPA] U.S. Environmental Protection Agency. 1992a. Estimating exposure to dioxin-like compounds. Washington DC: Office of Research and Development. EPA/600/6-88/005B.

[USEPA] U.S. Environmental Protection Agency. 1992b. Framework for ecological risk assessment. Washington DC: Risk Assessment Forum. EPA/630/R-92/001.

[USEPA] U.S. Environmental Protection Agency. 1992c. Water quality criteria. *Federal Register* 57(246):60911–60915.

[USEPA] U.S. Environmental Protection Agency. 1993s. Deposition of air pollutants to the Great Lakes: first report to Congress. Research Triangle Park NC: USEPA Office of Air Quality Planning and Standards. EPA-453/R-93-055. 89 p + append.

[USEPA] U.S. Environmental Protection Agency. 1993b. Interim report on data and methods for assessment of 2,3,7,8-tetrachlorodibenzo-*p*-dioxin risks to aquatic life and associated wildlife. Washington DC: USEPA Office of Research and Development. EPA/600/R-93/055.

[USEPA] U.S. Environmental Protection Agency. 1994a. Estimating exposure to dioxin-like compounds. Volumes I, II, and III. Washington DC: Office of Research and Development. External Review Draft. EPA/600/6-88/005Ca.

[USEPA] U.S. Environmental Protection Agency. 1994b. ECOSAR: Computer program and user's guide for estimating the ecotoxicity of industrial chemicals based on structure-activity relationships. Washington DC: USEPA Office of Pollution Prevention and Toxics. EPA-748-R-93-002.

[USEPA] U.S. Environmental Protection Agency. 1994c. USEPA environmental fate and ground water branch one liner data base. Version 3.04. Washington DC: USEPA Office of Pesticides and Toxic Substances, Office of Pesticide Programs, Environmental Fate and Effects Division.

[USEPA] U.S. Environmental Protection Agency. 1994d. National air quality and emissions trends report, 1993. Research Triangle Park NC: USEPA Office of Air Quality. EPA 454/R-94/026.

[USEPA] U.S. Environmental Protection Agency. 1994e. National water quality inventory: 1992 Report to Congress. Washington DC: USEPA Office of Water. EPA 841/R-94/001.

[USEPA] U.S. Environmental Protection Agency. 1995. Great Lakes water quality initiative technical support document for the procedure to determine bioaccumulation factors. Washington DC: USEPA Office of Water. EPA-820-B-95-005. 96 p + append.

[USEPA] U.S. Environmental Protection Agency. 1996. Proposed guidelines for ecological risk assessment; notice. *Federal Register* 61:47552–47631.

Valo RJ, Salkinoja-Salonen M. 1986. Microbial transformation of polychlorinated phenoxy phenols. *J Gen Appl Microbiol* 32:503–517.

van den Berg M, Craane LHJ, Sinnige T, van Mourik S, Dirksen S, Van der Gaag M, Lutke-Schipholt IJ, Spenkelink B, Brouwer A. 1994. Biochemical and toxic effects of polychlorinated biphenyls (PCBs), dibenzo-*p*-dioxins (PCDDs) and dibenzofurans (PCDFs) in the cormorant (*Phalacrocorax carbo*) after in ovo exposure. *Environ Toxicol Chem* 13:803–816.

van den Berg M, DeJongh J, Poiger H, Olson JR. 1994. The toxicokinetics and metabolism of polychlorinated dibenzo-*p*-dioxins (PCDDs) and dibenzofuran (PCDFs) and their relevance for toxicity. *CRC Rev Toxicol* 24:1–74.

Van den Heuvel MR, Munkittrick KR, Van Der Kraak GJ, McMaster ME, Portt CB, Servos MR, Dixon DG. 1994. Survey of receiving water environmental impacts associated with discharges from pulp mills. 1. Bioassay derived 2,3,7,8-tetrachlorodibenzo-*p*-dioxin toxic equivalent concentration in white sucker (*Catostomus commersoni*) in relation to biochemical indicators of impact. *Environ Toxicol Chem* 13:1117–1126.

Van Der Kraak GJ, Munkittrick KR, McMaster ME, Portt CB, Chang JP. 1992. Exposure to bleached kraft pulp mill effluent disrupts the pituitary-gonadal axis of white sucker a multiple sites. *Toxicol Appl Pharmacol* 115:224–233.

Van Loon WM, Boon JJ, de Groot B. 1993. Quantitative determination 32: of macromolecular chlorolignosulfonic acids in water by pyrolysis-gas chromatography/mass spectrometry with single-ion monitoring. *Analytical Chem* 65:1728–1735.

Van Loon WM, Boon JJ, de Jong RJ, de Groot B. 1993. Isolation of macromolecular chlorolignosulfonic acids and lignosulfonic acids from pulp mill effluents and the River Rhine using XAD-8 macroporous resin and ultrafiltration. *Environ Sci Technol* 27:332–343.

Veith GD, DeFoe DL, Bergstedt BV. 1979. Measuring and estimating the bioconcentration factor of chemicals in fish. *J Fish Res Board Can* 36:1040–1048.

Velleux M, Endicott D. 1994. Development of a mass balance model for estimating PCB export from the lower Fox River to Green Bay. *J Great Lakes Res* 20:416–434.

Voldner EC, Li Y-F. 1993. Global usage of toxaphene. *Chemosphere* 27:2073–2078.

Voldner EC, Li Y-F. 1995. Global usage of selected persistent organochlorines. *Sci Tot Environ* 160/161:201–210.

Voldner EC, Scholtz MT, Li Y-F, Slama C. 1994. Global usage and emissions of selected organochlorine pesticides. 5th Global Emissions Inventory Activities (GEIA) Workshop; 1994 Sep 5–9; Fuji-Yoshida, Japan.

Vonwesternhagen H, Dethlefsen V, Cameron P, Janssen D. 1987. Chlorinated hydrocarbon residues in gonads of marine fish and effects on reproduction. *Sarsia* 72:19–22.

Walker JD. 1993a. The TSCA interagency testing committee, 1977 to 1992: creation, structure, functions and contributions. In: Gorsuch JW, Dwyer, CG, Ingersoll CG, LaPoint TW, editors. Volume 2, Environmental toxicology and risk assessment. Philadelphia PA: American Soc for Testing and Materials (ASTM). STP 1216. p 451–509.

Walker JD. 1993b. Review of ecological effects and bioconcentration testing recommended by the TSCA Interagency Testing Committee and implemented by USEPA under the Toxic Substance Control Act: chemicals, and methods. In: Landis WG, Hughes JS, Lewis MA. Environmental toxicology and risk assessment. Philadelphia PA: American Soc for Testing and Materials (ASTM). p 92–115.

Walker MK, Cook PM, Batterman AR, Lothenbach DB, Berini C, Libal J, Hufnagle L, Peterson RE. 1994. Early life stage mortality associated with maternal transfer of 2,3,7,8-tetrachlorodibenzo-*p*-dioxin to lake trout oocytes. *Can J Fish Aquat Sci* 51:1410–1419.

Walker MK, Olson JR, Spitsbergen JM, Peterson RE. 1991. 2,3,7,8-tetrachlorodibenzo-*p*-dioxin (TCDD) toxicity during early life stage development of lake trout (*Salvelinus namaycush*). *Can J Fish Aquat Sci* 48:875–883.

Walker MK, Peterson RE. 1991. Potencies of polychlorinated dibenzo-para-dioxin, dibenzofuran, and biphenyl congeners, relative to 2,3,7,8-tetrachlorodibenzo-para-dioxin, for producing early life stage mortality in rainbow trout (*Oncorhynchus mykiss*). *Aquat Toxicol* 21:219–238.

Walter B, Ballschmiter K. 1991. Biohalogenation as a source of halogenated anisoles in air. *Chemosphere* 22:557–567.

Walter B, Ballschmiter K. 1992. Formation of C1/C2-bromo-/chloro-hydrocarbons by halperoxidase reactions. *Fresenius J Analytical Chem* 342:827–833.

Wania F, Mackay D. 1993a. Global fractionation and cold condensation of low volatility organochlorine compounds in polar regions. *Ambio* 22:10–18.

Wania F, Mackay D. 1993b. Modelling the global distribution of toxaphene: a discussion of feasibility and desirability. *Chemosphere* 27:2079–2094.

Wania F, Mackay D. 1994. A global distribution model for persistent organic chemicals. *Sci Tot Environ* 160/161:211–232.

Water Pollution Control Act. 1948. Ch 758, 62 Stat. 1155 (33 U.S.C. § 1251 et seq.), June 25, 1947.

Wester PG, van der Valk J. 1990. Tetrachlorobenzyltoluenes in eel from The Netherlands. *Bull Environ Contam Toxicol* 45:69–73.

White DH, King KA, Kolbe EJ, Williams JM. 1982. Parathion poisoning of wild geese in Texas. *J Wildlife Diseases* 18:389–391.

White DH, King KA, Mitchell CA, Hill EF, Lamont TG. 1979. Parathion causes secondary poisoning in a Laughing Gull breeding colony. *Bull Environ Contam Toxicol* 23:281–284.

White DH, Mitchell CA, Wynn LD, Flickinger EL, Kolbe EJ. 1982. Organophosphate insecticide poisoning of Canada geese in the Texas panhandle. *J Field Ornithology* 53:22–27.

Whitlock Jr JP. 1990. Genetic and molecular aspects of 2,3,7,8-tetrachlorodibenzo-*p*-dioxin action. *Annu Rev Pharmacol Toxicol* 30:251–277.

Whitlock Jr JP. 1993. Mechanistic aspects of dioxin action. *Chem Res Toxicol* 30:251–277.

Wild SR, Jones KC. 1991. Studies on the polynuclear aromatic hydrocarbon content of carrots (*Dacus carota*). *Chemosphere* 23:243–251.

Williams RR, Bidleman TF. 1978. Toxaphene degradation in estuarine sediments. *J Agric Food Chem* 26:280–282.

Wœrn F, Flodström S, Busk L, Kronevi T, Nordgren I, Ahlborg UG. 1991. Relative liver tumour promoting activity and toxicity of some polychlorinated dibenzo-*p*-dioxin and dibenzofuran congeners in female Sprague-Dawley rats. *Pharmacol Toxicol* 69:450–458.

Wolfe NL, Zepp RG, Gordon JA, Baughman GL, Cline DM. 1977. Kinetics of chemical degradation of malathion in water. *Environ Sci Technol* 11:88–93.

Wolfe NL, Zepp RG, Schlotzhauer P, Sink M. 1982. Transformation pathways of hexachlorocyclopentadiene in the aquatic environment. *Chemosphere* 11:91–101.

Wong AS, Crosby DG. 1978. Photolysis of pentachlorophenol in water. In: Rao RK, editor. Pentachlorophenol: chemistry, pharmacology and environmental toxicology. New York: Plenum Pr.

Wong S, Fournier M, Coderre D, Banska W, Krzystyniak K. 1992. Environmental immunotoxicology. In: Peakall D, editor. Animal biomarkers as pollution indicators. London UK: Chapman and Hall. p 167.

Wood SC, Karras JG, Holsapple MP. 1992. Integration of the human lymphocyte into immunotoxicological investigations. *Fundamental Appl Toxicol* 19:450–459.

Yamashita N, Tanabe S, Ludwig JP, Kurita H, Ludwig ME, Tatsukawa R. 1993. Embryonic abnormalities and organochlorine contamination in double-crested cormorants (*Phalacrocorax auritus*) and Caspian terns (*Hydropogne caspia*) from the upper Great Lakes. *Environ Pollut* 79:163–173.

Zabel EW, Cook PM, Peterson RE. 1995. Toxic equivalency factors of polychlorinated dibenzo-*p*-dioxin, dibenzofuran and biphenyl congeners based on early life stage mortality in rainbow trout (*Oncorynchus mykiss*). *Aquat Toxicol* 31:315–328.

Zeeman M. 1986. Modulation of the immune response in fish. *Veterinary Immunol Immunopathol* 12:235–241.

Zeeman M. 1995. Ecotoxicity testing and estimation methods developed under Section 5 of the Toxic Substances Control Act (TSCA). In: Rand G, editor. Fundamentals of aquatic toxicology, effects, environmental fate, and risk assessment. 2nd ed. Washington DC: Taylor and Francis. p 703–715.

Zeeman M. 1996. Comparative immunotoxicology and risk assessment. In: Stolen JS, Fletcher TC, Bayne CJ, Secombes CJ, Zelikoff JT, Twerdok LE, Anderson DP, editors. Modulators of immune responses: a phylogenetic approach. Fair Haven NY: SOS. p 317–329.

Zeeman M, Auer CM, Clements RG, Nabholtz JV, Boethling RS. 1995. USEPA regulatory perspectives on the use of QSAR for new and existing chemical evaluations. *SAR QSAR Environ Res* 3:179–202.

Zeeman M, Brindley WA. 1981. Effects of toxic agents upon fish immune systems: a review. In: Sharma RP, editor. Volume II, Immunologic considerations in toxicology. Boca Raton FL: CRC Pr. p 1–60.

Zeeman M, Clements RG, Nabholz JV, Johnson D, Kim A. 1993. SAR/QSAR ecological assessment at EPA/OPPT: ecotoxicity screening of the TSCA Inventory. Poster Paper 312A at the Society of Environmental Toxicology and Chemistry (SETAC), 14th Annual Meeting; 1993 Nov 14–17; Houston TX.

Zeeman M, Gilford J. 1993. Ecological hazard evaluation and risk assessment under USEPA's Toxic Substances Control Act (TSCA): an introduction. In: Landis WG, Hughes JS, Lewis MA, editors. Environmental toxicology and risk assessment. Philadelphia PA: American Soc for Testing and Materials (ASTM). p 7–21.

Zeeman M, Nabholz JV, Clements RG. 1993. The development of SAR/QSAR for use under the USEPA's Toxic Substance Control Act (TSCA): an introduction. In: Gorsuch JW, Dwyer FJ, Ingersoll CG, LaPoint TW, editors. Volume 2, Environmental toxicology and risk assessment. Philadelphia PA: American Soc for Testing and Materials (ASTM). STP 1216. p 523–539.

Zepp RG, Cline DM. 1977. Rates of direct photolysis in the aquatic environment. *Environ Sci Technol* 11:359–366.

Zepp RG, Schlotzhauer PF. 1983. Influence of algae on photolysis rates of chemicals in water. *Environ Sci Technol* 17:462–468.

Zepp RG, Wolfe NL, Baughman GL, Hollis RC. 1977. Singlet oxygen in natural waters. *Nature* 207:421.

Zepp RG, Wolfe NL, Gordon JA. 1976. Light-induced transformations of methoxychlor in aquatic systems. *J Agric Food Chem* 24:727–733.

Zika RG, Cooper WJ, editors. 1987. Photochemistry of environmental aquatic systems. Washington DC: American Chemical Soc (ACS). 288 p.

Index

A

B

C

E

M

Q

R

S

T

U